BUILDING THE CRITICAL ANTHROPOLOGY OF CLIMATE CHANGE

This book applies a critical perspective to anthropogenic climate change and the global socio-ecological crisis.

The book focuses on the critical anthropology of climate change by opening up a dialogue with the two main contending perspectives in the field, namely the cultural ecological and the cultural interpretive perspectives. Guided by these, the authors take a firm stance on the types of changes that are needed to sustain life on Earth as we know it. Within this framework, they explore issues of climate and social equity, the nature of the current era in Earth's geohistory, the perspectives of the elite polluters driving climate change, and the regrettable contributions of anthropologists and other scholars to climate change. Engaging with perspectives from sociology, political science, and the geography of climate change, the book explores various approaches to thinking about and responding to the existential threat of an ever-warming climate. In doing so, it lays the foundation for a brave new sustainable world that is socially just, highly democratic, and climatically safe for humans and other species.

This book will be of interest to researchers and students studying environmental anthropology, climate change, human geography, sociology, and political science.

Hans A. Baer is Principal Honorary Research Fellow, School of Social and Political Sciences, University of Melbourne, Australia.

Merrill Singer is Professor Emeritus in the Department of Anthropology at the University of Connecticut, USA.

ROUTLEDGE ENVIRONMENTAL ANTHROPOLOGY

Environmental Anthropology explores historic and present human-environment interactions, highlighting the link between human-caused environmental problems such as climate change, species extinction, and pollution, with the complex cultural, political, and economic systems that have created them. This series aims to contribute to the growing subfield by providing a comprehensive survey of contemporary topics in environmental anthropology, including food procurement, ethnobiology, spiritual ecology, resilience, non-human rights, architectural anthropology, industrialism, and education.

The *Routledge Environmental Anthropology* series welcomes submissions that combine strong academic theory with practical applications, and as such is relevant to a global readership of students, researchers, policy-makers, practitioners, and activists. Please contact Grace Harrison (Grace.Harrison@tandf.co.uk).

Ecological Epistemologies and Spiritualities in Brazilian Ecovillages
In the Labyrinth of an Environmental Anthropology
Luz Gonçalves Brito

Building the Critical Anthropology of Climate Change
Towards a Socio-Ecological Revolution
Hans A. Baer and Merrill Singer

For more information about this series, please visit: https://www.routledge.com/Routledge-Environmental-Anthropology/book-series/REA

BUILDING THE CRITICAL ANTHROPOLOGY OF CLIMATE CHANGE

Towards a Socio-Ecological Revolution

Hans A. Baer and Merrill Singer

Routledge
Taylor & Francis Group

LONDON AND NEW YORK

from Routledge

Designed cover image: FG Trade © iStock

First published 2025
by Routledge
4 Park Square, Milton Park, Abingdon, Oxon OX14 4RN

and by Routledge
605 Third Avenue, New York, NY 10158

Routledge is an imprint of the Taylor & Francis Group, an informa business

© 2025 Hans A. Baer and Merrill Singer

British Library Cataloguing-in-Publication Data
A catalogue record for this book is available from the British Library

Library of Congress Cataloging-in-Publication Data
Names: Baer, Hans A., author. | Singer, Merrill, author.
Title: Building the critical anthropology of climate change : towards a socio-ecological revolution / Hans A. Baer and Merrill Singer.
Description: Abingdon, Oxon ; New York, NY : Routledge, 2024. | Series: Routledge environmental anthropology | Includes bibliographical references and index.
Identifiers: LCCN 2024001246 (print) | LCCN 2024001247 (ebook) | ISBN 9781032745770 (hardback) | ISBN 9781032745763 (paperback) | ISBN 9781003469940 (ebook)
Subjects: LCSH: Climatic changes–Social aspects. | Human beings–Effect of climate on.
Classification: LCC QC903 .B145 2024 (print) | LCC QC903 (ebook) | DDC 304.2/8–dc23/eng/20240530
LC record available at https://lccn.loc.gov/2024001246
LC ebook record available at https://lccn.loc.gov/2024001247

ISBN: 978-1-032-74577-0 (hbk)
ISBN: 978-1-032-74576-3 (pbk)
ISBN: 978-1-003-46994-0 (ebk)

DOI: 10.4324/9781003469940

Typeset in Times New Roman
by Taylor & Francis Books

CONTENTS

List of Figures and Tables   *xi*
Acknowledgments   *xii*

Introduction   1

1 Climate Change, Climate Science, and Anthropology   6

*Case Study: The Unheard Voices in Climate Change
  Science 6*
Introduction to a Changing World 7
*The Misnamed Planet Earth: Disturbing the
  Anthropocentric Perspective 7*
*A Short History of Climate Science and the Recognition of
  Climate Change 10*
Why Anthropology Engages Climate Change 11
*Anthropogenic Impact on a Small Blue Dot: Encountering
  the Effects of Greenhouse Gases on Earth 14*
Cataloging Five Fundamental Planetary Changes 17
   Rising Temperatures 17
   Melting Ice Caps, Glaciers, Sea Ice, and Tundra 18
   Intensifying Wildfires and Hurricanes 22
   Hurricanes 23
   Species Shifts and Loss of Biodiversity 24
   Transgressing Planetary Boundaries 26
Conclusion 28

2 Conflicting Anthropological Perspectives: Cultural
   Ecological/Environmental Anthropological, Cultural
   Interpretive, and Critical Anthropological Perspectives of
   Climate Change   30

   Introduction 30
   The Cultural Ecological/Environmental Anthropological
     Perspective of Climate Change 31
        Continuing Research on Climate Change from an
        Environmental Anthropological Perspective 36
   The Cultural Interpretive Perspective of Climate
     Change 37
   The Critical Anthropology of Climate Change 45
   Conclusion: Towards an Integrated Critical Understanding
     of Climate Change 52

3 Anthropocene, Capitalocene, or Whatever? Rethinking Our
   Era of Climate Change Production   58

   Introduction 58
   The Anthropocene as a Contested Concept 59
        Early, Middle, and Late Anthropocene 60
        The Good Anthropocene 61
        The Patchy Anthropocene 62
   Dipesh Chakrabarty's Postcolonial Historical Take on the
     Anthropocene 63
   Debate Between Anthropocene and Capitalocene 65
   Alternatives to the Anthropocene and Capitalocene 68
   Conclusion 69

4 Social Inequality and Climate Change   73

   Case Study: Living at Risk in the Time of Climate
     Change 73
   The Human Creation of Inequality 76
   Introducing a Demonic Duo: The Interface of Climate
     Change and Inequality 77
   Elite Polluters and the Unequal Distribution of Power 78
   Welfare for the Wealthy (Polluters) and Blaming the
     Poor 82
   Resistance from Below: The Grassroots Fight Against
     Climate Change and its Drivers 86
   Conclusion 91

5 Planetary Health: A Critical Health Anthropological
   Perspective   95

   *Case Study: Electric Batteries and Environmental and
     Human Suffering 95*
   Why Planetary Health? 99
   The Environment and the Spread of Infectious Disease 102
   Environmental Violence 106
   Is Planetary Health a Colonial Strategy? 108
   Conclusion 115

6 Towards a Critical Anthropology of Climate Refugees   120

   Case Study: The Floods of Starvation 120
   *Anthropological Study of Refugees of War, Conflict, and
     Oppression 122*
   The Climate Migration Nexus 125
   Climate Migration and Health 128
   Island Nations at Growing Risk 130
   *Climate Change and High-Altitude Communities: The
     Andean Altiplano 133*
   Fleeing Drought and Rain in Kenya 134
   Super Flooding in Pakistan 135
   Conclusion 137

7 Can Ecological Modernization Contain Climate Change?
   How the Rich and Powerful Seek to Address the Ecological
   Crisis   142

   Case Study 142
   The Technical Fix: A Full or Partial Solution? 143
   Mainstream Ecological Modernization 143
   Ecomodernism 145
   Rich Proponents of Ecological Modernization 146
        Bill Gates 147
   *Other Wealthy Proponents of Ecological
     Modernization 149*
   *Ecological Modernization "Down Under": Greenwashing a
     Developed Country 151*
   Critique of Ecological Modernization 157
   Conclusion 160

8 The Scholarly Elephant in the Sky: How Can
   Anthropologists and Other Scholars Grapple with their
   Heavy Reliance on Flying in the Era of Climate Crisis?   164

   Case Study 164
   Staying Grounded 165
   Increase in Airplane Flights 166
   Academic Air Travel 168
   Air Travel on the Part of Anthropologists 169
   Flying on the Part of Other Academics 171
   Alternative Forms of Flying and Flying Far Less 173
   Teleconferencing 178
   Conclusion 179

9 Two Genres of the Climate Movement: Climate Action vs.
   Climate Justice   183

   Case Study 183
   Climate Movements 184
   Tendencies in the International Climate Movement 185
   The Climate Action Movement 186
        350.org 187
        School Strike 4 Climate 187
        Extinction Rebellion 189
   The Climate Justice Movement 191
   The Climate Movement in the Global South 195
        Brazil 195
        China 195
        Kiribati 195
        Sub-Saharan Africa 196
        India 196
   Transforming the Climate Movement into a Climate Justice
      Movement 197
   Finding Cracks in the System and Dilemmas Facing the
      Climate Movement as an Anti-Systemic Movement 199
   Conclusion 203

10 Towards a Critical Anthropology of the Future: Climate
    Change and Future Scenarios   208

    Case Study 208
    Playing with Time 209
    The Road to Dystopia 210

Mark Lynas 211
The Possibility of Eco-Authoritarian Regimes 211
The Potential for Climate Wars 214
Reflexive Modernization 217
Geoengineering 219
Limitations of Reflexive Modernization 220
Eco-Socialism as an Alternative World System 221
Democratic Socialism 222
Eco-Socialism 223
Democratic Eco-Socialism 224
*Radical Perspectives with which Eco-Socialism Can
Dialogue 225*
Eco-Anarchism 226
Eco-Feminism 228
De-Growth 229
Indigenous Voices 230
Half-Earth Socialism 230
Conclusion 231

11 Eco-Socialism as the Ultimate Climate Change Mitigation
Strategy   237
*Fighting Climate Change: Reformist Reforms and
Nonreformist Reforms 237*
Transitioning to Eco-Socialism 238
New Progressive Parties Designed to Achieve
Democratic Capture of the State 240
Emissions Taxes 242
Public Ownership of the Means of Production 243
Increasing Social Equality 244
Achieving a Sustainable Global Population 245
Workers' Democracy and Socialist Planning 245
Meaningful Work and Shortening the Work Week 246
The Need for a Steady-State Economy 248
Renewable Energy Sources, Energy Efficiency, and
Green Jobs 249
Sustainable Public Transportation and Travel 252
Sustainable Food Production and Forestry 255
Resisting the Culture of Consumption and Adopting
Sustainable and Meaningful Consumption 257
Sustainable Trade 258

Sustainable Settlement Patterns and Local
    Communities 258
    Demilitarization 260
Conclusion 261

Epilogue   267

Index   270

LIST OF FIGURES AND TABLES

Figures

2.1 Douglas's Grid-Group Model   41
2.2 The Grid-Group Model with Respect to Views of Nature   41

Table

9.1 Tendencies in the International Climate Movement   185

ACKNOWLEDGMENTS

The authors thank Lani Davison for reading and commenting on several chapters. We also acknowledge permission from the *Journal of Australian Political Economy* to incorporate some material from our article titled "Can ecological modernisation contain climate change? An eco-socialist perspective" which appeared in *No. 90* (Summer 2022/2023), pp. 75–91, in Chapter 7 of this book.

INTRODUCTION

Adam Smith, seen by some as the celebrated "Father of Capitalism," asserted that by their very nature people are driven by self-interest. As he colorfully phrased it in his magnum opus *The Wealth of Nations*, "[i]t is not from the benevolence of the butcher, the brewer, or the baker that we expect our dinner, but from their regard to their own interest" (Smith 1776). If this is at the core of our moral fiber, then capitalism, a system founded on the marriage of private ownership of the means of production and a taken-for-granted acceptance of the naturalness of self-interest, is not only effective in producing great wealth from the resources of the planet and human labor, but is also a system that closely harmonizes with our fundamental dispositions and characteristics as a species. In his book, *The War Against the Common: Dispossession and Resistance in the Making of Capitalism*, historian Ian Angus (2023), however, asks a provocative question: if capitalism is such a natural outcome of human nature, why were systematic violence and draconian laws necessary to establish it? If, as defenders of capitalism tell us, narrow self-interest and greed are among the primary motivations for human behavior, how is it, as anthropology teaches us, that for the vast majority of humans living on Earth the core values are cooperation and avoidance of social inequality? Something is out of kilter in what we have been told about ourselves. At heart, rethinking this issue—who we are as a species—underlies the core argument of this book about the need for a radical new pathway to a just and sustainable future. Simply put: your future.

For us, the authors, this book has been some 18 years in the making. During that time, we have watched, and protested, the ever-hastening arrival of the predicted catastrophe of anthropogenic climate change. The book is a timely sequel to the previous two editions of our book *The Anthropology of Climate Change: An Integrated Critical Perspective* (Baer and Singer 2014, 2018). Rather than updating the latter into a third edition, in this book we

DOI: 10.4324/9781003469940-1

have opted to focus on advancing understanding of the critical anthropology of climate change in light of our deep-felt sense, shared by many, that environmentally things are getting worse faster.

In the five years since we published our prior collaborative book in 2018, what has changed? First, the window of opportunity for achieving a livable and sustainable future has come far closer to closing. We now have, at the very most, a decade in which to save life on Earth as we know it. The 2023 United Nations Intergovernmental Panel on Climate Change finalized a synthesis report on what climate science now unequivocally understands about the dire climate crisis, which says that global average temperatures are estimated to rise by 1.5 °C (2.7 °F) above preindustrial levels sometime around the early 2030s, and that they will continue to rise further as we rely more heavily on coal, oil, and natural gas and generate an ever-thicker greenhouse blanket. Under the 2015 Paris Agreement, essentially all the states parties agreed to hold global warming at 1.5 °C (2.7 °F) because above that point the joint impacts of catastrophic heat waves, flooding, drought, crop failures, the spread of diseases, and species extinction will overcome the human ability to handle them.

The writing, inescapably, is on the wall or more aptly inscribed in the growing climate expense report. Every three weeks, the United States, for example, now experiences a climate shock event that causes at least US $1 billion in damage. Forty years ago, extreme weather episodes that cost (an inflation-adjusted) $1 billion happened on average only once every four months. As of November 8, 2023, there had been 25 climate disasters in that year alone with losses exceeding or surpassing $1 billion, thereby exceeding the 2022 level even without data for the last two months of the year.

Second, scientists can now say with much greater confidence than they could five years ago that the climate crisis makes rainstorms, hurricanes/cyclones, and wildfires more intense or more frequent, long-term droughts more severe, and heat waves more deadly. During the summer of 2023, for example, the US southwestern city of Phoenix, Arizona, endured a record 31 consecutive days of temperatures at or above 43.3 °C (110 °F), a withering heat wave that contributed to more than 500 heat-related deaths in Maricopa County. In July, across the country, a torrential rainstorm inundated sections of the state of Vermont under deadly floodwaters. In August, Maui, Hawaii, was scorched by a devasting wildfire just as Florida's Gulf Coast was being slammed by the second high-velocity hurricane/cyclone in two years. The list goes on: in the Pacific Northwest, a major heat wave during which temperatures reached a high of 49.6 °C (121.3 °F) caused intense forest fires. According to climate scientists, these events got so bad because of climate change and some would have been virtually impossible without global warming. The pattern is the same globally. To cite a notable example, in 2020, a prolonged heat wave occurred in one of the coldest places on Earth, triggering widespread wildfires. Temperatures in Verkhoyansk, a small Arctic town in Siberia, Russia, rose to a completely unprecedented 38 °C (104.4 °F). Again, climate scientists conclude that this event would have been almost impossible without global warming. Similarly, from North America to Europe to China, enormous portions of the Northern Hemisphere experienced an extreme drought in the summer of 2022, straining water resources, ruining crops, and priming the landscape for perilous wildfires. This event, climate scientists concluded, was made 20 times more likely by climate change. In short, climate scientists can now say without hesitation when and where rising planetary temperatures will increase the human toil of punishing weather events.

Third, while popular awareness of the harmful impacts of climate change are rising, conservative leaders continue to claim loudly but falsely that climate change is a hoax. Some cases in point: in 2023, the ultra-right-wing President of Argentina, Javier Milei, called scientific evidence of the climate crisis a "socialist lie." Also in 2023, the victory of Geert Wilders' far-right nationalist Partij voor de Vrijheid (Party for Freedom) in the Dutch elections has left European climate activists fearful of a drastic shift to more fossil fuel use and a rollback of climate policies in the Netherlands, which already has the highest European greenhouse gas emission levels per person. Hungarian Prime Minister Viktor Orban has dismissed European Union plans to tackle climate change as a "utopian fantasy." At the time of writing, in France, Marine Le Pen's far-right Rassemblement National (National Rally), which opposes most if not all of the European Commission's Green Deal, is on course to win the largest share of French votes ahead of in the European Parliament elections scheduled to take place in mid-2024. In Germany, opinion polls are showing a steady rise in popularity of the far-right populist party Alternative für Deutschland (Alternative for Germany), which denies that human activity is responsible for climate change. Meanwhile, the former president of the World Bank David Malpass, who was nominated to the post by the climate change denying Donald Trump when he was US President, was forced to step down after refusing publicly to acknowledge that the burning of fossil fuels is rapidly warming the planet. We now face a reverse "Chicken Little" dilemma: the skies, in a sense, are falling, but Chicken Little is denying that there is a problem.

Finally, a primarily youth-led grassroots global climate action movement, although still in its infancy, is gaining traction. There are now at least 300 youth climate groups spread across every inhabited continent. Many share an awareness that capitalism puts profits over human and planetary welfare and is the root cause of climate change. They are demanding much more than the minimalist reforms of climate policies and instead are calling for systematic structural change, including the dismantling of capitalism. Moreover, such youth groups are seeking climate and social justice. Notably, sub-Saharan Africa is the most active region for climate protest. These understandings of the nature of the climate crisis and its solutions align with the critical anthropological perspective on climate change detailed in this book.

In the chapters that follow, we bring together our thoughts on a range of entwined issues pertinent to climate change, including:

- Reviewing the many impacts and primary causes of climate change;
- Examining the three most commonly used theoretical perspectives that have emerged in the anthropology of climate change;
- Deconstructing the ever-more popular notion of the Anthropocene;
- Analyzing the fateful global intersection of ongoing climate change and widening social inequality;
- Investigating the eco-conscious concept embraced by a growing number of health researchers known as planetary health;
- Developing a critical anthropology of climate refugees;
- Critiquing the perspective known as ecological modernization that is embraced by advocates of a technological fix of the climate problem;
- Revealing the contradictions of climate damaging frequent air travel by academics, including anthropologists and climate scientists;
- Contrasting the climate action movement and the climate justice movement;
- Exploring three possible scenarios for the future of humanity in a time of rapid climate change;
- Demonstrating why eco-socialism constitutes the ultimate climate change mitigation strategy.

The issue of how people respond to climate change zigzags between the distant poles of hope and despair. It is in part the dread of despair that no doubt pushes some to cling to corrupt leaders who deny climate change and offer in its place a seemingly safe but far less democratic, and less just future. Alternately, there is the approach to hope voiced by youth climate activist Greta Thunberg:

> Hope is not something that is given to you. It is something you have to earn, to create. It cannot be gained passively from standing by passively and waiting for someone else to do something. It is taking action. It is stepping outside your comfort zone.
>
> *(Quoted in Bergstein 2022)*

We embrace this perspective knowing that we must fight for the kind of world we want to live in.

References

Angus, Ian. 2023. *The War Against the Commons: Dispossession and Resistance in the Making of Capitalism*. New York: Monthly Review Press

Baer, Hans A. and Singer, Merrill. 2014. *The Anthropology of Climate Change: An Integrated Perspective* (1st Edition). London: Routledge

Baer, Hans A. and Singer, Merrill. 2018. *The Anthropology of Climate Change: An Integrated Perspective* (2nd Edition). London: Routledge

Bergstein, Al. 2022. Greta Thunberg on hope. *The Olympic Peninsula Environmental News.* https://olyopen.com/2022/06/25/greta-thunberg-on-hope/

Smith, Adam. 1776. *An Inquiry into the Nature and Causes of the Wealth of Nations, Book 1.* London: W. Strahan

1

CLIMATE CHANGE, CLIMATE SCIENCE, AND ANTHROPOLOGY

Case Study: The Unheard Voices in Climate Change Science

Climate change has transitioned from an obscure concept to an everyday word for many people, although it is a topic about which there is neither equal knowledge nor agreed understanding. One place there is little debate about the certainty and broad outlines of climate change is among scientists around the globe who study climate for their livelihoods, and who often have intense passion about their work. Most commonly, when the news media reports on climate change it relies on climate scientists from the wealthy Global North countries, including those in Europe, the United States, and Japan. As a result, although the study of climate change is a global concern, climate scientists from the poorer countries of the Global South often are underrepresented in climate change reports and discussions in the mainstream news media.

Carbon Brief, a UK-based climate website, and the Oxford Climate Journalism Network, have responded to this gap by establishing the Global South Climate Database tool, enabling journalists to contact climate scientists from 107 different countries in Asia, Africa, Latin America, and the Pacific. Their expertise covers multiple issues, including local climate change impacts, tropical meteorology, mapping carbon sinks, and energy modeling.

Why is it important to hear from these experts? As highlighted by Navroz Dubash, a professor at the Centre for Policy Research in New Delhi, India, who studies energy policy and the climate crisis, there is a media bias towards the perspectives of highly developed nations that influences the type of stories that news editors accept for publication. In particular, Global North media outlets, especially those in the United States, tend to avoid reporting on the requests voiced in the Global South for wealthier, more developed nations to

DOI: 10.4324/9781003469940-2

cut their energy consumption—a growing demand from nations that are still developing their social and economic infrastructures and cannot afford to limit their comparatively low energy use. Moreover, in the time-crunched, deadline-driven world of journalism, reporters tend to call Global North experts they have always counted on because they know they are going to respond. However, climate research from scientists who live in the regions they study can offer broader, deeper, and more contextual reporting. Such on-the-ground scientists have greater in-depth knowledge of the issues at hand, both technical and social (Donovan 2023).

Introduction to a Changing World

Climate scientists, who for a long time toiled in cloistered academic settings and were rarely mentioned on the stage of public discussion about the world, have become quite visible and increasingly vocal as the wreckage done to vital global systems advances due to environment disruptions and planetary warming. This chapter reviews what they are saying about the current and future status of the atmosphere, the environment, and life on Earth during a time of rapid and worrisome change. In all honesty, it is not a pretty picture they are painting. Many temperature records were shattered in 2023, accompanied by intensifying heat waves, floods, and droughts that took lives, destroyed communities, and forced many to flee as homeless climate refugees. Climate scientists say far worse is yet to come as temperatures continue to rise. The world is now moving towards a horrifying 3 °C (5 °F) of intolerable global heating. While the impact will be worst in the poorer countries of the Global South, we are all now on the frontline of climate change.

We begin our broad-based examination of the current state of Earth's climate and how we got here—and how we might survive the mounting threats we face—with a bit of contextualization about human striving for understanding the world and the role of science in this process. On this foundation, we present a brief history of climate science, the multidisciplinary field that rang the alarm on the tsunami we call climate change, and detail the slow-developing anthropological role in climate science. We then take a closer look at what climate change is and why it must be central to our thinking and actions from this point forward.

The Misnamed Planet Earth: Disturbing the Anthropocentric Perspective

Scientists categorize the planets based on their surface composition: for example, ice planets are covered by an icy surface; desert planets have a rocky surface with little water; ocean planets are largely covered by water. Among English speakers, we have named (or misnamed) the planet we inhabit "Earth," although generally we are aware that a far higher proportion of our planet (71%) is covered by water rather than dry land. While water on Earth is found in many forms, including water vapor in the atmosphere, in rivers, ponds, and lakes, in icecaps and glaciers, in the ground as soil moisture, in underground aquifers, in humankind and other life forms, the oceans hold about 96.5% of all Earth's water. If we take this information into consideration, we might conclude that we live on planet Ocean and not on planet Earth. Our planet is the only celestial body known to have bodies of liquid water on its surface, although several exoplanets (planets outside of our solar system) have been found with the right conditions to support liquid water. Using mathematical modeling, in 2020 scientists at NASA reported that there is a high probability that exoplanets with oceans are common in our galaxy (Kazmierczak 2020).

What the story of our planet's naming teaches us is that humans have a very species-centric view of reality. While humans are highly dependent on water—most of us can only live about three days without it—the vast majority of people live on dry land, and it is the part of the world we know best. In effect, our consciousness of the world is influenced by the soil beneath our feet and the objects and species found on or near that surface.

While it is true that millions of people around the world are dependent on resources from oceans, lakes, and rivers, today only a small minority are like the Bajau who traditionally have spent most of their lives on boat flotillas navigating the waters of the Philippines, Malaysia, and Indonesia. This land-dwelling pattern has been the norm for all of human existence, and even the ocean voyagers who carried people to the most distant islands of the Pacific were seeking land. Being conscious does not just mean having awareness of the outside world. It means being aware of oneself within one's surroundings. This awareness emerged at a critical moment in human evolution, most likely in the land mass we call Africa, when we developed consciousness and began to live in a world constructed in our own minds. This is not to say that the physical world does not exist—of course it does. However, our understandings of it are filtered through our senses, brains, and cultural frameworks. This feature of human consciousness has led us to refer to and think of our planet as "Earth" and to overlook the actual wet nature of most of our environment.

This realization points to an underlying feature of human understanding of the world, a perspective that puts us at the center. As discussed below, overcoming our tendency to centralize ourselves, our social group, our species, even our planet, and the myths emergent from this human characteristic, has been a dominant theme of scientific research. Ironically, with climate change, the reverse is true: contemporary global warming is a product of human activities, although more through the behavior of some people than others.

Anthropocentrism is the belief that humans are the central or most important entity in the universe. It is characterized by an acceptance of human exceptionalism. From an anthropocentric perspective, humankind is separate from nature and superior to it, and other entities (plants, animals, rocks) are viewed as resources for humans to use. This perspective found voice in a novel written by bestselling American author Jonathan Foer (2010) in which he describes anthropocentrism as "the conviction that humans are the pinnacle of evolution, the appropriate yardstick by which to measure the lives of other animals, and the rightful owners of everything that lives."

Science, by contrast, dispels the notion that we are the center of everything. Instead, humans are but one species — however dominant at the moment — of the millions that have ever lived or are alive today. Indeed, the science of paleontology focuses on the time before the evolution of humans, which is, in fact, the vast majority of our planet's history. Hundreds of species of dinosaurs, for example, often (erroneously) said to have "ruled" Earth during the Mesozoic Era between 245 and 66 million years ago, are gone (although their bird descendants remain).

Long before the scientific understanding of extinct life forms developed, another critical displacement of the human-centered world had already occurred. Gazing upwards at night from what felt to them like a fixed reference point and witnessing the stars and planets move across the sky, the ancient Greeks embraced a geocentric view of the universe. This model put Earth at the center of the universe with the Sun and other planets revolving around it. Building on the ideas of others before him, Aristotle developed a detailed geocentric model. In various versions, the geocentric model was accepted as valid for almost 1,500 years. Ultimately, modern astronomy rejected it. At the beginning of the 16th century, Nicolaus Copernicus challenged the geocentric model and proposed that Earth and the other planets revolve around the Sun. The ultimate shift to a Sun-centered model began when Galileo Galilei studied the sky with a telescope in 1610.

While sounding somewhat similar to anthropocentrism, ethnocentrism refers to the assumption that one's own cultural traditions, society, or ethnicity are the standard by which to evaluate and judge the worth of other cultural life ways, social practices, beliefs, values, and people. Since this basis for judgment generally paints those deemed as outsiders in a negative light, some researchers also use the term to label the common belief that one's culture is superior to, more evolved, or more normal than all others. Anthropology views ethnocentrism as a form of tunnel vision because it generally leads to an inability to understand cultures that are different from one's own while producing various forms of bias, including nationalism, racism, and even sexism, and disability discrimination. Opposing and dispelling the myths of ethnocentrism has long formed the cornerstone of anthropological thinking.

More recently, anthropologists have been rethinking their approach to human research. Part of this change involves seeing a need to bridge the nature/culture divide that keeps our species largely separated from all other species on the planet—a form of human exceptionalism. This has given rise to a decentering research and writing approach known as multispecies ethnography (Pacini-Ketchabaw et al. 2016). Multispecies ethnography focuses on how various organisms' livelihoods, including our own, shape and, in turn, are shaped by crosscutting political, environmental, economic, and cultural forces. This approach begins with a recognition of the foolishness of human exceptionalism and leads to new insights about people and many other species, connections that were often overlooked in the past.

Anna Tsing (2015), for example, began studying mushrooms to imagine a human nature that shifted historically in tune with various webs of inter-species dependence. She searched the parklands of northern California for mushrooms and discovered a world of mutually flourishing rhizomic compa-nions in the mushroom world connected by underground shoots. Seeking to duplicate the "rhizomic sociality" of mushrooms, Tsing founded the Matsu-take Worlds Research Group, an ethnographic research team concerned with the matsutake mushroom. Ethnographically, following the matsutake mush-room through commodity chains in Europe, North America, and East Asia, this research group experimented with new modes of collaborative ethno-graphic research while studying multispecies relations. Along the way, these social scientists have contributed to destabilizing human-centrism.

As this account reveals, an ongoing impact of science has been revealing faulty self-centered aspects of the way we popularly construct the world around us. It is of note then that one of the common misunderstandings of climate change is the failure to recognize the dramatic level of impact that human behavior can have on the basic Earth systems that make human life on the planet possible. In the causation of climate change, we are the center.

A Short History of Climate Science and the Recognition of Climate Change

It is challenging to establish the precise origin of a science. In the case of climate change research, it is appropriate to go all the way back to the mid-1600s to the work of Flemish alchemist Jan Baptist van Helmost and his discovery that burning coal releases a certain amount of carbon dioxide into the air and that in concentration this gas is unhealthy to breathe. A century later, Joseph Black, a British medical student, discovered that limewater can be used as a carbon dioxide detector.

In 1824, Jean-Baptiste Joseph Fourier, a mathematician in the employment of Napoleon, determined how Earth remains comfortably warm through a study of the high-and-low cycles of temperature between day and night. He concluded that the Sun radiates energy, which arrives on Earth as visible light. This energy passes through the atmosphere relatively easily, and, in the process, warms the planet's surface. But Earth, in turn, radiates energy back into the atmosphere. Building on the work of William Herschel on the dif-ference in temperature between the colors in the visible spectrum, Fourier called this energy "dark heat." Today we know it as infra-red radiation. Invisible to the human eye, infra-red radiation can be detected as a sensation of warmth on the skin. Fourier found that infra-red radiation does not pass though the atmosphere as easily as light heat, which slows the loss of energy back to space. By way of analogy, he described the planet's atmosphere as functioning like the glass walls of a greenhouse.

In the 1850s, Irish physicist John Tyndall conducted a series of experiments to measure how well "dark heat" passes through different gases. His work showed that nitrogen and oxygen lack the ability to trap energy, but carbon dioxide, ozone, and water vapor can achieve this feat. Meanwhile, Eunice Newton Foot, an American scientist, discovered that carbon dioxide, ozone, and water vapor cause air to warm when exposed to sunlight. At the close of the 19th century, Swedish chemist Svante Arrhenius determined that burning coal can increase the amount of carbon dioxide in the atmosphere, which in turn warms the planet.

Although the work of many other climate scientists subsequently contributed to our current robust understanding of climate change and its potential for making Earth uninhabitable for humans and many other species, the scientists mentioned above laid the foundation for climate science, including the contribution of human activity to the warming of the planet, and the need for sweeping efforts to limit planetary warming. As this brief history suggests, neither climate science nor a basic scientific understanding of climate change is new. What is new is the level of detailed understanding of the complicated dynamics and components of climate change emanating from the ongoing day-to-day work of climate scientists. Sadly, the predictions made by climate scientists are increasingly coming to fruition.

Until relatively recently, however, the insights of climate change have been corrupted, confused, and blocked from public awareness by those who for economic or political reasons are threatened by climate change awareness. Most explicitly, and with increasing vehemence, climate change has been erroneously labeled a hoax by those who profit from greenhouse gas (GHG) production, namely elite polluters and their political defenders. At the same time, the catastrophic impacts of climate change have motivated the birth of popular movements around the world to counter the climate myths peddled by climate change deniers and to call attention to and to oppose business as usual anthropogenic climate disruption. Additionally, since the 1990s, scientific research on climate change and its effects on life, the environment, and society has extended to multiple other disciplines, including anthropology (Earle 2021; Sawyer 1972).

Why Anthropology Engages Climate Change

At first glance, climate change could be assumed to be an issue solely tackled by the natural sciences. After all, these disciplines, ranging from climatology to oceanography and from geophysics to biogeography, provide quantitative data on and physical evidence of the various expressions of climate change, measurements of alterations in the environment due to climate change, and understandings of the adverse impacts of climate change as they already exist or will exist in the future. Using evidence-based modeling (sets of mathematical equations based on the laws of physics, chemistry, and biology), natural sciences can also make predictions about future climate patterns and consequences of change—an action that has become increasingly accurate. A study of 17 climate models going back to the early 1970s, for example, found that most of the models did a good job of predicting temperatures in the decades ahead (Hausfather et al. 2020).

It is important to remember, however, that climate change is very much a human issue—one that is caused by humans—leading to diverse and unequally distributed consequences for humans, and is a challenge that ideally will be addressed and hopefully remedied by humans while there is still time. Anthropology as a field helps to develop understandings of how societies experience and understand climate change, and how cultural values and practices have influenced Earth's changing climate. Many of the key issues about why anthropologists are ever more involved in the study of climate change and in applied efforts to address it are found in the American Anthropological Association's Statement on Humanity and Climate Change (American Anthropological Association 2015).

According to this statement, climate change creates global threats that affect all aspects of human life, including our health, homes, livelihoods, infrastructures, and cultural systems, as well as our physical environments whether "natural" or, like cities, created by people. Threats of this high magnitude affect our stability, including our sense of cultural identity, our well-being, and our experience of everyday security. Moreover, climate change is not a future threat, it is causing death and suffering right now. Anthropologists, like others, predict that over time climate change will accelerate human migration, destabilize communities and nations, and exacerbate the spread of infectious diseases. The people anthropologists study in both the Global North and the Global South increasingly will feel the pressures ushered in by a warming world. Those who have for centuries directly depended on the immediate use of locally available natural resources, such as people living in high latitude/altitude areas, on low-lying islands, in coastal environments, and in various other at-risk biomes, will see their lives most dramatically disrupted in the short run. Moreover, the impacts will fall unevenly and with particular weight on those already affected by existing vulnerabilities, including children, the elderly, people who live with disabilities and restrictive health conditions, and those who do not have sufficient means to move or change their lives. The most vulnerable will be uprooted from their homes and communities and forced to move elsewhere. How they fare in these new places will also be a human issue highly influenced by the reception offered by their new neighbors (who will themselves be feeling the negative impacts of climate change). As the consequences of climate change intensify, public expenditures needed for emergency aid and restoration will escalate.

Tragically, these kinds of resources are not equally distributed within and across nations, further threatening the most vulnerable. As the discipline most clearly devoted to studying the human condition over both time and space, anthropology has the potential to provide important insights that can help to create workable solutions to mitigate the damaging impacts of climate change and to alleviate the suffering of those most severely affected.

An important subdiscipline of anthropology in some counties is archeology, which provides a window, however blurry at times, on the past. An examination of the archaeological record reveals diverse human adaptations and innovations in response to climate stresses over the millennia, providing evidence that is relevant to contemporary human decision-making. The existing evidence suggests that climate change played a prominent role in the formation of various civilizations, the occupation or abandonment of different regions of the planet over time, and the sudden collapse of major civilizations and Indigenous societies. The archaeological record further shows that maintaining diversity and flexibility increases resilience to stress in complex adaptive systems, and that successful adaptations incorporate principles of sustainability. One lesson of these findings is that it is important for there to be community involvement in crafting, determining, and adopting measures for adaptation, not solely global or national top-down approaches. Moreover, focusing only on reducing carbon and other GHG emissions will not be sufficient to successfully address climate change, as such an approach will fail to address systemic social, political, and economic causes. As David Victor (2015: 27) stresses, the Intergovernmental Panel on Climate Change, which was created to provide policymakers around the world with regular scientific assessments on climate change,

> [M]ust overhaul how it engages with the social sciences in particular ... Fields such as sociology, political science and anthropology are central to understanding how people and societies comprehend and respond to environmental changes, and are pivotal in making effective policies to cut emissions and collaborate across the globe.

Anthropology's entry into the field of climate change has been relatively slow until recently. When Margaret Mead (1980), long a leading figure in anthropology, was a Visiting Scholar at the Fogarty International Center, she persuaded the Center in 1975 to sponsor a conference that would explore strategies to contribute to a healthy atmosphere. Mead appears to have been the only anthropologist to attend the conference and perhaps the only social scientist too, in a meeting that primarily drew physical and natural scientists as well as public health experts. While at the time Mead did not directly encourage her colleagues to start working on climate change, her involvement in the conference foreshadowed the beginning of the anthropology of environmental change and ultimately of climate change.

During the 1990s, two cultural anthropologists, Steve Rayner (1991) and Mary Douglas (see Douglas et al. 1998), as well as two archaeologists, Carol Crumley (1994) and Brian Fagan (2000), began to lay the foundation for the anthropology of climate change. Since then, this field of anthropology has matured into a diverse and robust effort. Many anthropologists are now asking questions from a cultural ecological perspective while examining all facets of human-environment relations. The pace of this work accelerated during the first decade of the 21st century.

In terms of applied work, anthropologists have been looking at sustainability issues at two broad and quite distinct levels, both by participating in the formulation of environmental policies and by studying and becoming involved in sectors of the environmental movement, which advocates for social, technological, and economic changes leading to long-term sustainable practices.

Over time, greater numbers of anthropologists have become involved as observers and engaged scholars in applied initiatives, seeking to respond to environmental change at the local, regional, national, and global level. This requires us to work as advisors and partners with international climate regimes, national and state or provincial governments, community organizations, and environmental groups. Anthropologists and other social scientists are not seeking to become climate scientists. Natural scientists, similarly, are generally not in a good position to develop a detailed understanding of the ways social systems operate, or how they contribute to climate change. Efforts to examine and respond to the adverse impacts of human practice on nature and, conversely, of environmental degradation on humanity, must be a multidisciplinary initiative.

Anthropogenic Impact on a Small Blue Dot: Encountering the Effects of Greenhouse Gases on Earth

As noted, radical changes occurring on what astronomer Carl Sagan called a "small blue dot"—an image based on a picture of Earth taken at great distance by spacecraft Voyager—are the result of the buildup of heat-trapping gases in the atmosphere surrounding the planet. There are a number of such GHGs, and each can have different effects on Earth's warming.

The Global Warming Potential (GWP) system was developed to facilitate comparisons of the global warming impacts of different gases. Specifically, it is a measure of how much solar energy the emission of one ton of a gas will absorb over a given period of time (usually 100 years), relative to the emission of one ton of carbon dioxide. The larger the GWP, the more that a given gas warms Earth compared to carbon dioxide over that time period. GWPs provide a common unit of measurement that allows climate analysts to add up emission estimates of different gases (enabling development of a national GHG inventory), while helping policymakers to compare emissions reduction opportunities across economic sectors and relevant gases. The following are the most important GHGs:

- Carbon dioxide, by definition, has a GWP of one because it is the gas being used as the reference point. Carbon dioxide remains in the atmosphere for a very long time, from 300 to 1,000 years. The concentration of carbon dioxide in Earth's thin atmosphere is about 412 parts per million and rising. This represents a jump of almost 50% since the beginning of the Industrial Revolution. There has been an 11% increase in carbon dioxide since 2000 (Buis 2019).

- Methane, a powerful GHG, is estimated to have a 100-year GWP of between 27–30. Methane emitted today will remain in the atmosphere for about a decade on average, a much shorter time than carbon dioxide. But methane also retains much more energy than carbon dioxide. The net effect of the shorter lifetime and higher energy absorption is reflected in the gas's high GWP. The methane GWP, moreover, also accounts for some indirect climate effects of the gas, such as the fact that it is a precursor to the formation of ozone, which is itself a GHG. Methane comes from both natural sources and human activities, but it is estimated that 60% of today's methane emissions stem from human activities. The largest sources of methane currently are agriculture, fossil fuels, and decomposition of landfill waste (NASA 2022).

- Nitrous oxide (aka laughing gas) has a GWP 273 times that of carbon dioxide for a 100-year timescale. Nitrous oxide emitted today will remain in the atmosphere for about 114 years on average. Agriculture is one sector that produces atmospheric nitrous oxide. Nitrous oxide emissions can become part of a "climate feedback loop." As Earth warms, nitrous oxide emissions increase, since warmer and wetter conditions stimulate dentification, which is the microbial process of reducing nitrate and nitrite into gaseous forms of nitrogen. These additional nitrous oxide emissions further contribute to global warming, creating a vicious circle. Nitrous oxide is also a by-product of wastewater treatment, as a consequence of active sludge processes used to speed up waste decomposition (Brind'Amour and Lee 2022).

- Chlorofluorocarbons, hydrofluorocarbons, hydrochlorofluorocarbons, perfluorocarbons, and sulfur hexafluoride are sometimes called high-GWP gases because, for a given amount of mass, they trap substantially more heat than carbon dioxide. The GWPs for these fluorocarbons can be in the thousands or tens of thousands. These GHGs are emitted from a variety of household, commercial, and industrial applications and processes. Fluorinated gases typically are emitted in smaller quantities than other GHGs, but they are potent promoters of climate change (EPA 2023).

Beyond this set of GHGs, there are several other phenomena that cause global warming:

- Ground-level ozone, commonly called the "bad" ozone that forms close to Earth in the air we breathe. By contrast, stratospheric ozone, or "good" ozone, occurs naturally in the upper levels of the atmosphere, where it forms a shield that protects us from the Sun's harmful ultraviolet rays. Ground-level ozone is created through a chemical reaction involving emissions of nitrogen oxides and volatile organic compounds (VOCs) from automobiles, power plants, and other industrial and commercial sources in the presence of sunlight. Many VOCs are human-made chemicals that are used and produced in the manufacture of paints, pharmaceuticals, and refrigerants. Typically, they are employed as industrial solvents (e.g., trichloroethylene), fuel oxygenates (e.g., methyl tert-butyl ether) or by-products produced by chlorination in water treatment (e.g., chloroform). VOCs often are components of petroleum fuels, hydraulic fluids, paint thinners, and dry-cleaning agents. They also are widely used as ingredients in household products like paints, varnishes, and many cleaning, disinfecting, cosmetic, degreasing, and hobby products. All these products can release organic compounds while in use, and to some degree when they are stored. In addition to trapping heat, ground-level ozone is a pollutant that can cause respiratory health problems and damage crops and ecosystems.

- Water vapor is another GHG. Indeed, it is the most abundant one. Water vapor plays a key role in climate feedback because of its heat-trapping capacity. Warmer air is able to hold more moisture than cooler air. As a result, as GHG concentrations increase and global temperatures rise, the total amount of water vapor in the atmosphere also rises, further amplifying the warming effect. Increased water vapor does not *cause* global warming. Rather, it is a consequence of it. But increased water vapor in the atmosphere amplifies the warming caused by other GHGs.

- Aerosols in the atmosphere also can affect climate. Aerosols are microscopic (either solid or liquid) particles that are so small that instead of quickly falling to Earth's surface like larger particles do, they remain suspended in the air for days or even weeks. Human activities, like burning fossil fuels and biomass (e.g., soy used to make gasoline, wood, and animal manure), contribute to emissions of small particle aerosols, although some aerosols come from natural sources such as volcanoes and marine plankton. Unlike GHGs, the climate effects of aerosols are variable depending on what they are made of and where they are emitted. Depending on their color and other factors, aerosols can either absorb or reflect sunlight. When they absorb sunlight, they actually have a cooling effect on Earth's temperature.

- Black carbon, a kind of aerosol, is soot emitted from gas and diesel engines, coal-fired power plants, and other sources that burn fossil fuel, as well as from forest fires. When this soot falls onto and darkens the surface of snow and ice, it causes these normally reflective white surfaces to become powerful absorbers of infrared radiation, and hence impactful producers of global warming. Different types of soot contain different amounts of black carbon because the blacker the soot, the higher its efficiency as a warming agent. Fossil fuels and biofuel soot are blacker than soot from biomass, which generally is more of a brownish color. At the same time, black carbon melts the ice beneath it, contributing to the loss of land ice sheets and glaciers (and, hence, to sea level rise). Research indicates that black carbon may be responsible for more than 30% of recent warming in the Arctic (Shindell and Faluvegi 2009). When suspended in air, black carbon absorbs sunlight and generates heat in the atmosphere, which in turn warms the air sufficiently to affect regional cloud formation and precipitation patterns.

It is clear that multiple factors, including many human behaviors, especially activities shaped by corporate production, advertisement, and distribution, and government policy, lead to the use of various manufactured products that, in interaction with natural processes, produce climate change.

Cataloging Five Fundamental Planetary Changes

What are the key features and expressions of climate change? While there are many such expressions, here we focus on five of those that are having the gravest impacts.

Rising Temperatures

The term climate change is most commonly invoked to describe dramatic and enduring changes in temperature and the related consequences of this shift. These alterations could occur at the local, regional, or global level. An ice age, for example, involves extensive cooling, resulting in the expansion of continental and polar ice sheets and alpine glaciers. There have been at least five major ice ages in Earth's history. Outside of these periods, Earth appears to have been ice-free even in high latitude regions. By this definition, and as is generally agreed among climatologists, Earth currently is in a comparatively mild ice age which has been scientifically named the Holocene. We are, however, enduring a period of increasingly intense global warming, and this section is concerned with the nature, features, and magnitude of this change and the other planetary transformations it unleashes.

While noting a long slow rise in planetary temperatures since modern humans (*Homo sapiens*) first evolved, climate scientists point to the Industrial Revolution as marking a significant upward turning point in Earth's global temperature. Since the Industrial Revolution, which accelerated during the 1830s and 1840s in Britain, and quickly spread to other parts of the world, the global annual temperature has increased in total by a little more than 1 degree Celsius (2 °F). In the 100 years between 1880, the year that accurate recordkeeping started, and 1980, planetary temperatures increased every ten years by 0.07 °C (0.13 °F) on average (NASA Earth Observatory 2023). Since 1981, however, the rate of increase has more than doubled. Until recently, nine of the ten hottest years on record since 1880 occurred since 2005, and the five warmest years have all occurred since 2015. In the last few years, the planet has grown even hotter. During September 2023, global surface temperature was 1.44 °C (2.59 °F) above the 20th-century average of 15.0 °C (59.0 °F), making it the warmest September ever recorded. Further, September 2023 marked the 49th consecutive September and the 535th consecutive month with temperatures at least somewhat above the 20th-century average. In fact, the past ten Septembers (2014–2023) have been the warmest Septembers since recordkeeping began (National Oceanic and Atmospheric Administration 2023).

With rising temperatures have come killer heat waves. A heat wave in 1995 killed 700 people in Chicago. More than 70,000 Europeans perished in the summer of 2003, while 55,000 succumbed to heat in Russia in 2010. More recently, it is estimated that 1,400 people died across Oregon, Washington, and British Columbia during the 2021 heat dome, and about 60,000 lost their lives because of extreme heat across Western Europe in 2022. It is estimated that thousands more have died in the heat waves that have hit the Global South, where the lack of public health capacity and reporting obscures the toll. As the world heats up, it is expected that many places on the planet will become uninhabitable for humans.

Earth warmed to the highest temperature ever recorded when the average global temperature reached 17.18 °C (62.92 °F) during the Fourth of July celebrations in 2023 in the United States (Climate Reanalyzer n.d.). This might not sound very hot for summertime, but it is a measure for the whole surface of planet Earth, including regions experiencing winter. The high temperature in Phoenix, Arizona, that day was a hellish 42.2 °C (108 °F). Patients with heat stroke and burns from the scorching asphalt swamped local hospitals and the office of the city's medical examiner was forced to deploy trailer-sized coolers to store the bodies of heat stroke victims.

Surface temperatures on Earth actually mask the true scale of climate change, because the oceans have absorbed about 90% of the solar heat trapped by GHGs. Measurements that have been collected over the last six decades by oceanographic expeditions and networks of floating instruments show that from bottom to top, every layer of the oceans is heating up.

Melting Ice Caps, Glaciers, Sea Ice, and Tundra

Most (99.5%) of the permanent ice volume in the world is locked up as land-based ice sheets and glaciers. An ice sheet is a mass of crystalline ice that is more than 19,000 miles in circumference (50,000 square kilometers). As ice sheets extend to the coast and then over part of the ocean, they are called ice shelves. A mass of glacial ice covering less area than an ice sheet is called an ice cap, while a group of connected ice caps form an ice field. Making up ice fields, ice caps, and eventually ice sheets, are individual glaciers. A glacier is a large, perennial accumulation of ice, snow, rock, and sediment that originates on land and moves down a mountain slope under the influence of its own weight and the pull of gravity.

Today, there are only two ice sheets in the world: in Greenland and in Antarctica. During the previous glacial period, however, much of Earth was covered by ice sheets. These massive fields of ice form like other glaciers. Snow accumulates each winter then melts to some degree in the summer. The slightly melted snow gets harder and slowly compresses. In the process, the snow changes texture from a fluffy powder into a block of hard, round ice pellets. New snow the next winter buries the grainy snow. This causes the hard snow underneath to become even denser. At this point it is known as "firn." Over the years, layer upon layer of firn build on top of one other. When the ice grows thick enough—about 165 feet (50 meters)—the firn grains fuse into a huge mass of solid ice. The oldest glacier, located in Antarctica, may be almost 1,000,000 years old. By comparison, the age of the oldest glacier ice in Greenland is not much more than 100,000 years old. Ice sheets tend to be slightly dome-shaped and spread out from their centers. Scientists have observed that they behave plastically, or like a liquid. An ice sheet flows and slides over uneven surfaces until it covers everything lying in its path, including valleys, mountains, and sections of oceans.

With global warming, Earth is experiencing extensive and increasingly rapid ice melt. The Greenland ice sheet has been losing total mass for over 20 years. Recent estimates suggest that from 2012 to 2016 the Greenland ice sheet lost about 250 gigatonnes (1 gigatonne = 1 billion tons) per year of ice volume. The total mass balance of Greenland has been increasingly negative since 1995, and it is now equivalent to the global contribution to sea level rise from the simultaneous melting going on among glaciers and ice caps worldwide.

While 60% of ice mass loss in Greenland is through discharge across the front of the ice facing the ocean (as sea ice or melting into the ocean), 40% is from surface melt. Increases in surface melt (called ablation) are primarily responsible for the increasing melting in Greenland. These increases in surface melt and mass losses from Greenland are caused by rising winter and summer air temperatures. This ongoing process is leading to a significant drop in the Greenland ice sheet surface elevation, and a decline in total ice volume.

Ice discharge from the major outlet glaciers of the Greenland ice sheet has also increased, especially in western Greenland. This faster ice flow leads to outlet glaciers discharging more ice volume to the ocean than is replaced by new snow, resulting in the outlet glaciers also thinning.

Melt is also occurring in Antarctica. As global temperatures rise, seawater is eating away at the underbelly of the Antarctic ice sheet covering the ocean, forcing the grounding line to retreat, and speeding the decline of Antarctica's glaciers. Antarctica is losing ice mass at an average rate of about 150 billion tons per year. West Antarctica is the continent's largest contributor to global sea level rise and has enough ice to raise sea levels by more than 17 feet (5.3 meters). It is home to the Thwaites Glacier, also known as the "Doomsday Glacier" because its collapse—which could happen in a few years—would raise sea levels by several feet, flooding coastal communities and low-lying island nations around the world (Naughten et al. 2023).

Like ponds and lakes in northern parts of the planet, oceans develop a layer of ice on their surfaces during cold winter months. Because of its salinity, ocean water has a lower freezing point than fresh water. But if it is it is chilled to about –2 °C (28 °F), the salty liquid begins to solidify. Ice crystals form on the sea surface, and if the air is sufficiently cold, the crystals begin to expand, forming a slushy mix. Eventually, if the weather remains cold enough, a solid covering of ice thickens over time.

In the Arctic Ocean, the area covered by sea ice grows and shrinks over the course of each year. In the fall, as less sunlight reaches the Arctic and air temperatures begin to drop, additional sea ice forms. The total area covered by ice increases throughout the winter, usually reaching a maximum size in early March. Once spring arrives, bringing more sunlight and higher temperatures, the ice begins to melt, shrinking to its minimum extent by September.

In winter, sea ice covers about 7% of Earth's surface and about 12% of the world's ocean surfaces. Much of the world's sea ice is found in the Arctic, Antarctica, and the South Ocean, the southernmost waters of the world ocean. Due to the action of winds, currents, and temperature fluctuations, sea ice is very dynamic, leading to a wide variety of ice types and features. Sea ice that forms in the ocean can be contrasted with the chunks of ice that calve off from ice shelves or glaciers.

The bright, shiny surface of sea ice serves a role in maintaining cooler polar temperatures by reflecting back into space much of the sunlight that hits it. As the sea ice melts, its surface area shrinks, diminishing the size of the reflective surface and therefore causing Earth to absorb more of the Sun's heat. As the ice melts it lowers heat reflection causing more heat to be absorbed by Earth and further increasing the amount of melting ice in a feedback loop. Though the size of ice floes is affected by the seasons, even a small change in global temperature can significantly affect the amount of sea ice, causing a shrinkage of the reflective surface that keeps the ocean cool. As a result, sea ice melting sparks a cycle of ice shrinking and warming temperatures, a transformative potential that makes the polar regions the most susceptible locations for climate change on the planet.

Sea ice provides an ecosystem for various species, particularly polar bears whose environment is now being threatened as global warming causes the ice to melt. Moreover, sea ice affects the movement of ocean waters. During the freezing process, much of the salt in ocean water is squeezed out of the frozen crystal formations, though some remains frozen in the ice. This salt becomes trapped beneath the sea ice, creating a higher concentration of salt in the water beneath ice floes. This concentration of salt contributes to the salinized water's density and this cold, denser water sinks to the bottom of the ocean. This cold water then moves in a current along the ocean floor towards the equator, while warmer water on the ocean surface moves towards the poles. This natural process is referred to as "conveyor belt" motion and is a regularly occurring event. Movement of ocean currents is important for several reasons including that it helps to regulate global climate by counteracting the uneven distribution of solar heat reaching Earth's surface.

Satellite data have documented ongoing declines in sea ice since continuous satellite-based measurements began to be recorded in November 1978. The data show a trend of more ice melting away during summers and less new ice forming during winters. The Beaufort Gyre, a looping current north of Alaska, historically acted as a nursery for young ice, enabling it to thicken and grow. Since the beginning of the 20th century, however, summers in the southern portion of the gyre have been too warm for sea ice to survive.

Loss of sea ice also leads to a loss of habitat for walruses and seals, leading to disruptions in the food supply of Indigenous Arctic communities that depend on these animals for portions of their diets and their acquisitioning as a reaffirmation of their cultural identities. A longer ice-free season exposes Arctic communities to extreme erosion along the coast from pounding waves during winter storms. Indigenous homes and other structures that stand near the advancing erosion are at risk of collapsing into the ocean. This process has made whole communities vulnerable, leading some to uproot and move to new locations, an economically and emotionally difficult undertaking.

The coldest land-based biome is known as tundra. It is characterized by extremely low temperatures, low precipitation, poor soil nutrients, and abbreviated growing seasons (50 to 60 days). The average winter temperature is $-34\,°C$ $(-30\,°F)$. Tundra contains dead organic material in the form of nitrogen and phosphorus. It also consists of a layer of permanently frozen subsoil called permafrost. Arctic tundra is located in the northern hemisphere, encircling the north pole and extending south to the upper edge of coniferous forests of the subarctic region. Most of the world's permafrost is found in northern Russia, Canada, Alaska, Iceland, and Scandinavia.

Global warming is now rapidly melting permafrost, turning vast swaths intro mud, silt, and peat, and releasing massive amounts of climate-warming GHGs—including methane, carbon dioxide, and nitrous oxide —that have been stored in the permafrost for millennia. GHG release from the tundra occurs in two ways. As permafrost melts, once-dormant microorganisms begin to break down organic matter, allowing methane and carbon to be released in the atmosphere. Thawing can also open pathways for methane to rise up from reservoirs deep in the ground. Villages, towns, and even cities are built on tundra. When the permafrost melts, buildings break apart and even sink below the surface. Cracking and collapsing structures, for example, are a growing problem in places like Norilsk—a nickel-producing city of 177,000 people located 180 miles above the Arctic Circle in western Russia. 60% of all buildings in Norilsk have been damaged as a result of climate change shrinking the permafrost zone. More than 100 residential buildings, or about one-tenth of the housing stock, has had to be vacated because of damage from thawing permafrost. On May 29, 2020, a fuel-storage tank cracked open, spilling 21,000 tons of diesel fuel into nearby waterways and turning the Ambarnaya River a metallic red. It is now feared by many climatologists that as much as 2.5 million square miles of permafrost—40% of the world's total—could melt by the end of the 21st century.

Intensifying Wildfires and Hurricanes

Wildfires

Intense record-breaking forest fires during California's summer months have become a regular occurrence. Wildfires not only cause catastrophic environmental and socioeconomic damage, but they also have harmful consequences for health, human and nonhuman alike. Burned areas in northern and central California increased fivefold between 1996 and 2021, compared to between 1971 and 1995. Further, ten of the largest California wildfires have occurred in the last 20 years, five of which were in 2020 alone. The 2021 fire season in California included four blazes that ranked in the top 20 largest wildfires on record in the state.

Earth, Atmospheric, and Planetary Sciences, an international group of environmental researchers, developed a climate-driven computer model of burned area evolution in California and combined it with natural and historical climate simulations to assess the importance of human-caused climate change on burned areas in the state. The study found that nearly all the observed increase in burned areas over the past 50 years was due to human-caused climate change. It is estimated that from 1971 to 2021 anthropogenic climate change contributed to an increase of over 170% in burned areas, with an increase of over 320% in just five years from 1996 to 2021. Based on expected increases in global warming in the coming decades, a further jump in annual forest burned areas is expected, ranging from 3% to a worrisome 52% (Turco et al. 2023).

Wildfires like those seen in California are becoming more intense and more frequent, burning down whole communities and vast ecosystems. Recent years have witnessed record-breaking wildfires around the world, from Australia to the Arctic, and in both North and South America. Beginning in March 2023, and with increased intensity starting in June, Canada was impacted by a record-setting series of wildfires that affected all 13 provinces and territories. In 2023 Canada experienced its most severe wildfire season in recorded history when wildfires burned about 5% of all forested land in Canada. Smoke emitted from the wildfires caused air quality alerts and evacuations in Canada and in the United States. Similarly, in Australia, the 2019–2020 bushfire season was characterized by unusually intense, wide-spread, and long-burning wildfires. This was because of a climate change-produced increase in the frequency and severity of what scientists call "fire weather," periods with a high fire risk involving a combination of hotter temperatures, low humidity, low rainfall, and strong winds. Hundreds of fires burned, mainly in the southeast of the country, until May 2020. The most severe fires reached a peak between December 2019 and January 2020. Together, the fires burnt approximately 1,000 square miles (243,000 square kilometers), destroying over 3,000 buildings and killing at least 34 people (Green 2020). It is estimated that three billion terrestrial vertebrates, including many plants and animals from endangered species, such as the Kangaroo Island dunnart, a mouse-like marsupial, and the nightcapped oak tree, did not survive the catastrophe.

Hurricanes

Although they are called by many names depending on where you live on Earth (hurricanes, typhoons, cyclones), scientists refer to all these storms as tropical cyclones. This is because they are large rotating systems that need tropical conditions to form, and they mostly originate in the tropics. Techni-cally, hurricanes/cyclones are rotating storms that have winds of more than 74 miles per hour (about 120 kilometers per hour). Hurricanes/cyclones need four things to form and gain strength: warm ocean water; masses of warm air; low vertical wind shear (which is the change in wind speed and/or direc-tion as you travel upwards in the atmosphere); and a trigger set of thunder-storms. If a hurricane/cyclone moves close to or over land, it brings with it a devastating combination of very strong winds, drenching levels of rain, inrushing surges of ocean water onto the coast, and, in some cases, tornadoes.

Scientists have long predicted that climate change will lead to an increase in extreme rainfall events like hurricanes/cyclones. As surface temperatures rise, more liquid water evaporates from both the land and the ocean. Evaporation adds moisture to the air depending on the temperature. Warmer air can hold more moisture. The increased moisture in the air leads to more intense rain-fall during extreme events. Modeling studies project an increase on the order of 10%–15% for rainfall rates in the coming years. In a hurricane/cyclone, spiraling winds draw moist air towards the center, fueling the towering thun-derstorms that surround it. As the air continues to warm due to climate change, hurricanes/cyclones can hold more water vapor, producing more intense rainfall. Moreover, most computer models show that climate change brings a slight increase in hurricane/cyclone wind intensity. This change appears to be related to warming ocean temperatures and more moisture in the air, both of which fuel hurricane/cyclones. The global proportion of tro-pical cyclones that reach very intense levels (category 4 and 5) is projected to increase due to anthropogenic warming during the 21st century. This change includes a major increase in the destructive potential of each storm.

The southern and eastern coastal and adjacent inland areas of the United States have increasingly been put in harm's way because of hurricanes/cyclones. From 2012 to 2022, the six costliest hurricanes/cyclones on record for the United States occurred. Many of the most damaging hurricanes/cyclones to impact the United States in recent years have been notable for the speed at which they have intensified. For instance, Hurricane Maria (in 2017), the climate disaster with the highest death toll since 1980, and the fourth highest hurricane-related economic cost in the past four decades, strengthened from a tropical storm to a category 5 hurricane in just over 48 hours. Hurricanes Harvey (in 2017), Ian (in 2022), Sandy (in 2012), Ida (in 2021), and Irma (in 2017), the five other costliest US climate disasters in the last decade, all similarly strengthened rapidly, with most evolving from tropical storms to major hurricanes/cyclones in less than three days.

Human-induced climate change is causing more frequent and intense hurricanes/cyclones. For example, on the west coast of India, they originate in the eastern Arabian Sea. However, one of the strongest and most disastrous hurricanes/cyclones, called Cyclone Amphan, hit the eastern Indian states of Bengal and Odisha, as well as neighboring Bangladesh in May 2020. This catastrophic storm caused over 100 deaths and damage valued at US $13 billion. At its peak, this deadly cyclone had a maximum recorded wind speed of 150 miles per hour (240 kilometers per hour). In 2022, Hurricane Sandy was another especially extreme weather event. It formed in the Atlantic, gaining intensity and speed from the warm water as it passed into the Caribbean Sea. The storm first made landfall in Jamaica, Cuba, and the Bahamas, where its impact was devastating. It then continued northeast up the Atlantic coast and made landfall again near Atlantic City, New Jersey. In New York City, it caused a storm surge that flooded streets and the subway system, leading to extensive power outages. It was the largest Atlantic hurricane on record as measured by diameter, with tropical storm-force winds spanning 1,150 miles (1,850 km). The storm inflicted nearly $70 billion (in 2012 US dollars) in damage and killed 233 people across eight countries from the Caribbean to Canada.

Species Shifts and Loss of Biodiversity

The warmer weather brought on by climate change is reshaping the world of many plant and animal species. Plants are now blooming earlier or expanding into what historically were cooler locations, such as higher elevations, while retreating from places that have become too hot for them. This change in plant range is possible because of reproductive strategies. To reproduce, plants generate large quantities of seeds, which, with assistance from the wind and animals, spread in every direction. The vast majority land in inhospitable places and die, but some seeds survive and grow into new plants. Given enough time, this uneven "recruitment" (the birth of new individuals in a population) into new places can result in a change in the locations where a plant species is found. Working together, differences in mortality and recruitment are causing trees and other plant species to migrate to higher elevations, often in more northernly areas, and to places farther away from the tropics. However, Benjamin Freeman, a biologist at the Georgia Institute of Technology, coined the phrase "escalator to extinction" to describe the threat faced by mountaintop species that have adapted to cooler conditions and increasingly have nowhere to go as the climate becomes too warm.

Further, while many plant species are moving away from hot places, often they are not migrating quickly enough to escape rising temperatures. Various human activities, such as the construction of roads, dams, and cities, get in their way while destroying the natural areas on which species depend. These entities function as dispersal limitations that make it harder for both plants and animals to move. This can lead to species becoming "trapped" in places that have grown too hot for them.

Heat waves are important climatic extremes, not only on land but in the sea as well, and have devastating and long-term impacts on ocean ecosystems. Annual marine heat waves have increased by over 50% during the last century. Eight of the ten worst marine heat waves on record occurred since 2010. These events are causing massive die-offs of marine plants like kelp and seagrass, as well as of coral, fish, seabirds, and ocean-dwelling mammalian species.

Even if plants can shift home ranges fast enough to escape rising temperatures, they still may be at risk of extinction. Plants and animals depend on each other to survive. If some plant species migrate quickly but the animal species that they rely on have slower rates of migration or cannot migrate at all, the plants will have a difficult time surviving. Also, as species migrate to colder places, these areas can get crowded and the amount of available land to grow on will decrease. As Earth continues to warm, there may be no available cooler locations remaining for slower moving plant species to inhabit (Freeman et al. 2021).

In places that are already quite warm, like the Amazon rainforest, temperatures may become so hot that no new species can move in, leaving only a small group of plant species that favor high heat. Consequently, in some places, the total number of plant and animal species will decrease over time, leading to biotic attrition and significantly decreasing biodiversity. Given that tropical forests limit global warming, so will their loss accelerate the rate of planetary warming.

As suggested, animal species are also facing climate-driven migration challenges. For example, many duck species in the US Midwest have been shifting further north in the winter and are much more frequently overwintering there rather than migrating back south in the fall. A study by the National Audubon Society predicts massive reductions in future climate suitability of habitat for hundreds of North American birds as the climate continues to warm (Fritts 2021). Research also shows that animal predator species seemed to respond to climate change differently than their prey species. This is causing a mismatch between predators and the prey that they hunt in order to survive. In the ocean, species like the North American right whale teeter on the brink of extinction. There are only an estimated 336 individual right whales remaining in North America, the lowest count in 20 years. A warming ocean, coupled with a failure to decrease conflicts with humans (such as vessel strikes and entanglement in fishing gear), appears to be driving this species to extinction.

Similarly, Africa's national parks, home to thousands of wildlife species, including iconic species like lions, elephants, and water buffaloes, are increasingly threatened by climate-driven, below-average rainfall, as well as by new human-built infrastructure projects, stressing habitats and the species that depend on them. A prolonged drought in much of the continent's east, exacerbated by climate change and large-scale developments, including oil drilling and livestock grazing, are putting animal species at risk.

As a consequence of these kinds of changes, Earth is enduring a loss of biodiversity, the variety of life, in all its forms, from bacteria to entire ecosystems, such as forests and coral reefs. Generally, the more species that live in an area or ecosystem, the more biodiverse it is. This web of complexity creates healthy ecosystems and makes Earth the perfect place for us and all our co-habitants to live on. The biodiversity now found on the planet is the result of 4.5 billion years of evolution and it forms the web of life that humans depend on for food, clean water, medicine, a hospitable climate, and economic growth, among others. Up to one million species, however, are threatened with extinction because of climate change, many within a few decades.

Transgressing Planetary Boundaries

The planetary boundaries framework draws upon Earth system science. It identifies nine processes that are critical for maintaining the stability and resilience of the Earth system, which includes all the planet's interacting physical, chemical, and biological processes. All boundaries are now heavily disturbed by human activities. Humans have pushed Earth outside of the Holocene's window of environmental comfort and variability, giving rise to what has been proposed as the Anthropocene (human-influenced) epoch.

The planetary boundaries framework establishes limits to changes in the Earth system by identifying a scientifically based safe operating space for humanity that can safeguard both Earth's interglacial state and its resilience. Understanding how biosphere, anthroposphere, and geosphere processes interact with one another is a prerequisite for developing reliable projections of possible future Earth system trajectories. The nine boundaries all represent components of the Earth system critically affected by anthropogenic activities and relevant to Earth's overall state. For each of the boundaries, control variables are chosen to capture the most important anthropogenic influence at the planetary level of the boundary in focus. Boundary positions do not demarcate or predict singular threshold shifts in the Earth system state. They are placed at a level where the available evidence suggests that further perturbation of the individual process could potentially lead to systemic planetary change by altering and fundamentally reshaping the dynamics and spatiotemporal patterns of geosphere-biosphere interactions and their feedbacks.

The nine planetary boundaries defined by Richardson et al. (2023) are:

1. Biosphere integrity: the photosynthetic flow of energy and materials into the biosphere (Earth's biomes).
2. Climate change.
3. Novel entities: the anthropogenic introductions of synthetic chemicals and substances (e.g., microplastics, endocrine disruptors, and organic pollutants); radioactive materials, including nuclear waste and nuclear weapons; and human modified organisms and other direct human interventions in evolutionary processes.
4. Stratospheric ozone depletion: the anthropogenic release of gaseous halocarbon compounds from industry and other human activities into the atmosphere leading to long-lasting depletion of Earth's ozone layer.
5. Fresh water change across the entire water cycle over land.
6. Atmospheric aerosol loading (e.g., desert dust, soot from wildfires).
7. Ocean acidification: drop in the carbonate ion concentration in surface seawater caused by the absorption of carbon dioxide.
8. Land system change: deforestation in tropical, temperate, and boreal forests.
9. Biochemical flows: changes in available nitrogen and phosphorus, two elements that are fundamental building blocks of life.

Six planetary boundaries were found by Richardson and colleagues to be transgressed (climate change, biosphere integrity, land system change, freshwater change, aerosol loading, and biogeochemical flows). Some of these were known in 2015 to have been surpassed but the degree of transgression has increased since then. These findings confirm that humanity is today placing unprecedented pressure on the life-enabling Earth system. Perhaps most worrisome is that all the biosphere-related planetary boundary processes that offer resilience (i.e., capacity to limit disruptions) to the Earth system are now at or very close to a high-risk level of transgression. Several regional climate tipping points, relevant for stabilizing the global system, have already been or are close to being transgressed, thus weakening global resilience capacity. This implies already low and falling resilience precisely at the very point when planetary resilience is needed to cope with increasing anthropogenic disturbances. These findings should serve as a clarion wake-up call that Earth is on the verge of radical changes that will usher in devastating consequences for its planetary inhabitants, including humans.

Conclusion

On the foundation presented in this chapter on the causes and disparate expressions of climate change, the chapters that follow examine several anthropological and related approaches to discerning the nature and impacts of climate change, its adverse interactions with other harmful anthropogenic disruptions on Earth, conventional, minimalist-change thinking on the alleged mitigation of global warming, and the mounting conceptual and applied anthropological effort to create a truly sustainable climate and environment.

References

American Anthropological Association 2015. Statement on humanity and climate change https://s3.amazonaws.com/rdcms- aaa/files/production/public/FileDownloads/pd fs/cmtes/commissions/CCTF/upload/AAA-Statement-on-Humanity-and-Climate-Change.pdf

Brind'Amour, Molly and Lee, Nation. 2022. Laughing gas is no joke: The forgotten greenhouse gas. Environmental and Energy Study Institute https://www.eesi.org/arti cles/view/laughing-gas-is-no-joke-the-forgotten-greenhouse-gas#:~:text=As%20the% 20Earth%20gets%20warmer,warming%2C%20creating%20a%20vicious%20circle

Buis, Alan. 2019. The atmosphere: Getting a handle on carbon dioxide. NASA https:// climate.nasa.gov/news/2915/the-atmosphere-getting-a-handle-on-carbon-dioxide/#:~: text=Carbon%20dioxide%20is%20a%20different,timescale%20of%20many%20human %20lives

Climate Reanalyzer. n.d. Daily surface air temperature. Climate Change Institute, University of Maine. https://climatereanalyzer.org/clim/t2_daily/?dm_id=world

Crumley, Carol, ed. 1994. *Historical Ecology: Changing Knowledge and Landscapes.* Santa Fe, NM: American School of Social Research

Donovan, Robin. 2023. Climate journalism needs voices from the Global South. *Eos* https:// eos.org/features/climate-journalism-needs-voices-from-the-global-south#:~:text=Climate %20research%20from%20scientists%20who,of%20the%20issue%20at%20hand

Douglas, Mary, Gaspar, DesNey, Steven and Thompson, Michael. 1998. Human needs and wants. In *Human Choice and Climate Change*, Vol. 1: *The Societal Framework.* Columbus, OH: Batelle Press

Earle, Steven. 2021. *A Brief History of the Earth's Climate: Everyone's Guide to the Science of Climate Change.* Gabriola Island, BC: New Society Publishers

Ensler, Eve. 2008. Eve Ensler on Sarah Palin. *Gambit* https://www.nola.com/gam bit/news/the_latest/eve-ensler-on-sarah-palin/article_cd53ea dc-1f21-5992-94bb-b55a94 f40f30.html

Fagan, Brian. 2000. *The Little Ice Age: How Climate Made History, 1300–1850.* New York: Basic Books

Foer, Jonathan. 2010. *Eating Animals.* Boston, MA: Back Bay Books

Freeman, B.G., Song, Y., Feeley, K.J. and Zhu, K. 2021. Montane species track rising temperatures better in the tropics than in the temperate zone. *Ecology Letters* 24: 1697–1708. doi:10.1111/ele.13762

Fritts, Rachel. 2021. Ducks are moving north as winters warm. *Audubon* https://www. audubon.org/news/ducks-are-moving-north-winters-wa

Green, Matthew. 2020. Australia's massive fires could become routine, climate scientists warn. Reuters https://www.reuters.com/article/us-climate-change-australia-report/austra lias-massive-fires-could-become-routine-climate-scientists-warn-idUSKBN1ZD06W

Hausfather, Zeke, Drake, Henri, Abbott, Tristan, and Schmidt, Gavin. 2020. Evaluating the performance of past climate model projections. *Geophysical Research Letters* 47(1): e2019GL085378

Kazmierczak, Jeanette. 2020. Are planets with oceans common in the galaxy? It's likely, NASA scientists find. NASA https://www.nasa.gov/centers-and-facilities/goddard/a re-planets-with-oceans-common-in-the-galaxy-its-likely-nasa-scientists-find/

Mead, Margaret. 1980. Preface. In *The Atmosphere: Endangered and Endangering.* William W. Kellogg and Margaret Mead, eds. pp. xvii–xxii. Turnbridge Wells: Castle House Publications

NASA. 2022. Methane https://climate.nasa.gov/vital-signs/methane/

NASA Earth Observatory. 2023. World of change: Global temperatures https://ea rthobservatory.nasa.gov/world-of-change/global-temperatures

National Oceanic and Atmospheric Administration. 2023. Topping the charts: September was Earth's warmest September in 174 years https://www.noaa.gov/news/topping-cha rts-september-2023-was-earths-warmest-september-in-174-year-record#:~:text="Sep tember%202023%20was%20the%20fourth,174%20years%20of%20climate%20keeping

Naughten, Kaitlin, Holland, Paul, and De Rydt, Jan. 2023. Unavoidable future increase in West Antarctic ice-shelf melting over the twenty-first century. *Nature Climate Change* doi:10.1038/s41558-023-01818-x

Pacini-Ketchabaw, V., Taylor, A., and Blaise, M. 2016. De-centring the human in multi-species ethnographies. In *Posthuman Research Practices in Education.* C. Taylor and C. Hughes, eds. pp. 149–167. Basingstoke: Palgrave Macmillan

Rayner, Steve. 1991. A cultural perspective on the structure and implementation of global environmental agreements. *Evaluation Review* 15: 75–102

Richardson, Katherine, Steffen, Will, Lucht, Wolfgang, Bendtsen, Jorgen, Cornell, Sarah, Donges, Jonathan*et al.* (2023) Earth beyond six of nine planetary boundaries. *Science* 9(37) doi:10.1126/SCIADV.ADH2458

Sawyer, J. 1972. Man-made carbon dioxide and the "greenhouse" effect. *Nature* 239 (5366): 23–26 doi:10.1038/239023a0

Shindell, Drew and Faluvegi, Greg. 2009. Climate response to regional radiative forcing during the twentieth century. *Nature Geoscience* 2: 294–300

Tsing, Anna. 2015. *The Mushroom at the End of the World: On the Possibility of Life in the Capitalist World.* Princeton, NJ: Princeton University Press

Turco, Marco, Abatzoglou, John, Herrera, Sixto, Zhuang, Yizhou, Jerez, Sonia, Lucas, Donald, AghaKouchak, Amir, and Cvijanovic, Ivana. 2023. Anthropogenic climate change impacts exacerbate summer forest fires in California. *PNAS* 120(25): e2213815120 doi:10.1073/pnas.2213815120

United States Environmental Protection Agency (EPA). 2023. Overview of greenhouse gases https://www.epa.gov/ghgemissions/overview-greenhouse-gases.

Victor, David. 2015. Climate change: Embed the social sciences in climate policy. *Nature* 520: 27–29

2

CONFLICTING ANTHROPOLOGICAL PERSPECTIVES

Cultural Ecological/Environmental Anthropological, Cultural Interpretive, and Critical Anthropological Perspectives of Climate Change

Introduction

In this chapter, we examine the three most commonly used theoretical perspectives that have emerged in the anthropology of climate change that seek to provide a framework for understanding various aspects of the human–climate change nexus over the course of the past century or so. These are the cultural ecological/environmental, the cultural interpretive, and the critical anthropological perspectives of climate change. While some of the debate over theory in anthropology is the product of the fact that different theoretical frameworks raise different questions, and lead to answers that may be deemed unsatisfactory to those who see other questions as being of equal or greater importance, disagreement also arises from different understandings of what anthropology is and its role among the disciplines and within society.

In his effort to explore the issue of explanation in science, philosopher Karl Popper (2002: 37) defined theories as "universal statements" about the nature of the world we live in. Anthropology, especially the subfield of cultural anthropology, lacks an overarching and unifying theory of the order of Darwinian evolution and natural selection in biology and related sciences. Instead, the field has tended to construct often contested alternative explanatory frameworks that draw attention to certain issues, focus work on answering particular questions, and assist anthropologists in explaining or interpreting their research findings. In writing up the findings of research, the theoretical frameworks in use by anthropologists may be stated explicitly or they may be implied by the types of questions addressed as well as the types of questions not addressed by the researcher. While the anthropology of climate change has not yet generated intense internal theoretical debates (but see Singer 2023), it has embraced contrasting and potentially conflicted understandings of the human

DOI: 10.4324/9781003469940-3

condition and the human dimension in climate change, as discussed later in this chapter. The leading alternative anthropological frameworks for understanding climate change and society are presented below.

The Cultural Ecological/Environmental Anthropological Perspective of Climate Change

Cultural ecology or ecological anthropology examines human-environment relations and has a long history in anthropology, dating back 70 years ago to the work of Julian Steward (1955). It also draws on the field of human ecology, and while denying the simplistic environmental determinism of an earlier era, recognizes that environments play a vital role in shaping cultures and providing possibilities and setting constraints on what they are likely to do. For example, irrigation agriculture is much more likely to develop in areas with flowing rivers than in arid environments without immediate water sources (although intensive and expensive hydraulic systems are used in some places to move water to dry regions to support agricultural production). There is a strong tendency in cultural ecology to see culture as an adaptive mechanism, although certainly many cultural ecologists recognize that culture can exhibit environmentally maladaptive dimensions as well. Moreover, cultural ecology views each individual cultural system as having evolved in and adapted to specific landscapes, such as savannahs, tropical rainforests, or Arctic tundra environments. In other words, the cultural ecological perspective tends to view culture and its various techno-economic, social structural, ideological, and attitudinal components as expressing collective human engagement with challenges and opportunities in local environments and to ask questions directed at expanding understanding of the adaptive process. According to Milton (1996: 59), the "ecosystem approach developed by ecologists in the 1940s and 1950s, and adopted into anthropology in the 1960s, held no place for an understanding of culture in its narrower sense—people's thoughts, feelings and knowledge about the world." As this statement suggests, the framework of cultural ecology does not easily lend itself to asking questions—questions some anthropologists might see as important—such as how people's attitudes about the world shape their engagement with it, or the way in which knowledge about the environment is structured within a cultural system or dispersed within a society across gender or other social divisions. Cultural ecologists generally have been somewhat slow in fully coming to grips with the far-reaching impact of climate change on human societies.

Historical ecology has served as a precursor to the cultural ecology of climate change. The School of American Research (now known as the School of Advanced Research) in Santa Fe, New Mexico, held an advanced seminar on historical ecology in October 1990. Several of the chapters in the book *Historical Ecology* that resulted from the seminar touch on topics related to climate change. Archaeologist Carol Crumley (1994: 7), for example, defines historical ecology as the "practice of globally relevant archaeology, ethnohistory, ethnography, and related disciplines." She refers to global warming along with pollution, species extinctions, and massive disruptions of critical ecosystems as manifestations of environmental change that is "arguably the most pressing and potentially disastrous facing the global community" (Crumley 1994: 1). Furthermore, Crumley (1994: 8) maintains that historical ecology has the potential to address several questions related to needed global action on the global ecological crisis, including climate change. Based on ethnohistorical research on late prehistoric and early historic Europe, Crumley proposes an interesting hypothesis about the relationship between social structures and climate change:

> Periods of stable climate (whether hot or cold, wet or dry, or even consistently unpredictable) allow humans the opportunity to experiment and convey the results to succeeding generations in the form of successful strategies given to a particular set of conditions ... Periods of unpredictable climate require much greater flexibility on the part of human populations in their utilization of resources: also required is a much larger store of potentially useful information.
>
> *(1994: 192)*

While some anthropologists working on climate change draw heavily on cultural ecology, others draw heavily on political ecology, an approach that examines how and why economic structures and power relations drive environmental change in an increasingly interconnected human-built world.

Although in their Introduction to *Cultural Ecology* Mark Sutton and E.N. Anderson (2004: xiii) assert that "[o]ne of the goals of most cultural ecological work is to use the knowledge in an effort to stem global catastrophe," they only briefly mention global warming in passing, observing that "deforestation releases massive amounts of greenhouse gases as the trees are burned or allowed to decay. These gases have been a factor in global warming, although fossil fuels are a far more serious cause" (Sutton and Anderson 2004: 298). In our view, a more comprehensive approach involves assessment of pluralea interaction (Singer 2021) between anthropogenic deforestation and fossil fuel emission, as both of these human activities are changing the planet, often in tandem. Within the anthropology of climate change, there are various anthropologists who are asking the questions raised by a cultural ecological understanding.

As part of their extensive Arctic Climate Impact Assessment project, for example, Nuttall et al. examined the impact of climate change on subsistence patterns and adaptive strategies of Indigenous Arctic peoples in the past and present. They argue that "as the climate changes, the Indigenous peoples of the Arctic are facing special challenges and their abilities to harvest wildlife and food resources are already being tested" (Nuttall et al. 2004: 685). As this statement suggests, in questions about human-environment interaction, attention must be paid not only to changing human behaviors (e.g., adoption of new adaptive strategies), but to changing environments as well. In other words, human communities are not merely adapting to existing environment niches; indeed, the environment itself is a dynamic arena, and humans are increasingly critical players in forcing the directions of environmental change.

Ben Orlove (2005) examines climate variability in three frequently mentioned historical cases, namely the Mayan civilization of Mesoamerica, the Norse settlements in Greenland, and the US Dust Bowl. He asserts that while these three societies were remarkably different from each other, they all experienced climatic conditions that threatened their basic subsistence patterns. Orlove (2005: 596) seeks to blend the sociology of the future with comparative historical analysis: "The former offers broad concepts, particularly adaptation and mitigation, which can contribute to the urgent debates of the present. The latter permits us to trace the complex interactions of different elements within each society." Orlove believes that the possibility of migration provides a bridge between the sociology of the future and the comparative history of the past. Orlove, who is based at Columbia University, along with fellow social scientists Arun Agrawal, Maria Lemos, and Jesse Ribot, created the Initiative on Climate Adaptation Research and Understanding through the Social Sciences (ICARUS), a body that seeks "to bring together researchers and practitioners to address the growing need for social-scientific contributions to address climate change" (Agrawal et al. 2012: 330).

Focusing on three primary research areas, namely (1) theorizing key concepts such as vulnerability, adaptation, adaptive capacity and resilience; (2) understanding the causal structures of vulnerability, effects of adaptation, and the empirical referents of both adaptation and vulnerability at various scales; and (3) understanding and informing adaptation policy, ICARUS has organized a series of major conferences as well as smaller seminars (which they call writeshops), and related network-building activities. To date, ICARUS's founders argue, social scientists have made significant contributions to thinking about climate change. Included in these contributions are "integrated assessments of risks and costs; theorizing about vulnerability, adaptation, and mitigation; institutional analyses of climate mitigation at different scales; and the extent to which climate change and responses are likely to be equitable, just, or ethically acceptable" (Agrawal et al. 2012: 329).

Emilio Moran, one of the few anthropologists to have served on the Intergovernmental Panel on Climate Change (IPCC), refers to climate change in various places in his book *People and Nature* (Moran 2006). He stresses that the growing amount of carbon dioxide in the atmosphere threatens the climate system, coral reefs, and the Antarctic ice sheets (Moran 2006: 1). Moran (2006: 3) argues that both the North, or developed countries, and the South, or developing countries, have had a significant impact on nature, the former largely due to a high level of consumption, and the latter due to high population growth. Moran (2006: 21 and 23) also mentions the role of climate change in the spread of various vector-borne diseases, plant growth, and animal migration, and poses a critical question:

Once we begin to operate well above any recorded levels [of greenhouse gas emissions] not just for one but more many measurable parameters, the question has to be asked if we have begun to play a reckless game with the survival of our species on planet Earth. Do we recognize that business-as-usual threatens the end of life as we know it?

Moran (2006: 166) views our "global economy" in its present form as unsustainable and destructive of the "planet's future productive capacity" because it is exhausting Earth's "fisheries, its water, its soils, and a host of other resources" (Moran 2006: 171).

McElroy and Townsend (2009: 116–118), two anthropologists well known for their work in applying ecological models in medical anthropology, view the "traditional ecological knowledge" of the Inuit as an adaptive mechanism for reducing vulnerability to climate change which has impacted very dramatically upon their culture over the past several decades. Townsend (2011: 191) asserts that the strongest contribution of anthropology to the study of global climate change is likely to be its patient accumulation of hundreds of local-level, small-scale studies, both prehistoric and contemporary, of how human populations have adapted, or failed to adapt, to drastic changes in climate from melting polar ice to encroaching desert edge. These types of studies, based on ethnographic climate change response research, now constitute a dominant segment of the growing literature on the anthropology of climate change.

While utilizing the concept of adaptation in her own work on the Viluni Sakha, horse and cattle breeders in the Viliui regions of northeastern Siberia, Susan Crate (2008: 571) critiques what has become an excessive reliance on the concept in "boardrooms, living rooms, and government offices, often as a substitute for mitigation of the human-induced effects of global climate change" as well in international forums and even within the corridors and publications of the IPCC.

Crate and Nuttall (2009: 9) observe that much of the discourse on climate change treats adaptation as both a research and a policy priority, but query whether the "frames of adaptive capacity and resilience … are sufficient," in that the "ability to respond to climate change is severely constrained for many people around the globe." They go on to argue:

Resilience, both social and ecological, is a crucial aspect of the sustainability of local livelihoods and resource utilization, but we lack sufficient understandings of how societies build adaptive capacity in the face of change. Furthermore, we suspect that environmental and cultural change, far beyond the reach of restoration, is occurring. Combined with institutional and legal barriers to adaptation, the ability to respond to climate change is severely constrained for many people around the globe. Some of us feel we are in an emergency state as field researchers and struggle to design conceptual architecture sturdy enough to withstand the storminess of the intellectual and practical challenges before us.

(Crate and Nuttall 2009: 10–11)

As Wijkman and Rockström (2011: 42) aptly observe: "No one knows with any certainty how long Earth's capacity for resilience with regard to greenhouse gases will last" and when the climate system will reach tipping points that are irreversible. As Crate (2009: 147) adds, "we need not be overly confident in our research partners' [i.e., interlocutors] capacity to adapt." One of the reasons for the threat to resiliency wrought by climate change that Crate points out is the potential to cause relocation, which can result in a breakdown of locally situated environmental knowledge. As she emphasizes, climate change is forcing not just community adaptation and resilience, but also relocation of human, animal, and plant populations. Lost with those relocations are the intimate human-environment relationships that not only ground and substantiate Indigenous worldviews but also work to maintain and steward local landscapes. In some cases, moves also result in the loss of mythological symbols, meteorological orientation, and even the very totem and mainstay plants and animals that ground a culture, as discussed in Chapter 5). Various other anthropologists who have adopted a cultural ecological perspective also employ the notions of "risk" and "resilience" in their analysis of climate change. Mark Nuttall (2009: 298), in his work on the Greenland Inuit, argues that these Indigenous people exhibit ecological resilience which has allowed them over time to "adjust to climate variation and change, to move around, and to see and seize opportunities in the environment."

In some ways, the anthropology of climate change may be viewed as a subfield of ecological anthropology or environmental anthropology which examines the historic and current human-environment interactions (Kopnina and Shoreman-Ouimet 2017). Even prior to the emergence of the anthropology of climate change as a subfield of ecological anthropology, the latter subsumed concerns such as primate ecology, paleoecology, cultural ecology, ethno-ecology, ethno-ornithology, historical ecology, spiritual ecology, human behavioral ecology, and evolutionary ecology. While human ecology continues to exist as an interdisciplinary subfield, evidenced by the continuing existence of the journal *Human Ecology*, sometime around the beginning of the 21st century cultural ecology appears to have been superseded by environmental/ecological anthropology. Hornborg (2007: 9–10) observes:

Political ecology emerged in the 1970s as a reaction to the preoccupation of cultural ecology with local homeostatic processes and adaptation, and to its neglect of global political economy ... Recently, however, political ecology appears to have become so diverse in terms of theoretical perspectives and methodologies—for examples, the tensions between scientific and constructivist approaches—that it is no longer a coherent field of study.

Bearing these thoughts in mind, our own orientation to political ecology has been inspired by its neo-Marxian variant.

Continuing Research on Climate Change from an Environmental Anthropological Perspective

Finan and Rahman (2016) have applied the concepts of adaptation, vulnerability, and resilience in their examination of beels, small depressions situated at various distances from the coast which are connected through canals and small rivers, in southwest Bangladesh. Beels vary in size depending upon the season, and are used in boro rice cultivation during the wet months and as pasture for livestock during the dry months. During the monsoon season, when the beels fill up with water, they serve as a public fishing resource for nearby communities. Furthermore, a prawn export industry based on the beels has appeared over the course of the past three decades or so. Unfortunately, the possibility of sea level rise associated with climate change threatens the intricate system of boro rice cultivation and prawn production, in that it may contaminate the canals with salt water. The Bangladesh government responded to previous catastrophes by building a few thousand concrete shelters high above pilings. While the government is considering building massive earthworks to stem the rise in sea level, such a development may not prove sufficient to save a complex hydrology system.

Bangladesh is home to the Sundarbans, the world's largest mangrove forest which has long functioned as a vast wetland frontier, but which has been transformed into a climate hot spot. Jason Cons (2021) has examined what he terms the "ecology of capture" by which bandits, known as *dakats*, seek to eke out a living by extracting funds from fishermen by accosting them, and often kidnapping them for ransom money, given the deteriorating climatic conditions in the region. In an effort to make the Sundarbans safe for conservation and tourism, paramilitary forces have attempted to subdue the dakats, sometimes even killing them. The government has attempted to intervene by encouraging the dakats to surrender, meaning that the size of dakat groups have declined from 18–50 members to just 5–6 members (Cons 2021: 551).

Kirsten Hastrup (2009a) edited a volume titled *The Question of Resilience: Social Responses to Climate Change*, derived from a symposium held at the University of Copenhagen, Denmark, on a collaborative research project titled "Waterworlds, Natural Environmental Disaster, and Social Resilience in Anthropological Perspective." The Waterworlds project focused on climate change/water-related threats, namely "the melting ice, the rising seas, and the drying lands, emanating to communities around the world" (Hastrup 2009a: 9). In the opening chapter, Hastrup (2009b: 15) observes that the risks related to climate change are unevenly distributed around the world, leading to "new patterns of regional migration, political unrest, economic vulnerability, shifting resource bases, and a profound sense of risk affecting everyday life in many parts of the world." She argues that "[r]esilience ... is not simply a question of systemic (social or cultural) adaptation to external factors, but a constitutive element of any working society" (Hastrup 2009b: 28). Indeed, it is culture that bestows a high degree of resilience on societies, but it cannot be assumed that this capacity is limitless any more than the capacity of Earth to absorb anthropogenic disruptions is limitless.

In Chapter 2 of *The Question of Resilience*, Orlove and Caton (2009: 34) observe that water issues are linked to concerns about Earth's atmosphere or "airworlds," which started with concerns about air pollution in the 1960s and 1970s and continued with concerns about acid rain in the 1970s and 1980s, and subsequently concerns about climate change and ocean acidification. Both the airworlds and the waterworlds of Earth are subject to the entwined and adverse effects of human activity, an interaction that threatens the resilience of impacted communities. A few medical anthropologists who adopt a biocultural or medical ecological perspective have engaged the issue of climate change. Armelagos and Harper (2010: 303) observe that various infectious vector-borne diseases, such as dengue and chikungunya, are on the rise, partly due to global warming. They maintain that humanity is in the throes of a "third epidemiological transition in which we have the re-emergence of diseases thought to be near extinction" (e.g., tuberculosis) that is "occurring in an era of globalization, unprecedented urbanization, and global warming as we are approaching the end of an antibiotic era" (Armelagos and Harper 2010: 304). There is also concern that with Arctic melting long-frozen pathogens will reactivate and cause new human and/or animal infections. Overall, the cultural ecological perspective in the anthropology of climate change makes its mark in studies designed to answer questions about sociocultural responses to the climatic chaos of a warming planet. The perspective draws attention to the ways people collectively react and use their cultural systems to cope with new threats and develop new options in the face of climate change adversity.

The Cultural Interpretive perspective of Climate Change

Environmental anthropologists who employ a cultural interpretive perspective view culture as a set of "perceptions as well as interpretations" that situate humans within the environment (Milton 1996: 63). Most cultural interpretive or phenomenological examinations of climate change tend to focus on perceptions on the part of diverse peoples or "local knowledge" about climate change. Adherents of the cultural interpretive perspective observe that the "topic of 'climate change' encompasses people's perceptions and behavior based on the threat (or, in a few cases, the promise) of such change, as well as the causes, processes, and prospective impacts of the change itself" (Rayner and Malone 1998: xix). In Volume 1 of *Human Choice and Climate Change*, a chapter by anthropologist Mary Douglas et al. (1998: 196) "pin[s] responsibility for massive greenhouse gas emissions on human efforts to satisfy their wants." While the authors note the concern on the part of some environmentalists that "insatiable consumption habits" are contributing to "depletion of the globe's resources, including capacity to absorb atmospheric carbon," Douglas and her colleagues make no direct mention of the roots of consumerism in global capitalism. The implication is that unquenchable consumptive desires are inherent rather than cultural and are a product of a particular political economy. However, the authors do refer to the distinction that Agarwal and Narain (1991) made between survival emissions and luxury emissions.

While Douglas et al. (1998: 201) recognize that the "theory of wants will one day come to terms with the theory of society," they evade the issue that in the modern world, capitalism plays a major role in shaping people's wants, with nonstop, multimedia, ubiquitous advertising serving as one influential link between the production and consumption processes. Douglas et al. (1998: 202) acknowledge that the societal achievement of a clean environment will require that people "change their laws, and changing these will change the pattern of wealth and income distribution, and this will change the flows of goods and people on birthdays, anniversaries and weddings, retirements, funerals, sick visiting, and ordinary Sunday family gatherings." They employ the notion of sustainable development proposed by the World Commission on Environment and Development, perhaps better known as the Brundtland Commission, as providing a guideline for constructing a "theory of human needs" that could be used to determine "how the burden of reducing greenhouse gas emissions could be distributed in order to meet 'the needs of the present without compromising the ability of future generations to meet their own needs'" (Douglas et al. 1998: 204). However, the concept of sustainable development presupposes a complementarity between ongoing economic expansion and environmental sustainability, an assumption that has come under increasing scrutiny from a growing number of theorists, not only political ecologists or eco-Marxists, but also major environmental thinkers such as James Gustave Speth (2008).

Perhaps the earliest examples of an explicitly cultural interpretive analysis of climate change was a component of a study done by Willet Kempton (1991) and later Kempton et al. (1995). The latter research examined US environmental values by conducting semi-structured interviews with 43 informants (20 laypeople, 21 environmental specialists, and two pilot subjects) as well as surveys of 142 respondents (30 Earth First members, 27 Sierra Club members, 29 sawmill workers) from New Jersey and Maine (Kempton et al. 1995). The researchers investigated their informants' perceptions of three major local environmental changes, namely ozone depletion, species extinction, and global warming. Many laypeople, they found, confounded global warming as a subset or a consequence of ozone depletion.

Celeste Ray (2002) wrote what perhaps constitutes the first relatively broad overview of the anthropology of climate change, reflecting a particularly multidisciplinary anthropological approach that includes contributions from a geologist, an atmospheric physicist, two ecologists, an economist, and a political scientist. In terms of anthropology, she notes that archaeology, ecological anthropology, ethnohistory, and historical ecology, all contain the potential to contribute to the anthropology of climate change. Ray utilizes a cultural interpretive approach to climate change, noting the importance of "cultural models" in terms of shaping human responses to environmental crises. She asserts that cultural models of the environment "change over time with environmental changes, with population dynamics, with religious beliefs, and with new technologies" (Ray 2002: 85), which is surely not an exhaustive list of factors that shape cultural models. In terms of climate change mitigation, Ray (2002: 98) argues that effective approaches will "draw support from those of different cultural perspectives, socioeconomic classes, and educational and religious backgrounds." However, what she fails to acknowledge in her pluralist framework is that some players, such as the higher social classes and corporations, have much greater input into climate policies than people from the lower social classes or a wide diversity of subaltern groups, such as Indigenous people. Moreover, various sectors of the wealthiest social classes, those who in the early years of the 21st century have come to be known as the "1%," may have vested economic interests in particular carbon-emitting industries, a fiscal commitment to ever-expanded production, and a desire to keep using the environment as a free rubbish bin for the by-products and wastes of the industrial production process.

Carla Roncoli et al. (2009: 95), key figures in climate anthropology, observe that anthropologists have begun to explore the relationship between "local knowledge and climate phenomena." This perspective may have become the predominant one in the anthropology of climate change, given that it flows naturally from prior work that cultural anthropologists have done on small-scale societies or local communities where they tend to gather data on people's "emic" (insider) views, information that can be grouped under the rubric of local knowledge. Ironically, while local knowledge may recognize the reality of climate change, for larger segments of people in modern societies, culture may also serve to downplay or even deny that it is occurring. As Milton (1996: 138) observes:

> Our understanding of past climatic changes, in which successive ice ages have given way to warmer periods, helps to insulate us, to some extent, from the fear that we might ultimately be responsible for irreversibly changing the Earth's climate. The changes provoked by human activities seem less alarming when set in the context of the larger, "natural" pattern.

One of the concerns of the body of work we see as fitting a cultural interpretive perspective is that large-scale social and technological responses to climate change must be sensitive to local sociocultural configurations and local understandings of the environment. Celeste Ray (2002: 90), for example, argues that policies about climate change "must employ cross-cultural knowledge of benign environmental practices while maintaining consciousness of traditional family and community structures, labor division, and localized subsistence strategies."

A perusal of the arc of human history affirms that an obsessive focus on ever-expanding desire and consumption is not an inherent or automatic feature of the human condition or of human nature. Rather, contemporary modes of insatiable consumption are part and parcel of the larger system of global capitalism, an economic system that in its impulse for profit-making requires ongoing accumulation and non-stop expansion. Global capitalism fosters a worldwide, energy-gobbling machine. This global political economy ensures wealth and profits for the richest sectors of society, but in the process, because they are rated of lesser importance than profit-making, it sacrifices basic human needs for large sectors of humanity as well as environmental sustainability. In order to survive, capitalism must generate an artificial need, namely the need to unceasingly consume a wide array of commodities (sometimes just to show off, assert success, or reward oneself for small accomplishments), even potentially dangerous and lethal commodities.

Kay Milton (2008), a well-known environmental anthropologist, has sought to delineate a culture theory of climate change by drawing on the work of Mary Douglas. Douglas delineates two key variables of human social organization which she calls "grid" and "group," respectively. Grid "measures the freedom of choice people exercise," and group "measures their degree of collective allegiance" (Milton 2008: 40). Put otherwise, "grid" refers to the extent to which people's actions are constrained, and "group" refers to the extent to which people act for themselves as individuals or for the groups to which they belong. The relationship between grid and group results in a matrix depicted in Figure 2.1, which in turn produces different cultural orientations.

A hierarchical perspective within society occurs when "group allegiance is high and people's actions are relatively constrained" (Milton 2008: 40). An egalitarian perspective entails a "low level of constraint together with a high degree of group allegiance ... in which social justice and common interests are valued" (Milton 2008: 40), while an entrepreneurial perspective entails a "low level of constraint combined with a low degree of group allegiance" which contributes to individualistic ambitions such as is promoted in market economies (Milton 2008: 40). A fatalistic perspective occurs in a situation where "actions are tightly constrained and group allegiance is low," prompting people to feel powerless in the face of adversity.

In terms of perspectives about nature, Milton reconfigures the grid-group model as depicted in Figure 2.2 Milton (2008: 45) asserts that, in terms of

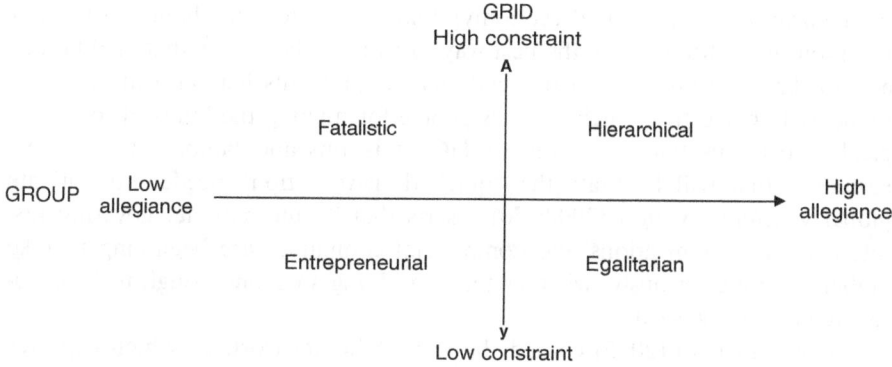

FIGURE 2.1 Douglas's Grid-Group Model

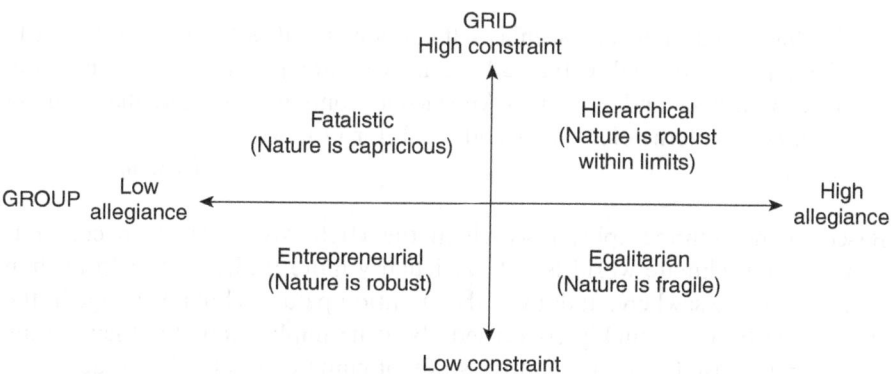

FIGURE 2.2 The Grid-Group Model with Respect to Views of Nature

global warming, multinational corporations and Western governments, which have tended to operate with an entrepreneurial perspective, have had to admit that nature is not as robust as they have believed to be the case (a belief that has rationalized an enormous level of environmental release of industrial toxins and greenhouse gases—GHGs), meaning that "increasing numbers of people are adopting the view that nature is fragile." A growing understanding of climate science and climate change has made growing numbers of people aware that their lifestyles are contributing to a historic ecological crisis. While many people have adopted a fatalistic attitude about global warming, or deny its reality, or at least the role that human activities play in creating it, Milton argues for an "expansive egalitarianism," one which attempts to merge egalitarianism and green values to "include future generations and/or non-human nature" (Milton 2008: 48). She argues that the debate around global warming has renewed the left–right divide. Milton (2008: 49), however, eschews the tendency on the part of leftist groups and commentators to "argue that a radical reorganization of [the global economy], and a fundamental change in the distribution of wealth, provide the best way forward." She asserts that egalitarians are not bringing about needed social change, and puts her faith in an entrepreneurial perspective which rewards people for making the kinds of technological innovations that will reduce GHG emissions and building hierarchical structures that will facilitate the "quick decisive action" needed to mitigate global warming. Milton (2008: 50) asserts that "some national governments, international organizations and commercial companies are beginning to take global warming seriously; whether they are doing so soon enough to be effective remains to be seen."

Kirsten Hastrup (2015) coined the term "climate worlds," which captures the reality in the modern world that people's knowledge about climate change is not simply situated in their local setting, but draws upon information they derive from the wider world. She asserts that:

> [W]hile ... cultures were seen as self-contained and rather fixed in the early days, today we realise that all worldviews are plastic and continuously incorporating—and locating—knowledge coming from elsewhere, along with new patterns of weather and wind, for instance.
>
> *(Hastrup 2015: 8)*

Based on her ethnographic research in the High Arctic, Hastrup contends that the Inuit climate world is not particularly more traditional or local than that of people elsewhere, and given the attention paid to climate change in the Arctic, the Inuit are highly concerned about its implications for their future. In making a case for a comparative study of climate worlds, she argues:

> It is one of the challenges facing anthropologists to show how located knowledge practices other than western science may move and affect not only present concerns but also future solutions. Anthropology can contribute vitally to the perception of humans, not only as destroyers but also creators of new possibilities, precisely because the burgeoning anthropological studies of how people deal with climate change show how all people are capable of integrating diverse forms of knowledge.
>
> *(Hastrup 2015: 151)*

Indeed, since June 2010 a Climate Worlds research team based in the Institute of Advanced Studies in Essen and the Graduate School in History and Sociology in Bielefeld, Germany, has been conducting interdisciplinary ethnographic research on the impact of climate change and how it is culturally framed in various parts of the world (Greschke 2015). The team has focused primarily on coastal areas, such as Ameland, a West Frisian island in the Netherlands; Churchill on Hudson's Bay in Manitoba, and Cape Verde in the North Atlantic, and is committed to long-term fieldwork, lasting approximately 20 months, at its various research sites. There is consider-able evidence that Indigenous peoples have been "particularly vocal about climate change as a threat to their traditional ways life represented in their cosmovision or traditional worldview," often translating into demands, particularly on the part of their leaders and representatives, that developed countries and the United Nations recognize the seriousness of climate change and adopt drastic climate change mitigation efforts (Eisenstadt and West 2017: 41). For example, a national survey of Ecuadorians revealed that respondents who subscribe to an Indigenous cosmovision are more likely to acknowledge the threats posed by climate change, particularly evident in Andean countries which are experiencing rapidly diminishing glaciers, than those respondents who do not subscribe to an Indigenous cosmovision. Indeed, the Ecuadorian constitution is the only constitution in the world that recognizes the "rights of nature," or grants that Mother Nature has rights just as do humans. Unfortunately, the Correa govern-ment decided in 2013 to discontinue its pledge to protect the biodiverse rainforest in Yasuni National Park by not allowing the drilling of oil on the condition that international donors, who only contribute approximately US $1 million per annum, compensate Ecuador for oil revenues, by leaving it in the ground (Eisenstadt and West 2017: 43).

Hannah Knox's (2020) *Thinking Like Climate* is a localized cultural inter-pretive analysis of examination of sustainability issues in Manchester, UK. She argues that the burgeoning anthropology of climate change has not engaged in a serious dialogue with studies of local weather as a "technologi-cal, infrastructural, political-economic phenomenon" (Knox 2020: 4). Never-theless, weather constitutes the basis upon which climate is conceptualized and experienced. Thus Knox (2020: 8) proposes that the anthropology of cli-mate change should think of climate as a "conceptual tool to assist an exploration of how the material dynamics of climate change—which have become known through the data, visualizations, and computer models"— are utilized by climate scientists. Much of her book draws on her observations and conversations over the course of eight years with government officials, climate scientists, and climate activists who are grappling with how to trans-form Manchester into a postindustrial and low-carbon city. Climate change became a local concern in 1994 at the Global Forum on Cities conference in Manchester, which city authorities hoped would propel Manchester into a center of global climate policy, an aspiration which fell short of the mark.

Manchester city council has had difficulty in addressing climate change in that a serious discussion about it interferes with its commitment to urban economic growth. However, the passage of the Climate Change Act in 2008, which committed the UK to reduce its carbon dioxide emissions by 80% from a 1990 baseline by 2050, compelled the county of Greater Manchester to make the same commitment in 2010. Although Manchester had initially committed to reducing carbon dioxide emissions by 41% from a 2005 baseline, in March 2019 it upped its target to become net zero carbon by 2038. Net zero carbon refers to the balance between the amount of GHG that is produced and the amount that is removed from the atmosphere. Kevin Anderson, the former director of the Tyndall Centre for Climate Change Research based in Manchester and a world-renowned climate scientist, appears to have played a role in firming up the city's targets in that he believes that the UK's emissions pathway is flawed. Furthermore, Knox (2020) observes that the official document *Principles of Tackling Climate Change in Manchester* seeks strategies to decouple economic growth from carbon dioxide emissions, a common ambition on the part of various advocates of green capitalism. In reality, this decoupling process has by and large thus far proven to be wishful thinking. Nevertheless, Manchester has developed a reputation for retrofitting its buildings to be more energy efficient and was chosen in 2011 as a pilot test site for a Green Deal program.

An even more recent cultural interpretive analysis of climate change is Aase Ravanneid's 2021 book *Perceptions of Climate Change in India*. For her fieldsite, she opted to focus on the village of Rana Majri situated in the area surrounding Changigarth, a city that was built to serve as the joint capital of the states of Punjab and Haryana in the 1950s. Although Ravanneid (2021: 155) found that by and large the villagers were aware that global warming was impacting their lives, given that the monsoon has become irregular and the Himalayan glaciers are melting, this "kind of awareness, however, failed to express itself in practices and conversations with government employees or scientists who worked (most notably) through the Haryana State Government to change—or assist the change of—Rani Majri into a 'progressed' condition." The villagers found greater solace in enduring the ravages of their hot, parched summers by engaging in a "dance of global warming" to ritually guarantee some semblance of modest harvests at their end (Ravanneid 2021: 142–160). Conversely, Schnegg (2021: 268) reports that pastoralists in the Fransfontein region of Namibia blamed deteriorating rainfall on German and later South African settlers who grabbed the best parcels of land for their commercial farms.

Shifting to the Global North, Dominic Boyer and Mark Vardy (2022) examine how residents in Houston, Texas, have come to define a "new normal" in the wake of three "500-year flood events," including Hurricane Harvey, within the space of 24 months (2015–2017). They delineate three forms of affective orientations that flood victims have clustered around as they attempt to cope with their precarious ontologies:

> Diluvial individualism that entails resignation to adopting individual mitigation strategies, such as elevating their new houses higher than previous ones, given that the flood victims view governments as being incapable of managing their security; hydraulic citizenship in which floodies (a term used for flooded homeowners who reside in areas that have been declared a federal disaster) proactively engage the "vast and

complex technological assemblage of water management, asserting rights, challenging responsible authorities, and arguing for political change; amphibious acceptance in which floodies resign themselves to the inevitability that eventually their coastal city will be inundated by the rising Gulf of Mexico, thus forcing them to become climate refugees by relocating inland.

(Boyer and Vardy 2022: 622)

From the cultural interpretive perspective, the questions that are raised concern how people culturally construct a meaningful world of nature and society, including their perceptions and understandings of the changing environment around them. Answering such questions will provide insight about how to best work with various peoples, both in responding to climate change and in the development of culturally grounded climate change policies. Failure to recognize that people are not blank slates, but have their own understandings of the world and of changes that are occurring within it, can lead to failed efforts to impose climate change adaptations that conflict with local cultural knowledge (as discussed in Chapter 5). This type of failure is not new, but has in fact long been a feature of externally initiated development or health promotion programs. Given the stakes, interpretively oriented anthropologists argue that it is vital that we do not repeat old errors. Conversely, it is important to recognize the limitations of ethnographic fieldwork which focuses on local perceptions of climate change. As Alf Hornborg (2013: 54) observes, some anthropologists miss the forest because of the trees by immersing themselves in "obscure representations of exotic, local particularities of experience," thus downplaying the macroscopic forces driving anthropogenic climate change in the modern era, a central concern of the critical anthropology of climate that we discuss next, and one to which we adhere.

The Critical Anthropology of Climate Change

The critical anthropology perspective on climate change is guided by an ecosocial perspective that is informed by three theoretical currents in and beyond anthropology: world system theory, and its particular application within an ethnographically informed anthropology; political ecology theory, with its understanding of the politicized nature of human interaction with the environment; and critical health anthropology, and its focus on the social determinants of health inequalities, experiences, and behaviors. In the contemporary period, the first of these is a perspective on the features of capitalism as a particular kind of economic and social system that transcends national boundaries and ties regions and countries together around the privatization of the means of production, a market-based distribution of goods and services, and the control of labor based on the buying and selling of it as a commodity. These features of capitalism, world systems theory argues, create two interlocked hierarchies. First, there is a hierarchical ordering of relationships among the countries of the world involving a tripartite division into economically dominant (and greatest GHG-producing) core countries, developing but subordinate semi-periphery countries (that rapidly become important GHG emitters as they strive for economic development), and the periphery countries with relatively weak (and potentially failing) governments and local economies focused on extracting and exporting raw materials from nature to core nations (and often receiving back the waste products of industrial production for local environmental disposal). It is people living in the peripheral zones of the world who produce the least amount of GHGs but who are beginning to suffer the most from the impact of emissions on global warming. The second hierarchy involves labor and the construction of social classes. As Wolf (1982: 354) observes, a market economy "creates a fiction that this buying and selling [of labor] is a symmetrical exchange between partners, but in fact the market transaction underwrites asymmetrical relationship between classes."

Consequently, in ways that parallel the climate change features of the world economic system, wealthier (labor-buying) classes benefit from the processes that lead to the environmental emission of GHGs (i.e., expanding production, growing markets), while the poorest laboring classes, through structures of environmental injustice and limited protective resources, suffer disproportionate harm from global warming. World systems theory has tended to generate the asking of macro-level questions such as to what degree do core countries need the periphery to remain underdeveloped and how do they contribute to their underdevelopment? And what causes the world system to change? In its uptake within anthropology, and the field's focus on on-the-ground sociocultural arrangements around the world and cross-cultural human subjectivity, research questions have been directed at the nature of the relationship of macro-level forces and structures and more micro-level social realities and actions, including issues of resistance, resilience, and emergent cultural heterogeneity in the face of the homogenizing influences of globalism.

Second, political ecology theory is a multidisciplinary perspective that developed out of the study of societies in environmental context, especially the cultural ecology initiative discussed above (e.g., Cole and Wolf 1974; Wolf 1972). This perspective draws attention to the fundamental importance of the interplay among political, economic, and environmental factors (Foster 1994; Roberts and Grimes 2002). In particular, political ecology moves beyond the cultural ecological approach by drawing attention to questions about the role of power in the unequal distribution of the costs and benefits of environmental change (e.g., across class, ethnic, or other social divisions), the ways this unequal distribution reinforces (or, in particular contexts, reduces) existing social and economic inequalities in society, and the political consequences of environmental changes.

Finally, critical health anthropology is a theoretical perspective on the nature and causes of human health and treatment systems. During the early years of medical anthropology's formation, explanations within the discipline tended to be narrowly focused on accounting for the nature of health-related beliefs and behaviors at the local level in terms of specific ecological conditions, cultural configurations, or psychological factors. While increasing the sum of knowledge about the nature and function of local medical models, the original perspectives in the field overlooked the wider causes and determinants of human decision-making and health-related behavior. Questioning the values of explanations that are limited to accounting for health-related issues in terms of the influence of human personalities, culturally constituted motivations and understandings, or even local ecological relationships, beginning in the 1980s medical anthropologists pointed to the importance of understudied "vertical links" that connect the local social group to larger and cross-cutting regional, national, and global processes (e.g., commodification and the capitalist market, the globalization of production and restructuring of labor forces, the spread of biomedicine as a reflection of the penetration of capitalism globally). Since it emerged, what came to be called critical medical anthropology and more recently the critical health anthropology has drawn attention to questions about the social origins of illness, such as the way poverty, discrimination, violence, and exposure to stress and violence create and sustain disease. Critical health anthropology is also concerned with the origins of dominant cultural constructions in health and with the role of social inequality in the structuring of healthcare practices and delivery.

Critical health anthropology, earlier termed critical medical anthropology, has focused increasingly on environmentally mediated inequalities in health, such as the role of polluting industries in causing disease and creating disparities of health across social classes, genders, or ethnic populations (Baer and Singer 2009; Singer 2011). Integrating components of these approaches, the critical anthropology of climate change asks questions about the relationship of the capitalist mode of production and planet sustainability, the role of power in the production and control (or non-control) of greenhouse gases and industrial pollution, the social origins and social impact of the climate change denier movement, the unequal and unjust distribution of health and other climate change effects, the contradictions of green capitalism, and local and wider social movements that have emerged in opposition to business-as-usual corporate environmental degradation. In short, the critical anthropology of climate change seeks to develop integrative understanding of the interface of power and social hierarchy in the anthropogenic making of climate change and other environmental disruptions, and the unequal health and social consequences, differential experiences, and responses that occur as a result of climate change turmoil. The critical anthropology of climate change seeks to be conversant with archaeological, cultural ecological, and cultural interpretive approaches to climate change, while emphasizing the limitations of analyses that fail to consider how local patterns connect to the world system and to the distribution and direct and hegemonic roles of power in society locally, regionally, and globally. Like critical anthropology in general, the critical anthropology of climate change is committed to an activist/ scholarly notion of praxis or the merger of theory and social action in which the two arenas of work feed each other.

Thus, like many others involved in the anthropology of climate change, regardless of whether they adopt an archaeological, cultural ecological, or cultural interpretive approach, the critical anthropology of climate change promotes an applied anthropology of climate change involving collaboration with local communities and other subordinated populations and social movements of climate change response. The critical anthropology of climate change asserts:

- Social systems do not last forever, whether at the local, regional, or global level.
- The capitalist world system or global capitalism has been around for about 500 years, but it has come to embody so many inherent contradictions that it must be transcended to ensure the survival of humanity and animal and plant life on a sustained basis.
- There is a need for an alternative global system, one that is committed to meeting people's basic needs, social equity and justice, democracy, and environmental sustainability.

Proposals for such an alternative system have come under various terms, including global democracy, Earth democracy, economic democracy, radical democracy, eco-anarchism, and eco-socialism, an issue that we explore in Chapters 10 and 11.

In sum, the critical anthropology of climate change posits that the roots of recent climate change largely lie within the context of the capitalist world system as a political economy committed to continual growth and expansion, heavy reliance on GHG-emitting fossil fuels, and global social inequality within and across nations. The critical perspective challenges the views of conventional economists such as Nicholas Stern (2007) who, while expressing alarm about the potential ravages of climate change and the need to take measures to mitigate it, nonetheless accept the premise that the global economy must continue to grow and that humans have an inherent need to consume, and to acquire more and more things. Panic sets in within these circles when retail sales drop, particularly during the mad (and highly profitable and GHG-producing) Christmas shopping season or during the recently invented Black Friday online shopping holiday.

Perhaps more than any other environmental crisis, anthropogenic climate change forces us to examine whether global capitalism needs to be transcended and whether humanity needs to begin to develop an alternative or, as some would see it, a democratic eco-socialist world system. From the critical perspective, climate change constitutes one of the most important social issues, indeed perhaps the most important issue, in that it is related to numerous other environmental challenges of the 21st century. The capitalist world system exhibits numerous contradictions, including (1) its emphasis on profitmaking, economic expansion, and the treadmill of production and consumption; (2) the growing socioeconomic gap between rich and poor both within nation-states and between nation-states; (3) the depletion of natural resources and environmental degradation, the most profound form of which is climate change; (4) population growth, which in large part is stimulated by ongoing poverty; and (5) the resource wars waged by various developed countries, particularly the United States, the United Kingdom, and Australia, in promoting the interests of multinational corporations. Climate change, perhaps more than any other environmental crisis, illustrates the unsustainability of the capitalist world system, but its impact does not occur in a vacuum; rather it is magnified by interactions with other eco-crises.

Since the late 1980s, climate regimes have emerged at the international, regional, provincial, state, and even the local level. The vast majority of climate regimes function within the parameters of green capitalism, a notion that capitalism, by adopting emissions trading schemes, various technological innovations, energy efficiency, recycling, and other practices, can be environmentally sustainable. In seeking to mitigate both environmental damage and climate change, a growing number of corporations, politicians, policymakers, research scientists and academics have become advocates of green capitalism, even climate capitalism—the idea that embracing efficiency and renewable energy sources, the curbing of waste, carbon markets, new technologies, and sustainable business plans is not only good for the planet, but can also be very profitable. While historically corporations have been resistant to the assertion on the part of environmental activists that many of their practices are environmentally destructive and also contribute to climate change, a growing number of corporations have begun to contend that they can achieve sustainable development while reducing their GHG emissions by engaging in a process of ecological modernization. Indeed, while eschewing climate denialism, many corporations have come to define climate change as a business opportunity.

Technological innovations, such as renewable sources of energy and energy efficiency, have an important role to play in climate change mitigation, but even they cannot contain climate change in the long run so long as they accept the capitalist imperative of continual economic growth. Consequently, the critical anthropology of climate change argues that green capitalism as well as climate capitalism exhibits essential limitations in terms of mitigating climate change. As a result, climate change compels us to engage in a serious assessment of alternatives to global capitalism. The critical anthropology of climate explores various social justice initiatives which, while not seeking to transcend global capitalism per se, attempt to make it both more socially just and environmentally sustainable, including in terms of climate change. In her ethnographic examination of the oil and gas industry in Weld County, Colorado, USA, for example, Mette High (2022) found that its financial experts envisioned the industry as having the potential to provide socioeconomic opportunities not only for capitalists but also for ordinary people. But, in the view of Mark, one of High's (2022: 749) interlocutors, oil is a gift from God that serves as the "basis for relations of care, and ultimately, human flourishing." Such thinking closely parallels the ideas of Tony Abbott, a former politician who served as prime minister of Australia and is a devout Catholic, who repeatedly stated that "coal is good for humanity" (quoted in Wilkinson 2020: 250).

The critical anthropology of climate change maintains that it is imperative to think outside of the socially allowable box and construct an alternative to global capitalism as the ultimate climate change mitigation strategy. Although historically many capitalist nations have emphasized individual freedoms, there has often been little tolerance for truly critical thought, as some ideas are rapidly dismissed as illegitimate and not worthy of serious consideration. This mechanism of social control is woven deeply into the fabric of capitalist society, finding expression overtly or more covertly across social institutions. For example, one of the ways researchers have brought to light subtle academic forms of hierarchicalization and legitimization is through the analysis of what is known as the politics of citation. In this form of academic politics, authors tend to be cited based on their relative status in a field of study or the stature of their home institutions, while texts that question reigning social mythologies about the capitalist world system are delegitimized (through academic snubbing). This pattern can even be seen, for example, in the emergent anthropology of climate change and its growing literature, in the failure in much of that literature to recognize the body of work done within a critical anthropology of climate change framework. The effort to examine the impact of climate change on humanity and how to mitigate it has to be an interdisciplinary one that involves collaboration among natural and social scientists, public health people, policy analysts, and humanists who are willing to collaborate with the climate justice movement and other anti-systemic movements. Going from the present capitalist world system, which has created and continues to generate anthropogenic climate change, to an alternative global political economy, however it is defined, will require much effort, and there are no guarantees that we will be able to create a more socially equitable and environmentally sustainable world.

In building a critical anthropology of climate change, Myanna Lahsen (2005a), an associate researcher in the Earth System Science Center at the Brazilian Institute for Space Research, has done critical ethnographic research on how climate models and atmospheric scientists deal with issues of certainty and uncertainty associated with general circular computerized models that seek to project possible global climatic changes emanating from GHG emissions. She has also written about US climate politics and discussed the role that "conservative and financial elites" have played in supporting campaigns to counter growing concerns among Americans about climate change (Lahsen 2005b), as well as how Northern countries dominate the framing of science, including climate science, that underpins international environmental and climate negotiations (Lahsen 2007).

In her commentary on Crate's (2008) seminal article, which asserts that theoretical perspectives are strategically well situated to interpret the impact of climate change on local populations, communicate information about this process, and even respond to climate change both in the field and at home as advisors to climate policy decision-makers and as advocates for the people they study, Lahsen (2008: 587) astutely observes:

> To truly enhance our effectiveness and overcome our marginality in scholarship and policy arenas related to global change research, we need to study all types of relevant "locals" and especially those populating institutions of power. That means overcoming our abated but continued aversion to study power brokers such as scientists, government decision makers, industry leaders, journalists, and financial elites, all of whom are much more important in shaping climate change and associated knowledge and policies than are the marginal populations we are accustomed to studying.

For several years, Thomas Hylland Eriksen and his collaborators have discussed how the world system has been overheating due to the drive for profits, economic growth, increased production and consumption, and high dependency on fossil fuels, all of which result in an increase in GHG emissions and ultimately contribute to anthropogenic climate change (Eriksen 2016). Eriksen does not use the term *overheating* to refer to global warming per se, although global warming is one of the consequences of overheating. Overheating refers to the speed by which economic, environmental, and social changes are occurring in the capitalist world system.

Eriksen (2018) conducted ethnographic research on overheating in Australia, a country with an economy heavily reliant on fossil fuels in a similar vein to his home country of Norway. Whereas Norway's economy is highly dependent on offshore oil extraction, Australia's economy is highly reliant on coal, coal seam gas, and natural gas, as well as other mineral resources, particularly iron ore and bauxite, having earned Australia the designation of being a "quarry nation." The industrial boomtown of Gladstone in Queensland constituted the focus for Eriksen's fieldwork, a community situated next to the iconic Great Barrier Reef, which has been undergoing coral bleaching due to the ravages of climate change, and also nearby coal and natural gas extraction sites, fossil fuels which in large part are exported to various parts of the world. Eriksen's research constitutes a case study of fossil fuel extraction in Australia, particularly the states of Queensland and New South Wales. The Australian extractive sector has been speeding up over the course of the 21st century.

Astrid Stensrud and Eriksen edited an anthology titled *Climate, Capitalism and Communities* which presents evidence from places as diverse as the Arctic,

Mongolia, and the South Pacific that testifies to the growing critical anthropological awareness of the link between global capitalism and climate change (Stensrud and Eriksen 2019). In another volume that explores efforts around the world to cool down the overheated world, not so much at the global level but at the local level, Susanna Hoffman, Eriksen, and Paulo Mendes, along with contributors in various countries, make another significant contribution to the critical anthropology of climate change (Hoffman et al. 2022). Their book is literally a *tour de force* focusing on impacts of overheating in various locations, including Northwest Namibia, the Eastern Himalayas of India, New Zealand, French Polynesia, Belém in northeast Brazil, Bangladesh, East Africa, the Austrian Alps, the Elbe River Valley near Dresden, Germany, Native American communities of the American Southwest, the remote mountains of Portugal, the Isthmus of Tehuantepec in Oaxaca, Mexico, and the state of Georgia in the United States, along with diverse efforts to cool down the ecological and climatic systems of these locales. Included is a chapter depicting disasters consequent to climate change in Colorado and Louisiana. As the various chapters reveal, thus far efforts to cool down the climatic and ecological systems have met with mixed results at best.

Conversely, various subalterns at various levels around the world are challenging emissions-intensive industries and countries. For example, a Peruvian farmer challenged RWE, a German energy company which does not have operations in Peru, in a German court which "saw the possibility for a precedent under German law if evidence could be found to prove a causal link between RWE's emissions and the glacial lake flood risk to Saul's house in Peru" (Walker-Crawford 2023: 91–92). South Pacific Islanders "have played active roles as initiators, motivators, and shapers of global climate change policy regimes" instead of resigning themselves to being doomed to find refuge from their low-lying island homes as sea levels rise (Kirsch 2020: 830). Anna Szolucha (2021) reports that in northwest England, some residents have formed anti-fracking groups to counter the British state's failure to adequately regulate a burgeoning natural gas industry. These grassroots initiatives signal the potential for people to organize and challenge elite polluters as well as fatalistic thinking about the inevitability of disastrous climate change.

Conclusion: Towards an Integrated Critical Understanding of Climate Change

As discussed in this chapter, the anthropology of climate change has produced three alternative perspectives, and with them differing sets of questions that anthropological researchers bring to their encounter with climate change issues. While we tend to favor a critical perspective which draws on world systems theory, we acknowledge that this approach, as Orr et al. (2015: 161) assert, may appear "overdrawn because it seemed to localize all agency at the level of global actors and none at the level of local communities." But, in fact, we believe that critical anthropologists need to examine closely the climate movement and other localized and Indigenous responses to climate change, as well as global capitalism and social inequality, around the world (Baer 2022; Burgmann and Baer 2012; Connor 2016; Singer 2018). While alternative perspectives can lead to debate and disagreement, this is not an inherently negative occurrence, as even heated discussion can produce new insights, raise new questions, and result in productive synergies.

Bearing these thoughts in mind, the effort to critically examine and respond to the adverse impacts of climate change on humanity and the ecosystem must be a multidisciplinary effort. It entails collaboration between climate scientists, Earth system scientists, energy analysts, and physical geographers, on the one hand, and social scientists, including anthropologists, archaeologists, sociologists, political scientists, and human geographers, on the other hand (Baer and Singer 2018). The reality is that natural scientists and mainstream economists tend to dominate much of the discourse on climate change, as is evidenced by the composition of the IPCC. Wijkman and Rockström (2011: 26–27) advocate moving beyond the "disciplinary status quo" characteristic of the sciences and universities and towards emphasis on "more integrated and problem-solving programmes." Collaboration serves to bring the strengths from various disciplines to what is a monumental but undeniably vital task: understanding and responding effectively to climate change. Climate change research needs to move beyond research centers and universities. It needs to collaborate with communities and Indigenous peoples, particularly those that are being adversely impacted by climate change, as well as nongovernmental organizations, progressive political parties, women's groups, and climate action groups that are pushing for effective climate change mitigation strategies informed by a strong sense of social and climate justice.

In contrast to applied and practicing anthropology as a whole, the application of the anthropology of climate change, or engaged climate anthropology, is in its infancy and is still struggling to have an impact on climate change policies and practices around the world. Anthropologists need to contribute in various ways by participating in a growing but still disparate climate justice movement, which is particularly strong in the Global South. This differs from the narrower climate movement, prevalent in the Global North, which tends to focus on technological solutions, particularly renewable energy sources as the major means for decarbonization, while sometimes downplaying social justice issues. Just as anthropology went through an effort to reinvent itself in the late 1960s and early 1970s (Hymes 1972), it needs to once again reinvent itself by moving beyond the political impasses of particularism and postmodernism and shift towards becoming a more ethically, socially just, localized, postcolonial, and ecologically sustainable endeavor while maintaining a global vision that recognizes that all human beings, particularly those in the Global South whom many of us continue to study, face the threat of catastrophic climate change if emissions are not drastically reduced over the next few decades. In terms of the anthropology of climate change, we view the critical anthropology of climate change as part of the ongoing process of reinventing anthropology.

Various engaged climate anthropologists argue for the strengths that anthropology brings to the study of and response to climate change as a dual social and environmental issue. However, from the perspective of critically engaged climate anthropology, issues of social justice and the unsustainability of the global capitalism are of fundamental importance. Given that many climate scientists argue that humanity faces the prospect of a 4 °C (7.2 °F) world by 2100 if emissions do not rapidly decline (which they did temporarily during the COVID-19 pandemic owing to reduced industrial production, motor vehicle driving, and air flights), it behoves engaged climate anthropologists to engage in a larger project, namely the anthropology of the future, which prompts us to be a part of a growing global effort to envision alternatives to the existing world system and to find ways to achieve a world based on social justice, environmental sustainability, and a safe climate.

References

Agarwal, Anil and Narain, Sunita. 1991. *Global Warming in an Unequal World*. New Delhi: Centre for Science and Environment

Agrawal, Arun, Maria Lemos, Ben Orlove, and Ribot, Jesse. 2012. Cool heads for a hot world: Social sciences under a changing sky. *Global Environmental Change* 22: 329–331

Armelagos, George J. and Harper, Kristin N. 2010. Emerging new diseases, urbanization, and globalization in a time of global warming. In *The Blackwell Companion in Medical Sociology*. William C. Cockerham, ed. pp. 291–311. Malden, MA: Wiley-Blackwell

Baer, Hans A. and Singer, Merrill. 2009. *Global Warming and the Political Ecology of Health: Emerging Crises and Systemic Solutions*. Walnut Creek, CA: Left Coast Press

Baer, Hans A. and Singer, Merrill. 2018. *The Anthropology of Climate Change: An Integrated Critical Perspective* (2nd edition). London: Routledge

Baer, Hans A. 2022. *Climate Change and Capitalism in Australia: An Eco-Socialist Vision for the Future*. London: Routledge

Boyer, Dominic and Vardy, Mark. 2022. Flooded city: Affects of (slow) catastrophe in post-Harvey Houston. *Current Anthropology* 63: 615–636

Burgmann, Verity and Baer, Hans A. 2012. *Climate Politics and the Climate Movement in Australia*. Melbourne: Melbourne University Press

Cole, John and Wolf, Eric. 1974. *The Hidden Frontier: Ecology and Ethnicity in an Alpine Valley*. Berkeley: University of California Press

Connor, Linda H. 2016. *Climate Change and Anthropos: Planet, People and Places*. London: Earthscan

Cons, Jason. 2021. Ecologies of capture in Bangladesh's Sundarbans. *American Ethnologist* 48: 245–259

Crate, Susan A. 2008. Gone the bull of winter? Grappling with the cultural implications of anthropology's role(s) in global climate change. *Current Anthropology* 49: 569–595

Crate, Susan A. 2009. Gone the bull of winter? Contemporary climate change's cultural implications in northeastern Siberia, Russia. In *Anthropology and Climate*

Change: From Actions to Social Transformations. Susan A. Crate and Mark Nuttall, eds. pp. 139–152. Walnut Creek, CA: Left Coast Press

Crate, Susan and Nuttall, Mark. 2009. Introduction: Anthropology and climate change. In *Anthropology and Climate Change.* Susan A. Crate and Mark Nuttall, eds. Pp. 9–36. Walnut Creek, CA: Left Coast Press

Crumley, Carol, ed. 1994. *Historical Ecology: Changing Knowledge and Changing Landscapes.* Santa Fe, NM: School of American Research Press

Douglas, Mary, Gaspar, Des, Ney, Steven, and Thompson, Michael. 1998. Human needs and wants. In *Human Choice and Climate Change,* Vol. 1: *The Societal Framework.* Steve Rayner and Elizabeth L. Malone, eds. pp. 195–263. Columbus, OH: Batelle Press

Eisenstadt, Todd A. and Jones West, Karleen. 2017. Indigenous belief systems, science, and resource extraction: Climate change and attitudes in Ecuador. *Global Environmental Politics* 17: 40–68

Eriksen, Thomas Hylland. 2016. *Overheating: An Anthropology of Accelerated Change.* London: Pluto Press

Eriksen, Thomas Hylland. 2018. *Boomtown: Runaway Globalisation on the Queensland Coast.* London: Pluto Press

Finan, Timoth J. and Rahman, Md. Ashiqur. 2016. Storm warnings: An anthropological focus on community resilience in the face of climate change in southern Bangladesh. In *Anthropologists and Climate Change: From Actions to Transformations* (2nd edn). Susan A. Crate and Mark Nuttall, eds. pp. 172–185. London: Routledge

Foster, John Bellamy. 1994. *The Vulnerable Planet: A Short Economic History of the Environment.* New York: Monthly Review Press

Greschke, Heike. 2015. The social facts of climate change: An ethnographic approach. In *Grounding Climate Change.* Heike Greschke and J. Tischler, eds. pp. 121–138. Dordrecht: Springer Science + Business Media

Hastrup, Kirsten, ed. 2009a. *The Question of Relevance: Social Responses to Climate Change.* Copenhagen: Det Kongelige Danske Videnskabrnes Selkskab

Hastrup, Kirsten, ed. 2009b. Waterworlds: Framing the question of social relevance. In *The Question of Relevance: Social Responses to Climate Change.* Kirsten Hastrup, ed. pp. 11–30. Copenhagen: Det Kongelige Danske Videnskabrnes Selkskab

Hastrup, Kirsten. 2015. Comparing climate worlds: Theorising across ethnographic fields. In *Grounding Global Climate Change.* H. Greschke and J. Tishler, eds. pp. 139–154. Dordrecht: Springer Science + Business Media

High, Mette. 2022. Utopias of oil: Private equity and entrepreneurial ambition in the U.S. oil and gas industry. *Cultural Anthropology* 37: 738–763

Hoffman, Susanna M., Eriksen, Thomas Hylland, and Mendes, Paulo, eds. 2022. *Cooling Down: Local Responses to Global Climate Change.* New York: Berghahn

Hornborg, Alf. 2007. Conceptualizing socioecological systems. In *The World System and the Earth System.* Alf Hornborg and Caroline Crumley, eds. pp. 1–11. Walnut Creek, CA: AltaMira Press

Hornborg, Alf. 2013. The fossil interlude: Euro-American power and the return of the physiocrats. In *Cultures of Energy: Power, Practices, Technologies.* Sarah Strauss, Stephanie Rupp, and Thomas Love, eds. pp. 41–59. Walnut Creek, CA: Left Coast Press

Hymes, Dell, ed. 1972. *Reinventing Anthropology: Radical Anthropology of a New Generation.* Ann Arbor: University of Michigan Press

Kempton, Willet. 1991. Lay perspectives on global climate change. *Global Environmental Change* 1(3): 183–208

Kempton, Willet, Boster, James S., and Hartley, Jennifer A. 1995. *Environmental Values in American Culture.* Cambridge, MA: MIT Press

Kirsch, Stuart. 2020. Why Pacific islanders stopped worrying about the Apocalypse and starting fighting climate change. *American Anthropologist* 122: 827–839

Knox, Hannah. 2020. *Thinking Like a Climate: Governing a City in Times of Environmental Change.* Durham, NC: Duke University Press

Kopnina, Helen and Shoreman-Ouimet, Eleanor. 2017. An introduction to environmental anthropology. In *Routledge Handbook of Environmental Anthropology.* Helen Kopnina and Eleanor Shoreman-Ouimet, eds. pp. 3–9. London: Routledge

Lahsen, Myanna. 2005a. Seductive simulations? Uncertainty distribution around climate models. *Social Studies of Science* 35(6): 895–922

Lahsen, Myanna. 2005b. Technocracy, democracy, and U.S. politics: The need for demarcations. *Science, Technology & Human Values* 30(1): 137–169

Lahsen, Myanna. 2007. Trust through participation? Problems of knowledge in climate decision making. In *The Social Construction of Climate Change: Power, Norms, and Discourses.* Mary E. Pettenger, ed. pp. 173–196. Aldershot: Ashgate

Lahsen, Myanna. 2008. Commentary on "Gone the bull of winter? Grappling with the cultural implications of and anthropology's role in global climate change" by Susan A. Crate. *Current Anthropology* 18: 204–219

McElroy, Ann and Townsend, Patricia K. 2009. *Medical Anthropology in Ecological Perspective* (5th edn). Boulder, CO: Westview Press

Milton, Kay. 1996. *Environmentalism and Culture Theory: Exploring the Role of Anthropology in Environmental Discourse.* London: Routledge

Milton, Kay. 2008. Climate change and culture theory: The need to understand ourselves. In *The Impact of Global Warming on the Environment and Human Societies.* PASI Research Paper No. 1. Hans A. Baer, ed. Pp. 39–52. Melbourne: School of Philosophy, Anthropology, and Social Inquiry, University of Melbourne

Moran, Emilio F. 2006. *People and Nature: An Introduction to Human Ecological Relations.* Malden, MA: Blackwell

Nuttall, Mark. 2009. Living in a world of movement: Human resilience to environmental instability in Greenland. In *Anthropology and Climate Change: From Action to Social Transformation.* Susan A. Crate and Mark Nuttall, eds. pp. 292–310. Walnut Creek, CA: AltaMira Press

Nuttall, Mark, Berkes, Fikret, Forbes, Bruce, Kofina, Gary, Vlassov, Tatrina, and Wenzel, George. 2004. *In Impacts of a Warming Arctic: Arctic Climate Impact Assessment.* Carolyn Symon, Lelani Arris, and Bill Heal, eds. pp. 649–660. Cambridge: Cambridge University Press

O'Reilly, Jessica, Isenhour, Cindy, McElwee, Pamela, and Orlove, Ben. 2020. Climate change: Expanding anthropological possibilities. *Annual Review of Anthropology* 49: 13–39

Orlove, Ben. 2005. Human adaptation to climate change: A review of three historical cases and some general perspectives. *Environmental Science & Policy* 8: 589–600

Orlove, Ben and Caton, Steven C. 2009. Water as an object of anthropological inquiry. In *The Question of Resilience: Social Response to Climate Change.* Kirsten Hastrup, ed. Pp. 31–47. Copenhagen: Det Kongelige Danske Videnskabernes Selskab

Orr, Yancey, Lansing, J. Stephen and Dove, Michael R. 2015. Environmental anthropology: Systemic perspectives. *Annual Review of Anthropology* 44: 153–168

Popper, Karl. 2002. *The Logic of Scientific Discovery.* London: Routledge

Ravanneid, Aase. 2021. *Perceptions of Climate Change in India.* London: Routledge

Ray, Celeste. 2002. Cultural paradigms: An anthropological perspective on climate change. In *Global Climate Change.* Sharon L. Spray and Karen L. McGlothin, eds. pp. 81–100. Lanham, MD: Rowman & Littlefield

Raynor, Steve and Malone, Eleanor. 1998. Why study human choice and climate change. In *Human Choice and Climate Change*, Vol. 1: *The Societal Framework*. Steve Rayner and Elizabeth L. Malone, eds. pp. xiii–xlii. Columbus, OH: Batelle Press

Roberts, J. Timmons and Grimes, Peter E. 2002. World-system theory and the environment: Toward a new synthesis. In *Sociological Theory and the Environment: Classical Foundations, Contemporary Insights*. Riley Dunlap, Frederick Buttel, Peter Dickens, and August Gijswift, eds. pp. 167–198. Lanham, MD: Rowman & Littlefield

Rockström, Johan. 2011. Science's role and responsibility. In *Bankrupting Nature: Denying Our Planetary Boundaries*. Anders Wijkman and Johan Rockström, pp. 19–35. London: Earthscan

Roncoli, Carla, Crane, Todd, and Orlove, Ben. 2009. Fishing climate change in cultural anthropology. In *Anthropology and Climate Change: From Action to Social Transformation*. Susan A. Crate and Mark Nuttall, eds. pp. 87–115. Walnut Creek, CA: Left Coast Press

Schnegg, Michael. 2021. Ontologies of climate change: Reconciling indigenous and scientific explanations for the lack of rain in Namibia. *American Ethnologist* 48: 26–273

Singer, Merrill. 2011. Down cancer alley: The lived experience of health and environmental suffering in Louisiana's chemical corridor. *Medical Anthropology Quarterly* 25: 141–163

Singer, Merrill. 2018. *Climate Change and Social Inequality: The Health and Social Costs of Global Warming*. Abingdon: Routledge

Singer, Merrill. 2021. *EcoCrises Interaction: Human Health and the Changing Environment*. Hoboken, NJ: Wiley

Singer, Merrill. 2023. Climate change and retreatist anthropology. *Public Anthropology Blog*. https://publicanthropologist.cmi.no/2023/10/13/climate-change-and-retreatist-anthropology/

Speth, James Gustav. 2008. *The Bridge at the Edge of the World: Capitalism, the Environment, and Crossing from Crisis to Sustainability*. New York: Oxford University Press

Stern, Nicholas. 2007. *The Economics of Climate Change: The Stern Review*. Cambridge: Cambridge University Press

Stensrud, Astrid B. and Eriksen, Thomas Hylland, eds. 2019. *Climate, Capitalism and Communities: An Anthropology of Overheating*. London: Pluto Press

Steward, Julian. 1955. *Theory of Culture Change: The Methodology of Multilinear Evolution*. Urbana: University of Illinois Press

Sutton, Mark Q. and Anderson, E.N. 2004. *Introduction to Cultural Ecology*. Walnut Creek, CA: AltaMira Press

Szolucha, Anna. 2021. Watching fracking: Public engagement in postindustrial Britain. *American Ethnologist* 49: 77–91

Townsend, Patricia. 2011. The ecology of disease and health. In *A Companion to Medical Anthropology*. Merrill Singer and Pamela I. Erickson, eds. pp. 181–195. Malden, MA: Wiley-Blackwell

Walker-Crawford, Noah. 2023. Climate change in the courtroom: An anthropologist of neighborly relations. *Anthropology Theory* 23: 76–99

Wijkman, Anders and Rockström, Johan. 2011. *Bankrupting Nature: Denying our Planetary Boundaries*. London: Earthscan

Wilkinson, Marian. 2020. *The Carbon Club: How a Network of Influential Climate Sceptic Politician and Business Leaders Fought to Control Australia's Climate Policy*. Sydney: Allen & Unwin

Wolf, Eric. 1972. Ownership and political ecology. *Anthropological Quarterly* 45(3): 201–205

Wolf, Eric. 1982. Ownership and political ecology. In *Europe and the People without History*. Berkeley: University of California Press

3

ANTHROPOCENE, CAPITALOCENE, OR WHATEVER?

Rethinking Our Era of Climate Change Production

Introduction

Since its advent in the late 19th century, anthropology has in one capacity or other recognized the human-nature nexus, although other social science disciplines also have done so, particularly geography. Ironically, over the past decade or so, geologists and other earth scientists have adopted the notion of the *Anthropocene*, a term which bears close resemblance to anthropology. Andrew Mathews (2020: 67) argues that the Anthropocene is prompting anthropologists to engage in "experiments and collaboration beyond anthropology." He adds that the Anthropocene is a "term that inspires collaboration among natural scientists, social scientists, humanities scholars, and artists" (Mathews 2020: 73). In his Distinguished Lecture at the American Anthropological Association annual meeting in December 2014, Bruno Latour (2017) asserted that geologists had presented anthropologists with a gift by positioning the *anthropos* to center stage as the main geological force shaping the face of the planet Earth and thus dubbing this new epoch the Anthropocene. He explained:

> In other words, to designate the present period as that of the Anthropocene is to tell the other disciplines that the task of joining "physical" and "cultural anthropology" (I purposely use labels pertaining to the past of your discipline) is not your exclusive undertaking. Suddenly, without you having even asked for help, hundreds of subfields are also busy doing it. Everybody, it seems, is now converging on the same problem, ready to make the same mistakes and to live through the same traumatic experience that the discipline of anthropology as a whole had lived through since the beginning of the nineteenth century, namely, how to get bones and divinities to fit together.
>
> *(Latour 2017: 37)*

DOI: 10.4324/9781003469940-4

While a whole industry has developed around the notion of the Anthropocene with the term frequently being used in publications and the mass media, almost a decade later, it is not clear whether it best encapsulates an analysis of the human-nature nexus. However, anthropologist Alf Hornborg (2019: 8) maintains:

> In sum, the discourse on the Anthropocene has (1) strengthened environ-mental arguments, (2) challenged the conventional distinction between Society and Nature, (3) suggested that humans are a unique form of life, and (4) intensified the debate on climate justice.

Bearing this thought in mind, this chapter explores various notions of the Anthropocene, including the Early Anthropocene epoch dating back to the human invention of fire, the Middle Anthropocene epoch dating back to development of agriculture, and the Late Anthropocene epoch dating back to the Industrial Revolution dependency on fossil fuels, and the advent of the Great Acceleration in the wake of World War II, with the development of nuclear weaponry and technology. We also explore two political ecological perspectives on the Anthropocene, one that prefers to grapple with the con-cept, despite its deficiencies, and the other which seeks to transcend it by adopting the concept of the *Capitalocene*, particularly given that the late Anthropocene has coincided with the development of capitalism as a regional system initially situated in Western Europe into a hegemonic world political-economic system. In this chapter we present another alternative to the Anthropocene conception based on the criticisms of Indigenous peoples.

The Anthropocene as a Contested Concept

The term Anthropocene reportedly was "introduced in the 1980s by biologist Eugene Stoermer, with whom [Paul] Crutzen published a discussion on the meaning of the Anthropocene in 2000" (Provenzale 2023: 167). Crutzen, a Dutch atmospheric chemist, and Stoermer, a marine biologist, suggest that the geological age in which humans are living should be renamed the Anthropocene ("Age of People") (Crutzen and Stoermer 2000). They argue that for the past 150 years it is human activity that has had the most impact on shaping the bio-geological environments of Earth and its climate, more so than natural forces. Crutzen and Stoermer maintain that the Anthropocene began in the later part of the 18th century, when analyses of air trapped in polar ice showed the beginning of growing global concentrations of carbon dioxide and methane. This period also coincides with James Watt's design of the steam engine in 1784.

In reality, Provenzale (2023: 168) maintains:

> The idea of an era in which the atmospheric effect has become dominant is not a new one. As early as the end of the nineteenth century, Italian geologist and paleontologist Antonio Stoppani proposed the expression

the "Anthropozoic era" to indicate precisely that new and dominant human intervention.

More recently, Crutzen and other scholars have dated the beginning of the Anthropocene to around 1945, marked by the creation of the atomic bomb and the onset of incredible global economic growth.

Early, Middle, and Late Anthropocene

However, US paleoclimatologist William Ruddiman (2005) pushed the beginnings of the Anthropocene further back in his recognition that carbon dioxide emissions began slowly to increase as humans started to clear the land in their shift from foraging to farming about 10,000 years ago in places such as China, India, and the Americas, and that methane emissions began to increase around 5,000 years ago as various populations started to irrigate for rice production and to rear livestock. He maintains that this earlier Anthropocene changed the normal oscillation between glacial and interglacial periods, with the last glacial period being the fourth one. Thus, the Anthropocene has delayed the onset of yet another glacial period. Conversely, some might argue that we are we still in a receding glacial period given that there is ice on Earth at the poles and on mountains around the world.

However, some scholars push the advent of the Anthropocene even further back. Steffin et al. (2007: 614) assert: "The mastery of fire by our ancestors provided humankind with a powerful monopolistic tool unavailable to other species, that put us firmly on the long path towards the Anthropocene ... Fire allowed humans to migrate to harsh climatic zones." In the tripartite typology of the Anthropocene, geologist Andrew Glikson (2013: 91) pushes the Anthropocene even further back with the Early Anthropocene, starting roughly 2 million years ago with the discovery of fire, the Middle Anthropocene starting with the development of "extensive grain farming" in the Near East, and finally the Late Anthropocene commencing with the "onset of combustion of fossil fuels," beginning with coal. However, there is evidence that the Neanderthals were burning coal in Europe over 50,000 years ago (Pettit 2011). In terms of the modern era, anthropologist Leslie C. White (2008: 118) referred to coal as the "king, or the father, of the Fuel Revolution," which served as part and parcel of "modern capitalist culture," a fact that suggests to various scholars that the Late Anthropocene might be more appropriately termed the Capitalocene. Lewis and Maslin (2015) argue that European colonization of the America may mark the advent of the Anthropocene, suggested 1610 as a starting date, an event that resulted in the transfer of plant and animal species between the Old and New Worlds.

Geographer Frans Berkhout (2014: 156) maintains that social scientists can play a major role in predicting and forecasting of the future, including Anthropocene futures, despite their reluctance to venture into this endeavor.

Indeed, reflecting on the first Grand Challenge Symposia organized by the Smithsonian Institution in 2012, anthropologist Shirley Fiske (2012) emphasized that "the meaning of the Anthropocene is ethical and moral—how do we want the future to look and what can we do with the knowledge we have?" Aside from human intentions, it is clear that, barring a global catastrophe (e. g., a major meteor impact, all-out nuclear war, an unstoppable infectious disease pandemic, or worst-case global warming), humanity will likely remain a major environmental force for the foreseeable future. Despite his recognition of the limitations to all future scenarios, Berkhout (2014: 156–158) delineates some possible Anthropocene futures:

- The possibility of peak oil can be expected to result in a drive for non-oil energy sources, something that has already happened with the fracking of shale gas and the gradual turn to renewable energy sources, particularly solar and wind energy.
- The adverse impact of climate change on agricultural production can be expected to "generate the search for new, more diversified but intensified global food production systems."
- A diminishing availability of phosphorous for agriculture may stimulate low-phosphorous forms of agricultural production.
- New planetary risks will be very unevenly distributed, creating "winner" and "loser" scenarios around the world, at least in the short run.
- In other words, "[s]hort of a real cataclysm, it is likely that 'good' and 'bad' Anthropocenes will continue to exist side-by-side."

In essence, much of the discussion of winners and losers in assessing the impact of climate change on human societies closely resembles neoclassical microeconomic or neoliberal discourse. For example, Ward (2010: 196) asserts:

Among the winners and losers will be locales that today are too cold to be desirable for year-round dwelling. Through geographic accident, most such places are in the Northern Hemisphere. The biggest victors will be Canada, Alaska, Greenland, Russia, and Scandinavia, and in the Southern Hemisphere, Argentina most of all. Perhaps future world power will not be relocated to the countries in the Southern Hemisphere, as is often predicted, but stay concentrated, if not redistributed, in the Northern Hemisphere.

The Good Anthropocene

While recognizing that both the planet and humanity face the danger of ecological disaster, the notion of the Good Anthropocene asserts that "we can save ourselves and planet with technological fixes and other strategies that provide opportunities for industry, science, and technology" (Sklair 2020: 178). Latour (2012) expressed an ambivalent view of the Anthropocene but sometimes seem to accept the notion of the Good Anthropocene. He lamented that Green politics "has succeeded in leaving citizens nothing but a gloomy asceticism, a terror of trespassing Nature, and a diffidence toward industry, innovation, technology, and science" (Latour 2012: 4). Latour (2012: 5) called for management of the environment rather than seeking to keep it in a pristine state, and asserted that nature "needs our constant care, our undivided attention, our costly instruments, our hundreds of thousands of scientists, our huge institutions, our careful funding." He became intrigued with ecomodernism having read and reviewed Ted Nordhaus and Michael Schellenger's (2007) book *Break Through* and later became a fellow at the Breakthrough Institute, the principal ecomodernist think tank. Despite Latour's earlier flirtation with ecomodernism, he now expresses reservations about it, referring to it as a "monster":

> A great technical fix which will allow the addicted to behave just as before, except now he or she will go on with the benefit of a high tech product and the happy support of his or her physician, mother and significant other. In other words, "ecomodernism" seems to me another version of "having one's cake and eating it too."
>
> *(2015: 2020)*

The Patchy Anthropocene

Various anthropologists have come to grapple with what they term the *patchy Anthropocene*. Borrowing the term *patch* from landscape ecology—where it is used to refer to the fact that ecosystems contain an unevenly distributed mixture of organisms and other features—Tsing et al. (2019: S186) maintain the patchy Anthropocene illustrates the "uneven conditions of more-than-livability in landscapes increasingly dominated by industrial forces," such as in the case of coffee production. They propose five ways to illustrate the patchy Anthropocene concept, noting that landscapes show us the existence of Anthropocene patches (Tsing et al. 2019: S187–S188):

The first proposed way entails "*Noticing landscapes shows us Anthropocene patches*" (Tsing et al. 2019: S187; emphasis in the original). They argue that humans have always reshaped landscape structures such as forests, cities, and plantations, but since the advent of the Anthropocene landscape structures have been shaped and disturbed as a result of European colonialism, imperialism, global capitalism, and industry.

The second proposed way states: "*Two kinds of landscape structures are key to the anthropogenic disturbances that we call Anthropocene: 'modular simplifications' and 'feral proliferations'*" (Tsing et al. 2019: S187; emphasis in the original). They delineate dominant landscape forms in the contemporary world, namely modular simplified forms and feral proliferations. A modular simplified form could be a plantation or a mineral extraction site. A coffee plantation, for instance, could invite feral proliferations, such as avian viruses capable of infecting both humans and other birds.

The third proposed way states: *"Systems are thought experiments with which to make sense of structures"* (Tsing et al. 2019: S187; emphasis in the original). Three kinds of system-thinking illuminate the patchy Anthropocene: ecological models, nonsecular cosmologies, and political economies. Earth systems and climate models have served to depict the impact of humans on nature during the Anthropocene. Nonsecular cosmologies depict spirits as entities that may intervene into the human-nature nexus in a multiplicity of settings, such as chicken farms, scientific laboratories, or even airports.

The fourth proposed way asks: *"Can we acknowledge catastrophe while engaging possibility?"* (Tsing et al. 2019: S188; emphasis in the original). In contrast to the "doomism" often associated with the Anthropocene, can humans find light at the end of the tunnel, in the way that ecomodernists seek to do by proposing technological solutions as part of a Good Anthropocene? While Tsing et al. do not suggest the possibility of a socio-ecological revolution, can ecological and climatic crises provoke an environmental proletariat to rise from the ashes and create an alternative world system based upon social justice, deep democracy, and environmental sustainability? To paraphrase Gramsci, it is a matter of pessimism of the intellect, optimism of the will.

The fifth proposed way views that *"patches" are sites for knowing intersectional inequalities among humans* (Tsing et al. 2019: S188; emphasis in the original). Patches illustrate histories of genocide, displacement, exploitation, oppression, as well as ecocide, particularly exemplified over the course of the past several hundred years of capitalist history. Tsing et al. (2019) assert that the patchy Anthropocene brings social justice-based analysis into Anthropocene studies.

Dipesh Chakrabarty's Postcolonial Historical Take on the Anthropocene

In *The Climate of History in a Planetary Age*, Dipesh Chakrabarty (2021: 1), an eminent postcolonial historian, asserts that the COVID-19 pandemic, the rise of authoritarian, racist, and xenophobic regimes around the world, discussions about climate change, fossil fuels, renewable energy, water shortages, biodiversity loss, the Anthropocene, etc. "signal that something is amiss with our planet and that this may have to do with human actions." Chakrabarty (2021: 26–43) revisits his four theses first delineated in a widely circulated article, "The climate of history," published in 2009:

- Thesis 1: Anthropocentric explanations of climate change spell the collapse of the humanist distinction between Natural History and Human History.

- Thesis 2: The idea of the Anthropocene, the new geological epoch when humans exist as a geological force, severely qualifies humanist histories of modernity/globalization.
- Thesis 3: The geological hypothesis regarding the Anthropocene requires us to put global histories of capital in conversation with the species history of humans.
- Thesis 4: The crosshatching of species history and the history of capital is a process of probing the limits of historical understanding.

Chakrabarty asserts that anthropogenic global warming illustrates the collision of three histories, namely the history of the Earth systems; the history of living beings, including humans, on the planet; and the more recent history of industrial civilization or capitalism. While admitting that climate change raises serious moral and political issues, he argues that even a "more prosperous and just world made up of the same number of people as today" could be one in which the "climate crisis could be worse" (Chakrabarty 2021: 57). This assertion appears to view the poor as collateral damage and overlooks the fact that a more just world would inevitably have to be a postcapitalist one in which there would be a more even playing field in which there would be no large distinctions in access to resources among humans. Furthermore, it would be imperative that such a world would entail a radical decarbonization agenda, in contrast to earlier socialist-oriented states, such as the Soviet Union and the German Democratic Republic; and it would entail a weaning away from coal, petroleum, and natural gas in both the Global North and the Global South as quickly as possible.

Chakrabarty (2021: 67) says that the climate change literature reconfigures an older debate on anthropocentrism and nonanthropocentrism. The climate crisis demonstrates the "planet's otherness" (Chakrabarty 2021: 67) and that humans are latecomers to Earth who function in a "position of passing guests" (Chakrabarty 2021: 67) or as a mere blip in cosmic time. Chakrabarty (2021: 68) asserts that Earth system science, a product of the Cold War and the race to space, entails the conjuncture of three histories: "the history of planet, the history of life on planet, and the history of a globe made by logics of empires, capital, and technology." The globe is a sociohistorical construction and a by-product of globalization, not only in terms of its land surface but also its skies and waters, a process which has resulted in anthropogenic global warming. Conversely, Chakrabarty reports that planetary science tells us that global warming has occurred both on Earth—in the distant past chiefly owing to natural causes—and on other planets. Many Earth scientists fear that anthropogenic global warming may herald the sixth Great Extinction.

For Chakrabarty (2021: 86), whereas the global is a human-centered process, the planetary "discloses vast processes of unhuman dimensions." He argues that the planetary crisis has prompted important insights from both posthumanists who query the nature/culture dualism and some Marxists who want to refer to the Anthropocene as the Capitalocene. At any rate, he observes that the "climate crisis concerns the balance of all terrestrial life on planet" (Chakrabarty 2021: 128).

With climate denialism still rampant around the world, Chakrabarty identifies two principal approaches to mitigating climate change: (1) green capitalism, entailing a rapid shift to renewable energy coupled with market mechanisms; and (2) some form of postcapitalism. In reality, the former is hegemonic, while the latter is marginal but appears to be on the ascendency, at least in terms of advocacy, an issue that we explore in subsequent chapters. Chakrabarty (2021: 180) asserts that climate change defies the "ontic certainty of earth that humans have enjoyed through Holocene epoch and perhaps for longer." He maintains that the notion of the Anthropocene recognizes that humans have been interfering with processes that make the planet habitable for complex life forms, including themselves. However, the notion of the Anthropocene tends to downplay the fact that certain actors, such as wealthy multinational corporations, have contributed much more to this interference, something that Marx recognized in his assertion that capitalism is in a metabolic rift with nature (Foster 2000).

While Chakrabarty does not wish to take sides in the debate about the pros and cons of geoengineering as a viable climate mitigation strategy, he does observe that geoengineering champions "belong as a rule to sciences that are ahistorical in their analytical approach—such as physics and chemistry" (Chakrabarty 2021: 182). In the postscript to his book titled "The global reveals the planetary" he engages with the ecomodernist writings of Bruno Latour. Ironically, Crutzen (2006) suggested that failure to reduce greenhouse gas (GHG) emissions might warrant geoengineering, a stance that has been given more consideration in policy and research areas, including on the part of the Breakthrough Institute.

Debate Between Anthropocene and Capitalocene

Aside from a heated debate as to the origin point of the Anthropocene concept, various political ecologists, eco-socialists, and neo-Marxist scholars have asserted that the term Anthropocene is misleading for a number of reasons, including that it inadvertently—even if unintentionally—implies that all humans have been equally complicit in contributing to a rise of GHG emissions, anthropogenic climate change, and an ecological crisis (Malm and Hornborg 2014).

In this section, we examine two books, one edited by Jason Moore (2016) titled *Anthropocene* or *Capitalocene?*, and the other authored by Ian Angus (2016) titled *Facing the Anthropocene*. What is distinct about the two books reviewed in this chapter is that by and large their respective contributors adopt political, ecological, or perhaps more precisely, eco-Marxist perspectives in addressing the Anthropocene concept. However, on the one hand, Jason Moore (2016) and most of his collaborators, and Ian Angus (2016), on the other hand, adopt different stances on the utility of the Anthropocene concept, with the former group preferring the concept of the Capitalocene and the latter preferring to work with the concept of the Anthropocene.

Anthropologist Alf Hornborg (2015: 57) reports:

> "Capitalocene" was coined by Andreas Malm at a seminar in Lund in 2009. It usefully emphasises the role of "capitalism" in generating trans- formations of the biosphere, but might raise the objection that various forms of capital accumulation had caused ecological degradation, albeit before the Industrial Revolution.

The contributors to *Anthropocene or Capitalocene?* generally prefer Capita- locene to Anthropocene. Moore (2016) in his chapter asserts that the defor- estation of vast regions that began in Europe under medieval feudalism sped up under capitalism, particularly between 1450 and 1750 when the greatest landscape revolution in human history occurred on both sides of the Atlantic. The Capitalocene as a world ecology system ramped up reliance on cheap labor and cheap energy, particularly in the form of fossil fuels. In a similar vein, Elmar Altvater (2016), who is well known for his notion of fossil capit- alism, argues that while it is impossible to give an exact date for the beginning of the Anthropocene, its onset probably occurred between the beginning of European modernity in Braudel's long 16th century and the industrial-fossil revolution of the second half of the 18th century. He asserts that with the Industrial Revolution, owners of the capitalist mode of production converted nature into a capital asset. In her chapter Eileen Crist (2016) argues that while Anthropocene exponents have understandable misgivings about too disruptive a climate, too much manmade nitrogen, or too little biodiversity, the Anthropocene discourse calls us to the high road of becoming good managers of the standing reserve, a position that is compatible with a green capitalist agenda, one which she rejects. Daniel Hartley (2016) raises an additional problematic in noting that as a way of talking about geological changes the Anthropocene discourse is relatively harmless. Danger arises, however, when geologists enter the political arena, calling for collective eco- logical intervention on the basis of the Anthropocene. In delineating various conceptual problems associated with the Anthropocene discourse, he notes that *Anthropos*, or the human, is never clearly defined, but relies on vague, ahistorical terms such as the *human enterprise*, whereas a historical concep- tion of humanity, in contrast, would see humans as internally differentiated and constantly developing through contradictions of power and (re)produc- tion. Christian Parenti (2016) implores critical scholars wishing to elucidate the Capitalocene to theoretically grapple with the state, which has historically played a role in seizing portions of the Earth's surface, measuring it, control- ling it militarily, legally, and scientifically, and developing communication and transportation networks upon it.

On the whole, *Anthropocene or Capitalocene?* constitutes a noble effort that moves the Anthropocene discourse to a new level, one that identifies global capitalism as the elephant in the room when it comes to the ecological and climatic crisis and the need to transcend it with a more sustainable world system, although the parameters of an alternative are not explicitly defined. The deeper question is how to shift earth scientists to recognize that the Capitalocene describes human impacts upon the Earth's geology better than the term Anthropocene.

In contrast to contributors to Moore's anthology, Ian Angus (2016) in *Facing the Anthropocene* advocates working with proponents of the Anthropocene, at least the more progressive ones. He argues that after remaining the exclusive domain of Earth science specialists for nearly a decade, the Anthropocene concept has been adopted by social scientists, humanists, and journalists and has become the focus of three academic journals. Angus (2016: 38) also views the Anthropocene concept as a useful organizing tool for eco-socialists, arguing:

> So the challenge for socialists is not to proclaim the revolution from every street corner, but rather to unite the broadest possible range of people, socialists or not, who agree that the climate vandals must be stopped. We need to work with everyone who is willing to join in fighting climate change in general, and the fossil fuel industry specifically (216?). He explains that in 2004 a team of earth scientists associated with the International Geosphere-Biosphere Program (IGBP) compiled a report titled Global Change and the Earth System: A Planet Under Pressure (Steffen et al. 2004) which includes B24 graphs—twelve showing historical trends in human activity (GDP growth, population, energy consumption, water use, etc.) and twelve showing physical changes in the Earth System (atmospheric carbon dioxide, ozone depletion, species extinctions, loss of forests, etc.) over 250 years.

Angus (2016: 218–221; emphasis in the original) asserts that the report makes an invaluable contribution to broad understandings of the Earth system in that it reveals that every trend line showed gradual growth from 1750 and a sharp upturn from about 1950. Angus maintains that eco-socialist movements must adopt four principles in order to be successful in their efforts in the Anthropocene:

- We must be pluralist and open to differing views within the green left.
- We must constantly extend our analysis and program in the light of changing political circumstances and scientific knowledge.
- *We must be internationalist and anti-imperialist.*
- We must actively participate in and build environmental struggles, large and small.

While his position is one that certainly aligns with eco-socialist and green left thought, he is fully cognizant that most leftists prefer the term Capitalocene to Anthropocene. Angus (2016: 232) maintains that substituting the latter for the former is a "category mistake." He asserts that capitalism is a 600-year-old social and economic system, while the Anthropocene is a 60-year-old Earth system epoch that will continue long after capitalism is a distant memory (Angus 2016: 232).

While wishing to retain the Anthropocene concept, John Bellamy Foster and Brett Clark (2021) seek to refine the Anthropocene concept by viewing the *Capitalinian* and the *Communian* as sub-types. They object to substituting Capitalocene for Anthropocene because the latter is "deeply embedded in natural science" (Foster and Clark 2021: 5). For them, the Capitalinian conforms to the historical period which many environmental historians view as having commenced around 1950, in the aftermath of World War II, the rise of multinational corporations, and unleashing of the process of decolonization and global development, thus constituting the first phase of the Anthropocene (Foster and Clark 2021: 2). While the Anthropocene is a by-product of capitalism, it would endure even if capitalism were to be transcended through a "Great Climacteric" marking a transition to a more sustainable world system.

Foster and Clark (2021: 8) associate the Capitalinian with the Great Acceleration of global monopoly capitalism and the advent of nuclear weapons technology in 1950s, "resulting in an age of planetary ecological crisis." This was also an era when various radioactive elements entered in lifeforms and when plastics and synthetic petrochemicals became major components of the world economy. For Foster and Clark (2021: 14), the Communian will hopefully be the next Anthropocene age—barring an Anthropocene human extinction—when the great mass of humanity "reaffirm[s] communal relations with earth."

Alternatives to the Anthropocene and Capitalocene

Beyond the debate as to whether the Anthropocene or the Capitalocene is the better term to describe the human-nature nexus at least in the modern era, a slew of competing terms have been proposed. These has included the *Chthulucene*, an awkward term proposed by Donna Haraway (2016) to refer to an epoch in the making and consisting of ongoing multispecies stories and practices of becoming-with in times that remain at stake, in precarious times, in which the world is not finished; the *Necrocene* referring to an "epoch in which capitalism has produced a Sixth Extinction while hoping to invent new corpses upon which to feast, such as it did with coal and then oil and then natural gas" (McBrien 2016); the *Plantationocene* referring to the "devastating transformation of diverse kinds of human-tended farms, pastures, and forests into extractive and enclosed plantations, relying on slave labor and other forms of exploited, alienated, and usually spatially transported labor"

(Haraway 2016: 162); and *Polemocene* as a designator for the Anthropocene as an "Age of War" (Bonneuil and Fressoz 2017). Finally, James Lovelock (2019), a proponent of the Gaia hypothesis which views Earth systems as constituting an interactive superorganism, proposed a successor to the Anthropocene called the *Novocene* that would see the advent of intelligent robotic agents or cyborgs which would exhibit thought patterns and intelligence greatly exceeding those of humans. In contrast to humans, he maintains that cyborgs driven by artificial intelligence would save both the planet and humanity from catastrophic climate change because their super-intelligence would recognize the danger of global warming and enable them to take the steps necessary to mitigate it. All these terms and various others have not served as longstanding replacements for either the Anthropocene or Capitalocene.

Conclusion

Spangenberg (2022: 367) expresses reservations about utilizing the term Capitalocene which implies a:

[C]all to arms against capitalism as a necessary condition to end the socio-dynamic unavoidably transgressing the planetary boundaries. Therefore, the question is if first capitalism has to be overcome, ending this economic dynamic and its impacts (or to be able to rein it in in a second step). If so, that would be bad news—while the resistance against capitalism has been going on for more than 170 years or so now with no end in sight, we have just one, at best two decades, to implement a "Great Transition" to safeguard liveable environmental conditions for ourselves and future generations.

Conversely, O'Lear et al. (2022: 3) argue:

Unlike the Anthropocene, the Capitalocene necessarily extends to the dominance and functioning of the neoliberal state and multilateral governance institution system, which works to manage human-environment relation within a global capitalist framework … More specifically, we suggest that the predominantly technomanagerial approaches to disasters pursued within the system reveal the tensions in addressing the causes of environmental change and disasters under capitalism. We argue that through an engagement with the Capitalocene, environmental politics could further contribute to a more nuanced, critical understandings of disasters and their making that in ways that foreground their in/justice implications.

As for ourselves, while the Anthropocene concept may foster dialogue between earth scientists and other scholars, we remain skeptical whether the concept explains anything that we have not known for a long time, namely that human activities, particularly those that fall under the rubric of global capitalism, have had and continue to have a profound impact on the Earth system. After all, while some scholars have embraced the Anthropocene concept or some variant of it with gusto, others have remained relatively indifferent to it. The deeper question is how do we humans learn to live in harmony with each other and with nature as we push further and further into an age characterized by social disruption and ecological and climatic crises. In our experience, while many earth scientists, particularly climate scientists, are aware that humans are in a metabolic rift with nature, something that Marx and Engels observed (Foster 2000), most of them are not willing to recognize global capitalism as the principal driver of the ecological and climatic crisis in the modern era.

References

Altvater, Elmar. 2016. The Capitalocene, or, geoengineering against capitalism's planetary boundaries. In *Anthropocene or Capitalocene: Nature, History, and the Crisis of Capitalism*. Jason W. Moore, ed. pp. 138–152. Oakland, CA: PM Press

Angus, Ian. 2016. *Facing the Anthropocene: Fossil Capitalism and the Crisis of the Earth System*. New York: Monthly Review Press

Berkhout, Frans. 2014. Anthropocene Futures. *The Anthropocene Review* 1(2): 154–159

Bonneuil, Christophe and Fressoz, Jean-Baptiste. 2017. *The Shock of the Anthropocene*. London: Verso

Chakrabarty, Dipesh. 2021. *The Climate of History in a Planetary Age*. Chicago: University of Chicago Press

Crist, Eileen. 2016. On the poverty of nomenclature. In *Anthropocene or Capitalocene: Nature, History, and the Crisis of Capitalism*. Jason W. Moore, ed. pp. 14–33. Oakland, CA: PM Press

Crutzen, Paul J. 2006. Albedo enhancement by stratospheric sulfur injections: A contribution to resolve a policy dilemma. *Climatic Change* 77: 211–219

Crutzen, Paul J. and Stoermer, E.F. 2000. The Anthropocene. *Global Change Newsletter* 41: 17–18

Fiske, Shirley 2012. Global climate change from the bottom up. In *Applying Anthropology in the Global Village*. Christina Wasson, Mary Odell Butler, and Jacqueline Copeland-Carlston, eds. pp. 143–172. Walnut Creek, CA: AltaMira Press

Foster, John Bellamy. 2000. *Marx's Ecology: Materialism and Nature*. New York: Monthly Review Press

Foster, John Bellamy and Clark, Brett. 2021. The Capitalinian: The first geological age of the Anthropocene. *Monthly Review*, September: 1–16

Glikson, Andrew. 2013. Fire and human evolution: The deep-time blueprints of the Anthropocene. *Anthropocene* 3: 89–92

Haraway, Donna J. 2016. *Staying with the Trouble: Making Kin to the Chthulucene*. Durham, NC: Duke University Press

Hartley, Daniel. 2016. Anthropocene, Capitalocene, and the problem of culture. In *Anthropocene or Capitalocene: Nature, History, and the Crisis of Capitalism*. Jason W. Moore, ed. pp. 154–165. Oakland, CA: PM Press

Hornborg, Alf. 2015. The political economy of the technocene: Uncovering ecologically unequal exchange in the world system. In *The Anthropocene and the Global Environmental Crisis: Rethinking Modernity in the New Epoch*. Clive Hamilton, Christophe Bonneuil, and Francois Gemenne, eds. pp. 57–69. London: Routledge

Hornborg, Alf. 2019. Colonialism, in the Anthropocene: The political ecology of the money-energy-technology complex. *Journal of Human Rights and the Environment* 10: 7–21

Latour, Bruno. 2012. Love your monsters: Why we must care for our technologies as we do our children. Feb. 14. https://www.bruno-latour.fr/sites

Latour, Bruno. 2015. Fifty shades of green. *Environmental Humanities* 7: 219–225

Latour, Bruno. 2017. Anthropology at the time of the Anthropocene: A personal view of what is to be studied. In *The Anthropology of Sustainability*. M. Brightman and J. Lewis, eds. pp. 35–49. New York: Palgrave Macmillan

Lewis, S.L. and Maslin, M.A. 2015. Defining the Anthropocene. *Nature* 519: 171–180

Lovelock, James. 2019. *Novacene*. London: Penguin

Malm, Andreas and Hornborg, Alf. 2014. The geology of mankind? A critique of the Anthropocene narrative. *Anthropocene Review* 1(1): 62–69

Mathews, Andrew. 2020. Anthropology and the Anthropocene: Criticisms, experiments, and collaborations. *Annual Review of Anthropology* 49: 67–82

McBrien, Justin. 2016. Accumulating extinction: Planetary catastrophism in the Necrocene. In *Anthropocene or Capitalocene: Nature, History, and the Crisis of Capitalism*. Jason W. Moore, ed. pp. 116–137. Oakland, CA: PM Press

McNeill, John R. and Engelke, P. 2014. *The Great Acceleration: An Environmental History of the Anthropocene since 1945*. Cambridge, MA: Belknap Press of Harvard University Press

Moore, Jason W. 2016. The rise of cheap nature. In *Anthropocene or Capitalocene: Nature, History, and the Crisis of Capitalism*. Jason W. Moore, ed. pp. 78–115. Oakland, CA: PM Press

Nordhaus, Ted and Shellenberger, Michael. 2007. *Break Through: From the Death of Environmentalism to the Politics of Possibility*. Boston, MA: Houghton Mifflin

O'Lear, Shannon, Masse, Francis, Dickinson, Hannah, and Duffy, Rosaleen. 2022. Disaster making in the Capitalocene. *Global Environmental Politics* 22(3): 2–11

Parenti, Christian. 2016. Environmental making in the Capitalocene: Political ecology of the state. In *Anthropocene or Capitalocene: Nature, History, and the Crisis of Capitalism*. Jason W. Moore, ed. pp. 78–115. Oakland, CA: PM Press

Pettit, P. 2011. The rise of modern humans. In *The Human Past* (3rd edn). C. Scarre, ed. Pp. 127–173. London: Thames and Hudson

Provenzale, Antonello. 2023. *History of Climate Change: From the Earth's Origins to the Anthropocene* (trans. Alice Kilgarriff). London: Polity

Ruddiman, William F. 2005. *Plows, Plagues and Petroleum: How Humans Took Control of Climate*. Princeton, NJ: Princeton University Press

Sklair, Leslie. 2020. Introduction to the special issues on world-systems analysis and the Anthropocene. *Journal of World-Systems Research* 26(2): 175–182

Spangenberg, Joachim H. 2022. Inside the Anthropocene-Populo-Consumo-Capitalocene. *Anthropocene Science* 1: 358–374

Steffen W., Sanderson, A., Tyson, P.D., Jäger, J., Matson, P.A., Moore, B. III, Oldfield, F., Richardson, K., Schellnhuber, H.J., Turner, B.L., and Wasson, R. J. 2004. *Global Change and the Earth System: A Planet Under Pressure*. Heidelberg: Springer-Verlag

Steffin, Will, Crutzen, Paul, and McNeill, John R. 2007. The Anthropocene; Are humans now overwhelmingly the great forces of nature? *Ambio* 36: 614–621

Tsing, Anna Lowenhaupt, Mathews, Andrew S., and Bibandt, Nils. 2019. Patchy Anthropocene: Landscape structure, multispecies history, and the retooling of anthropology. *Current Anthropology* (Introduction to Supplement 20) 60: S186–S197

Ward, Peter. 2010. *The Flooded Earth: Our Future in a World without Ice Caps*. New York: Basic Books

White, Leslie A. 2008. *Modern Capitalist Culture*. Burton J. Brown, Benjamin Urish, and Robert L.Carneiro, eds. Walnut Creek, CA: AltaMira Press

4

SOCIAL INEQUALITY AND CLIMATE CHANGE

Case Study: Living at Risk in the Time of Climate Change

In Dhaka, the capital of Bangladesh, poor populations often live in slum dwellings on low-lying land and are at high risk of flooding due to intense monsoon rain and ocean rise. Moreover, high population density makes the poor of Dhaka highly vulnerable to climate change. Climate change impacts Dhaka primarily in two ways: through recurring floods and congested drainage and through heat stress and related conditions. Most of the poor urban dwellers in Dhaka live on the worst-quality land in the city, such as on the edges of ravines, on flood-prone embankments, on slopes susceptible to mudslide or collapse during monsoon rains or in densely packed areas which are especially vulnerable to the adverse effects of climate change. Although they live on the edge, with little support or services provided by the government, the slum dwellers are critical for the daily functioning of Dhaka because they keep the economy going. Through their hard work in an array of low-paying jobs, they provide most of the necessary services needed by wealthier city dwellers (Khan 2010).

Over the last decade, Dhaka has become one of the fastest growing cities in the world. Today, the city's population is estimated at over 20 million, a figure that is projected to keep rising. The United Nations (Hasan and Macdonald 2021) estimates that about four million people live in the city's 5,000 or more slum areas, places that are continually fed by an influx of migrants from the countryside, many of them driven from their homes by extreme weather. In the past, migrants would come to Dhaka, find a way to earn some money, and then return home to their rural villages. But as the effects of climate change have increased, more people are staying in Dhaka's slums permanently despite the many hardships they encounter (e.g., inadequate and poor-quality housing, overcrowding, lack of electricity, poor sanitation).

DOI: 10.4324/9781003469940-5

Bangladesh's rapid urbanization has not been met with needed infra-structure improvements and environmental protections, which has further aggravated daily challenges. No mass transit system exists, nor are there pedestrian sidewalks, and the country's power generating capacity struggles to cover just 35% of the population. Also, children living in urban slums in Dhaka often have poor access to schooling and, if they can attend, they must go to poorer quality schools than students from middle-class households.

The already poor economic conditions of the urban impoverished house-holds in Dhaka become even worse during floods. During "normal" times when the city is not flooded, the average monthly income is only about US $116–$232 (or about $5.80 a day per family). During and after flooding, 30% of families lose their jobs or have their number of work hours cut, changes that befall female workers in particular. At these times, daily family incomes fall to a meager $3.91 per day (Akther and Mokbul 2022). Yet the prices of basic necessities are rising. All daily goods like rice, eggs, and vegetables and gas for cooking are becoming more expensive with global inflation.

Dhaka's poor have elevated rates of many chronic diseases, including heart diseases, respiratory diseases, and diabetes. Children suffer regularly from diar-rhea. During and after flooding, they suffer from additional health complica-tions like severe injuries and various water-borne skin and gastrointestinal diseases. Likewise, research has found that because of climatic change, as disease susceptibility increases, families suffer a decreased ability to adjust to the new daily challenges they face. This diminishes their ability ever to escape from pov-erty (Rahman 2008). Rahman also found that while more than 60% of the Dhaka families he studied could afford three meals daily in "normal" times, during times of intense flooding 60% could afford only a single meal per day. Similar findings have been reported for other areas in the Global South including India (Pandey et al. 2017) and Iran (Keshavarz et al. 2017).

Interviews with people living in extreme poverty in Dhaka reveal the level of suffering people must endure:

> If we are not extreme poor, then who is? Why do I work as a housemaid, why my husband pulls a rickshaw in scorching heat and in rain? Ever wondered how does it feel to work like a servant or slave? I return home in the evening and my husband returns at around 10 pm or even 11 pm. Our children stay at home on their own. It's a shame that I cook good food for other people but can't afford good food for my girls.
>
> *(Housemaid, quoted in Kamruzzaman 2021)*

> Of course, we are extreme poor. Let me tell you why. We live in a small room in a slum, when it rains, we cannot sleep because there are a few holes in the roof. Often there is not enough food for us. When I am sick the situation becomes worse.
>
> *(Rickshaw-puller, quoted in Kamruzzaman 2021)*

Typical in many ways of the lives of people living in Dhaka in the time of global warming is Parul Akter. She was driven from her rural home with her family by rising floodwaters. They found a tiny shack in the capital, which became a home for all six family members. But their ramshackle dwelling abuts a lake and when it rains water comes sloshing in. Only the bed, which they were able to raise with bricks, remains dry. Says Parul, "This room is all we have, so we need to stay here no matter what happens" (quoted in McPherson 2015).

Akther and Mokbul (2022) compared two slum areas in Dhaka with different average levels of household income. On the one hand, in Mirpur slum, 26.5% of households had very low-level incomes, while in Rail line slum, around 75% of households were living with low to very low incomes. On the other hand, in Rail line slum, only 6.5% of households had the highest incomes found in Dhaka slums, while in Mirpur slum about one-third of households were at this level. These differences reflect households' ability to recover from intense flooding, with the people living in Mirpur exhibiting greater capacity and a faster return to their pre-storm standard of living. This finding highlights the tight connection between climate change and poverty; while climate change exacerbates pre-existing poverty, the level of poverty helps to determine human ability to bounce back from extreme climate events.

While conditions are harsh in Dhaka's slums, about 80% of the households in Akther and Mokbul's 2022 study reported that they had supportive social networks with other families living in the slums. Families sought to assist each other as best they could. People watched over each other's children and frail family members during working hours, exchanged essential daily goods when available, and provided financial help to one another. Often these social networks held together during flooding with connected families fleeing in a group for shelter in a flood-free area such as along roadsides or in less impacted slums. The outcomes of this study match the results found in other Global South cities, including Lagos, Nigeria (Adelekan 2010), and several urban areas in Pakistan (Rana and Routray 2018).

It is estimated by some researchers that by the end of the 21st century more than a quarter of Bangladesh will be inundated with water, leaving 15 million people displaced—the equivalent of the inhabitants of New York, Los Angeles, and Chicago combined. It is likely that coastal cities like Dhaka will bear the brunt of climate change-related disasters, including floods, storms, and hurricanes/cyclones.

Because of the obstacles they face, Dhaka's poor endure a vicious cycle involving (a) a comparatively high level of exposure to the adverse effects of climate change; (b) increased susceptibility to the structural and health damages caused by climate change; and (c) decreased ability to cope and recover from the damage suffered putting them at greater risk of future assaults from climate change.

The Human Creation of Inequality

In the New Testament of the Bible, Jesus, while eating at the home of Simon, says: "The poor you will always have with you" (Mark 14:7). Whatever Jesus intended by this remark, from a historic perspective, anthropology has firmly established that there have not always been wealth divides between the rich and the poor. For most of our history—95%—humans have sustained themselves by foraging, involving the gathering of plants and shore-dwelling creatures and the hunting of wild animals in their natural environment. This way of life supported a high level of social and economic equality. There was no private property or ownable real estate, so nature was open to everyone as a place to acquire food and other resources. Bands were mobile in search of resources, taking with them only what they could carry on their backs. People were not rich by contemporary standards, nor were they poor relative to anyone else in their small band or in other nearby or distant groups. They could meet their caloric needs, clothe their bodies, and fashion simple dwellings. Sahlins (1968) called them the original affluent societies, not because their possessions were great (they weren't) but because all their material needs—including security—were almost always met. They were not overworked brutes; rather, they rested when they got tired and slept as much as they needed, and still had time for gathering together to tell stories, teach the young about the environment and its inhabitants (and their behaviors), dance, sing, gossip, decorate their implements and bodies, care for the sick, and engage in healing rituals. Based on various ethnographies of foragers, Sahlins estimated the average work week to be only 15–20 hours. There were no "bosses" or police, social life was kin-oriented, and social control relied solely on collective peer pressure. If someone found even this to be insufferable, they "voted with their feet" and moved to another band in which they had relatives.

The rise of systematic inequality was linked to the emergence of political centralization and the institutionalization of power. The relatively rapid transition—once it began—away from egalitarianism occurred when some members of society were able to gain control of social mechanisms that gave them leverage over productive resources. This movement could only occur after societies could produce a surplus beyond their daily needs and usually this required food production (agriculture and animal husbandry) and settled living patterns. Control over surpluses appears to be the foundation upon which inequality and hierarchy grew.

One force pushing societal movement in this direction may have been climate change. One hypothesis is that drying conditions forced groups to find new subsistence strategies like the domestication of plants and animals. Alternately, the warming of the planet that occurred as global ice sheets melted and carbon dioxide and rainfall levels increased enabled a rise in planetary verdancy and greater control of food sources. Notably, strategies for plant-intensive resource use appeared with the Holocene amelioration of global air temperature. While this began in various local areas around the globe (with site-specific arrays of plants and animals), agriculture and animal husbandry became dominant in all but marginal areas (where foraging or simple horticulture continued).

Through these radical changes, cities and state-level political formations were born and inequality became the norm of human life. The full transition into the capitalist mode of production—which is firmly built on inequality and a powerful economic driver of climate change and which is still the dominant economic system in the world—occurred with the colonial expansion that brought all societies under the yoke of dominant powers. Borrowing from Aristotle, European Christians viewed the Indigenous peoples they encountered in other lands as "natural slaves" who God intended to be subservient to white Europeans. This additional radical change pressed inequality onto all sectors of the planet (e.g., Indigenous peoples who lived on the lands that following colonial penetration became the United States or Australia).

Introducing a Demonic Duo: The Interface of Climate Change and Inequality

Climate change and social inequality are intimately connected because (a) most greenhouse gas (GHG) emissions are produced by the wealthiest nations and the mechanisms of their production and release profit the wealthiest, namely the most elite sections of those nations; (b) it is the world's two billion poorest people, with the fewest resources, who are most directly feeling the adverse effects of climate change and who are the most challenged in being able to manage them; (c) the prevailing narrative about adapting to climate change excludes the world's poorest people; and (d) the predicament of the poor is a product of their historic hierarchical relationships with wealthier sectors of society.

In short, inequality and global warming are inherently related: the poor and other marginalized groups suffer the highest consequences of anthropocentric climate change, while planetary warming promotes impoverishment. In the poorest economies, a sizeable part of the population directly depends on environment-dependent subsistence strategies that may be the most impacted by climate change, including subsistence agriculture, forestry, and fishing. Rising temperatures are exacerbating preexisting disparities in access to clean water and affordable food. Indeed, climate change threatens to set back any progress that has been made in reducing poverty by collapsing poverty eradication efforts worldwide, and this is especially true in the poorest regions and among the poorest people. A report published by the World Bank (2020) estimates that an additional 68–135 million people may be pushed into poverty by 2030 because of climate change. Research by the International Monetary Fund (IMF) (see Guivarch et al. 2021) indicates that if the worst projections of future economic damage from climate change being discussed in the scientific literature hold true, climate change would cause inequality between countries to rise significantly. According to the IMF, the effects of climate change also risk exacerbating inequality within countries. Consequently, the IMF views mitigating climate change as a necessary condition for sustainably improving living standards around the world. At the same time, it maintains that distributive and procedural justice intended to enhance social and economic equity must be at the forefront of every component of environmental and climate-related policymaking. In planning, development, and implementation, the effort to reduce GHG emissions, asserts the IMF, must be at the service of broader objectives of development, such as poverty and inequality reduction, the creation of decent accessible jobs, improvement of air quality, and the improvement of public health.

The link between poverty and climate change is a primary factor in the desperate efforts by Guatemalan families to be accepted in the United States. Guatemala is one of the most densely populated and poorest countries in Central America. More than half of its population lives in rural areas, and of that 70% live in poverty. Three and a half million Guatemalans were classified as food insecure in 2021. The Guatemalan economy is the fifth poorest economy in Latin America and the Caribbean region with persistently high rates of poverty and inequality.

Further, Guatemala is heavily affected by climate and weather events; its poorer populations are particularly vulnerable. The country ranks ninth in the world for level of risk to the effects of climate change. Frequent droughts tied to global warming in what is known as the Dry Corridor have significantly damaged agriculture, which employs more than 30% of Guatemalan workers (Bermeo et al. 2022). Rising rates of family migration from Guatemala to the United States are associated with rural poverty and agricultural stress linked to climate change. There is a strong link between being born in a rural area in Guatemala and migrating as part of a family unit to the United States.

Elite Polluters and the Unequal Distribution of Power

The difference between the GHG emissions of the rich and the poor within countries is now greater than the differences in emissions between countries. This fact points to the ever-expanding divide between a group of rich, elite polluters around the globe and everyone else. The term elite polluter was coined to call attention to the existence of people with lifestyles, investments, and positions in major corporations that not only bear little relation to those of the majority of people but who are responsible for polluting the planet and driving climate change. In terms of lifestyle, elite polluters commonly drive (often multiple) large, expensive cars, own multiple homes, usually travel by air between them and elsewhere around the globe, eat diets rich in costly and imported foods, and routinely purchase new clothes and imported luxury goods. There is a concentration of elite polluter investments in polluting industries like those involved in fossil fuel production. Elite polluters also often hold executive offices in and sit on the board of directors of fossil fuel companies.

Michael K. Wirth, for example, has been CEO and chairman of the board of Chevron Corporation since 2018. One of the primary successors of robber baron John D. Rockefeller, who founded the original Standard Oil Company, Chevron Corporation was founded in 1879 and since then has grown to become one of the world's largest multinational oil and gas industry companies with business operations in 22 countries. Wirth was born in 1960 and graduated from the University of Colorado Boulder. He has had a long career at Chevron, first joining the company in 1982 as a design engineer. Over his many years of employment with the company, Wirth has been responsible for overseeing supply and trading, pipeline and power operating units, business development, corporate strategy, and corporate governance. Since May 2023, Wirth has been Chevron's single largest individual shareholder, owning 1,504,813 shares of the company valued at over US $218 million. He has served on the board of directors for GS Caltex Corporation and Caltex Australia Limited and has served on boards for Catalyst, the American Petroleum Institute, and the International Business Council of the World Economic Forum. Under Wirth's direction, Chevron is alleged to have pursued a variety of greenwashing tactics intended to downplay and conceal the impact of the company's activities on the environment. A coalition of environmental groups filed a Federal Trade Commission complaint against Chevron 2021, stating that under Wirth the company had misled the public by claiming responsibility solely for carbon emissions associated with refining and transporting oil, not the total emissions created by the product it sells (Wright et al. 2021).

The adverse environmental impacts of fossil fuel companies like Chevron have been growing rapidly. Such companies and their products have released more GHG emissions in the last 28 years than in the 237 years prior to 1988. Of the thousands of companies in the world, just 100—including many fossil fuel companies—are responsible for 71% of the global GHG emissions that cause global warming, according to *The Carbon Majors Database*, a report published by the Carbon Disclosure Project (CDP 2017). CDP, a non-profit organization focused on global disclosure of vital information, points out that since 1988 more than half of the world's industrial emissions can be traced back to just 25 state companies and entities.

The CDP report also reveals that 32% of emissions come from public investor-owned companies, making their major investors elite polluters. Over half (52%) of all global industrial GHGs emitted since the start of the Industrial Revolution can be traced to these 100 fossil fuel producers. Notes Pedro Faria, Technical Director at CDP, "In particular, the report shows that investors in fossil fuel companies own a great legacy of almost a third of all industrial GHG emissions, and carry influence over one-fifth of the world's industrial GHG emissions today" (CDP 2017).

Richard Heede (2019a) of the US-based Climate Accountability Institute, one of the world's leading authorities on the role of big oil companies in escalating the contemporary climate emergency, has evaluated the fuel resources that global corporations have extracted from the ground, and the subsequent GHG emissions these fossil fuel producers are responsible for since 1965. This is the year in which the disastrous environmental impact of fossil fuels was known by both industry leaders and politicians.

Chevron topped the list of the eight investor-owned corporations, followed closely by Exxon, BP, and Shell. Together these four global corporations are responsible for over 10% of the world's carbon emissions since 1965. Twelve of the top 20 companies are state-owned and together their extractions are responsible for 20% of total emissions during the same period. The leading state-owned polluter is Saudi Aramco, which alone produced over 4% the global total. Nine of the top producers are government-run industries in countries such as the People's Republic of China, the Russian Federation, and Poland. The research also classified the 90 entities according to type of fossil fuel extracted and marketed. There are 56 oil and natural gas companies, and 37 coal producers on the list. Moreover, the study shows that many of the worst offenders are investor-owned companies that are household names around the world. These corporations spend billions of dollars on lobbying governments for lenient favorable policies while portraying themselves as environmentally responsible businesses.

According to Heede,

> [T]hese companies and their products [which] are substantially responsible for the climate emergency, have collectively delayed national and global action for decades, and can no longer hide behind the smokescreen that consumers are the responsible parties ... Oil, gas, and coal executives derail progress and offer platitudes when their vast capital, technical expertise, and moral obligation should enable rather than thwart the shift to a low-carbon future.
>
> *(Taylor and Watts 2019)*

Michael Mann, one of the world's leading climate scientists, said Heede's study made it clear that:

> The great tragedy of the climate crisis is that seven and a half billion people must pay the price – in the form of a degraded planet – so that a couple of dozen polluting interests can continue to make record profits. It

is a great moral failing of our political system that we have allowed this to happen.

(*Taylor and Watts 2019*)

Another study (Taylor and Watts 2019) found that the five largest stock market-listed (i.e., investor-owned) oil and gas companies spend nearly US $200 million annually lobbying governments to delay, control, or block policies intended to fight climate change.

Edward Collins noted that the

[o]il majors' climate branding sounds increasingly hollow and their credibility is on the line. They publicly support climate action while lobbying against binding policy. They advocate low-carbon solutions [in public statements] but [their investments in climate action] are dwarfed by spending on expanding their fossil fuel business.

(*Laville 2019*)

During 2019, oil and gas companies' expenditure on lobbying increased to US $115 billion, while only $4.5 million was directed at low-carbon projects.

In assessing the role of elite producers of GHGs and resulting climate change, three key issues emerge: (1) the ways social elites treat the environment as a limitless resource and dumping ground for the by-products of production and use, including the release of heat-retaining gases; (2) the ways elite polluters impose the costs of climate change on poor and working people; and (3) the nature of elite strategies to blame the poor for their suffering—including from climate change—while silencing their grievances by deploying tropes of personal responsibility, the innate goodness of economic growth, and the needs of national security, all of which, we argue, are forms of structural violence. Exposed by examination of this form of obscured violence are the ways in which the bodies, actions, communities, and social environments of the poor and marginalized people are intimately related to macroeconomic structures and power inequalities. In this relationship, the poor are treated as if they have "expendable bodies" and their deaths are attributed to natural causes, poor habits, inferior values, and even "God's will."

Notes British climate accountant Nicholas Stern,

Climate change is a result of the greatest market failure the world has seen. The evidence on the seriousness of the risks from inaction or delayed action is now overwhelming. We risk damages on a scale larger than the two world wars of the last century. The problem is global and the response must be a collaboration on a global scale.

(*Benjamin 2007*)

According to Richard Heede (2019b),

We need to eliminate subsidies and regulatory preferences, and to price carbon so as to "internalise" the vast costs of climate damages now mostly paid by people who did not cause the problem, such as today's farmers and tomorrow's children.

Welfare for the Wealthy (Polluters) and Blaming the Poor

Climate change deniers commonly complain about the evils of big government. With the support of the dominant political parties in the United States and elsewhere, however, fossil fuel companies continue to be awarded billions of dollars in federal production and exploration subsidies and other fiscal benefits by governments (welfare for the wealthy polluters, sometimes called crony capitalism). This helps to explains why they routinely spend enormous amounts of money on lobbying politicians and even, when possible, have a direct hand in writing the actual language of new policies at the invitation of those politicians who are the biggest recipients of energy sector campaign donations. This approach is exemplified by the American Petroleum Institute, a large, well-funded national trade association that promotes the oil and gas industry, which, among many other similar efforts over many years, developed a secret Global Climate Science Communications Plan. Although not intended for public viewing, the plan was allegedly leaked by someone at one of the Institute's member organizations. The plan was drawn up just months after the Kyoto Protocol, an international GHG reduction treaty, was signed in December 1997. The Institute's plan was developed by representatives from Exxon, Chevron, and Southern Company and from conservative and libertarian organizations. It includes a multimillion-dollar, multi-year budget and detailed strategies to seed doubt about climate change in the public policy arena. Target audiences are detailed in the plan and include media, policy-makers, and even teachers. The objective of this initiative is clearly stated:

> Victory will be achieved when average citizens understand uncertainties in climate science; recognition of uncertainties becomes part of the "conventional wisdom"... and when those who support the international climate agreement, the Kyoto Protocol, are seen as being "out of touch with reality."
>
> *(DeSmog 2020)*

Further, the American Petroleum Institute spent US $127 million on direct lobbying of politicians on behalf of fossil fuel companies between 1998 and 2022. The Institute recognized that almost all politicians are in an almost constant search for campaign dollars. They have been able to bend this need into a well-heeled system to persuade governments to be fossil fuel-friendly (Open Secrets 2023).

Also of note, the American Petroleum Institute was well aware of the risks of climate change as early as 1982 (Banerjee 2016). A Columbia University report titled *Climate Models and CO2 Warming, A Selective Review and Summary*, commissioned by the Institute, warned that global warming "can have serious consequences for man's comfort and survival." Written by Alan Oppenheim and William L. Donn of Columbia's Lamont-Doherty Geological Observatory, the report was intended to monitor and share among members emerging climate research. Exxon ran the most ambitious of the corporate programs to assess the relationship of GHG to global warming, but other oil companies had their own smaller projects that chiefly focused on climate modeling.

The task force specifically commissioned the report to gain a better understand how new climate models were being developed by climate scientists. The authors of the report recognized there is still some way to go in developing accurate models, but understood that the consequences of increasing GHGs in the atmosphere are serious for humanity. They wrote, "since patterns of aridity and rainfall can change, the height of the sea level can increase considerably and the world food supply can be affected" (Banerjee 2016). Rather than heed this warning—which was right on target—the Institute buried it and continued to promote expanded fossil fuel production.

In a particularly sinister move, the American Petroleum Institute launched its Explore Offshore project in June 2018. Its stated goal is to convince Hispanic and black communities to support the Trump administration's proposed expansion of offshore drilling. These, of course, are the communities in the US first and hardest hit by global warming, having the fewest resources to cope with the disruptions of climate change. Erik Milito, the American Petroleum Institute's director of Upstream and Industry Operations explained that: "We want to build support in minority communities because the message that increasing the supply of affordable energy and good paying jobs will resonate" (quoted in Volcovici 2018).

Generally, government pronouncements adopt altruistic language to assert that the work of elected politicians is dedicated to serving the public interest and the public good. In the United States, however, both the federal and state governments annually provide large subsidies, grants, tax credits, loan guarantees, bailouts, joint ventures, royalty waivers, corporate bond buybacks and price supports to the super-wealthy and the companies under their control, including many involved in fossil fuels. As part of this process, government actions and the spending of tax dollars help to keep wealth and power regularly flowing into the hands of the polluting elite. For example, the federal law known as Intangible Drilling Costs Deduction (26 US Code § 263) allows oil companies to deduct a majority of the costs incurred from drilling new wells in the United States. It is estimated that eliminating tax breaks for intangible drilling costs would have generated US $1.59 billion in revenue in 2017, or $13 billion over the next ten years (Coleman and Dietz 2019).

Similarly, 26 US Code § 613 offered a significant tax break to oil companies. The law allowed these companies to use an accounting method called depletion that works much like depreciation, allowing businesses to deduct a certain amount from their taxable income to reflect the declining level of production from an oil reserve over time. The depletion approach, however, is not based on capital costs. Instead, it enables companies to take deductions at levels than can exceed capital costs. It is estimated that eliminating depletion for coal, oil and gas would have generated US $12.9 billion between 2017 and 2028 (Coleman and Dietz 2019).

In fact, there is a long history to this pattern of taking money from the pockets of taxpayers to line the pockets of the energy industry. Numerous energy subsidies have been written into the law in the US tax code with the stated goal of promoting the production of abundant fossil energy. Often the circumstances that were relevant at the time older subsidy laws were passed (helping startup companies) no longer exist, but the money keeps on flowing. In today's market, the domestic fossil fuel industries (primarily comprising companies involved in coal, oil and gas production and distribution) are quite mature and are generally highly profitable. Chevron Corporation, for example, reported a net income of US $36.5 billion in 2022, a significant increase compared to the previous year. This oil producer's adjusted net profit for 2022 exceeded its previous record set in 2011 by about $10 billion (Valle 2023). Additionally, various cleaner and renewable alternative fuels exist, which increasingly are price-competitive with traditional fossil fuels.

As suggested, behind the business-as-usual drive for new production by elite fossil fuel producers and emitters of GHGs is the often suppressed or discounted issue of social inequality. Turning again to the case of Chevron, researchers have described a consistent campaign of deception (Center for Climate Integrity 2021). A Chevron subsidiary, Pittsburg and Midway Coal Mining, for example, was a member of the Information Council for the Environment, a fossil industry front group formed in 1991 to "reposition global warming as theory (not fact)." Also known as Informed Citizens for the Environment, or ICE, the Council mounted a public relations campaign run by Simmons Advertising to test climate change obfuscation ideas with target audiences in Fargo, North Dakota, Flagstaff, Arizona, and Bowling Green, Kentucky. This initiative served as a testing ground for examining the effectiveness of a proposed climate denial campaign for possible use nationwide to sway public opinion and to shut down public support of legislation to fight climate change. One product of this campaign was a set of advertisements in the *New York Times* and other influential publications with messages like: "Who told you the earth was warming ... Chicken Little?" and "Doomsday is cancelled. Again." One Council advertisement claimed that there is "no hard evidence [that climate change] is occurring." These ads also told readers how to get more information about the "reality" of climate change (that it was not happening, presented no threat, and was a made-up idea to financially benefit those who were beginning to warn about climate disruptions). The Council's campaign also involved the use of radio commercials read by ultra-conservative media influencer Rush Limbaugh. In these radio commercials Limbaugh asserted that the "facts don't jibe" with the "theory" of climate change and related effects like the melting of the polar ice caps, and instead places like Minneapolis have "actually gotten colder." He urged listeners to call the Council to obtain the real "facts." Additionally, the Council's campaign included a nationwide public relations tour, as well as mailers that were sent to people's homes.

In its public advertisements, Chevon announced that "Renewable energy is vital to our planet," and "At Chevron, we're investing millions in solar and biofuel technologies." In January 2013, however, employees of Chevron's renewable power group, whose mission was to launch large, profitable clean energy projects, were invited by the company to a dinner at San Francisco's upscale Sens restaurant to acknowledge the first full year of the group's operations. To the surprise of attendees, despite its financial achievements and the team's role in helping to set up more than a half-a-dozen solar and geo-thermal projects capable of powering at least 65,000 homes, they were told by their managers that funding for the effort was ending. Staffers were informed that they would have to find jobs elsewhere. Accordingly, Robert Redlinger, who was director of Chevron's renewable energy projects until he left the company in 2010, told a reporter:

> When you have a very successful and profitable core oil and gas business, it can be quite difficult to justify investing in renewables. It requires significant commitment at the most senior levels of management. I didn't perceive that kind of commitment from Chevron during my time with the firm.
>
> *(Upton 2014)*

Between 2010 and 2018, Chevron invested only 0.2% of its annual capital expenditure on the development of low-carbon energy. These same expenditures said to advance production of clean energy sources disingenuously included funds to promote natural gas as a means to reduce GHG emissions, despite the fact that natural gas is a significant source of carbon dioxide and methane emissions.

These efforts by Chevron and other fossil fuel industry companies have continued over the years. During a series of closed-door meetings, industry representatives argued for a significant roll back of the landmark climate change rule implemented by the US Environmental Protection Agency during the Barack Obama administration (Colman 2017). In fact, the Obama administration implemented a number of programs intended to address the impact of climate change, which, once elected, President Donald Trump claimed harmed American businesses and rescinded them. At a media event organized to celebrate the signing of his executive order, Trump said: "With today's executive action, I am taking historic steps to lift the restrictions on American energy, to reverse government intrusion and to cancel job-killing regulations" (Park 2017). During the Trump Administration, between 2017 and 2019, the US government provided an average of US $7.6 billion annually in subsidies to the fossil fuel industry. Approximately $2.5 billion of this public funding went to support exploration, production, and transportation, while a notable $4.8 billion went to subsidize an expansion of the consumption of GHG-producing coal, oil and gas across the country. The federal government also began to use new ways to reduce costs for the fossil fuel industry, such as threatening private banks with the loss of public subsidies if they did not continue financing oil and gas projects.

In the period after the COVID-19 pandemic struck and significantly slowed the economy, as part of its fiscal stimulus package the US government allocated nearly US $100 billion to support energy companies. 72% of this huge windfall went to the fossil fuel industry. The US provided more financial support to fossil fuels than any other country, guaranteeing that its economic recovery would be powered by oil and gas. In effect, the federal government channeled a huge chunk of taxpayers' money to an industry known to produce toxic air pollution during a respiratory pandemic—likely made worse by air pollution exposure—that by 2020 had cost nearly 1,150,000 American lives and pushed thousands more into bankruptcy (Colenbrander and Picciariello 2020).

Beyond the United States, government fossil fuel subsidies also are increasing. The IMF reports that an estimated 6.5% of global gross domestic product (GDP) (US $5.2 trillion) was spent on fossil fuel subsidies in 2017, a half-trillion-dollar increase since 2015. The largest subsidizers outside of the United States are China ($1.4 trillion in 2015) and Russia ($551 billion). According to the Fund, "fossil fuels account for 85% of all global subsidies," and reducing these subsidies "would have lowered global carbon emissions by 28% and fossil fuel air pollution deaths by 46%, and increased government revenue by 3.8% of GDP." An Overseas Development Institute study found that subsidies for coal-fired power increased almost threefold, to $47.3 billion per year, in the period from 2014 to 2017 (Coleman and Dietz 2019).

Resistance from Below: The Grassroots Fight Against Climate Change and its Drivers

In spite of the powerful elite forces aligned against them, poor and other marginalized people, including Indigenous groups, are not passive beings mindlessly manipulated by the global capitalist system. While social inequality may bring them many diseases, the damaging forces of climate change, and the polluting of their land, water, and air, those subject to both overt and covert suppression are never quietly acquiescent. They have developed local understandings of their circumstances, sought means of coping with the risks they face, and have banded together to challenge those causing their suffering in both small and larger ways.

In 1985, anthropologist James Scott published a book titled *Weapons of the Weak* based on two years of ethnographic study of the effects of the Green Revolution in rural Malaysia. The Green Revolution was an effort initiated by wealthy Global North counties to increase crop yields in the farming areas of the Global South. Central to the campaign was an attempt to convince small farmers, like those growing rice in Malaysia, to adopt new technologies like high-yield cereals, dwarf wheat and rice, and very heavy use of chemical fertilizers and pesticides. The new agricultural technologies often bypassed the poorest small farmers. This was especially common in areas in which there was inequitable land distribution as well as insecure land ownership and tenancy rights, in places with poorly developed input, credit, and output markets, and in locales in which there were policies enacted that discriminated against smallholders, such as subsidies for mechanization research and extension. The onset of the Green Revolution, however, eliminated two-thirds of the wage-earning jobs of small landholders and landless workers. Slow growth in the nonfarm economy blocked it from being able to absorb the rising numbers of rural unemployed or underused workers. Many Malaysians were wary of the introduced changes and resisted adopting them. Scott found that they used a variety of resistance strategies, including foot-dragging, dissimulation, desertion, false compliance, pilfering, feigned ignorance, and acts of sabotage. Given the disparities in the power of the change promoters and those who resisted, he called these strategies "weapons of the weak." In a subsequent book, Scott (1987) described how subordinate groups create a secret discourse that expresses a critique of those in power that is literally spoken behind the backs of the dominant. Rather than quietly obey, they act daily to give vent to their outrage in words and deeds.

Sometimes resistance to domination and exploitation mushrooms into full scale revolt. Six such uprisings among peasants around the world (Russia, Mexico, China, Algeria, Cuba, and Vietnam) have been documented and analyzed by Eric Wolf (1969). Wolf argued that North Atlantic capitalism exacerbated existing tensions and exploitation in peasant communities, leading to revolutions that challenged enduring injustices. He viewed these uprisings as reactions to adverse changes introduced by capitalism. Quoting dramatist, poet, and philosopher Bertolt Brecht (quoted in Benjamin 2005), he highlights that "it is not communism that is radical, it is capitalism." In his historic analysis, Wolf found that an important factor determining the peasantry's revolutionary potential is their physical proximity to centers of state power. He noted (1969: 293) that "frontier areas quite often show a tendency to rebel against the central authorities, regardless of whether they are inhabited by peasants or not." He also suggests that somewhat better-off peasants tend to be quicker to challenge authorities than their poorer and landless counterparts. The social existence of very poor peasants often is too precarious to initiate risky revolutionary action, although they will join in revolt if they feel protected by wealthier co-revolutionaries.

Similar movements have erupted among Indigenous groups. A current example is taking place in Papua New Guinea. Emmanuel Peni is a leader of an Indigenous campaign called Save the Sepik that has targeted a proposed massive open-face mining operation. Debris tailing (i.e., mine waste) from the mine will be dumped along the Frieda River, a tributary of the Sepik River. The level of potential ecological damage and the harmful effects on human health and wellbeing in the area are expected to be huge. Says Peni, "The Sepik River is us,' and we are the Sepik River, there is no distinction ... It's really hard to explain the connection, it's so diverse in terms of how people explain the connection to the river" (quoted in Tan 2023).

A joint statement from the clan leaders of 25 villages along the Sepik River, entitled the Supreme Sukudimi Declaration (SSD), has become a rallying cry for the campaign. Sukudimi is one of the names for the region's river god. The declaration, published by the Save the Sepik campaign in May 2020, called for a total ban of the mine, asserting that "[t]he Sepik River is not ours. We are only vessels of the Sepik Spirit that dwells to protect it. We will guard it with our lives." Specifically, the SSD declares:

> We call for a total ban on the Frieda River Mine. We respect our Sepik River and call on our [national] leaders to give it the same respect and uphold policies that protect it and promote our cultural heritage.
> We assert the value of our traditional economy, which promotes self-reliance amongst our people and communities, and we are opposed to actions and policies which encourage the dependency of our people on others, including the state.
> We are opposed to any extractive development that sees increased transportation on our river system. We are opposed to any mine tailings facility that could negatively impact the Sepik River. We oppose all foreign programs, bribes and inducements to bring about customary land registration in the Sepik region.
>
> *(Save the Sepik 2020)*

In addition to addressing scientific and ecological concerns over the mine, the SSD acknowledges the voices of ancestors and the spirits of the river, land, plants, and animals. Anthropologist Christiane Falck (2023) argues that the fight of Sepik people, who align themselves with ancestral spirits against the mining project and the national government of Papua New Guinea, can be understood as a conflict between different ontologies or ways of being-in-the-world.

What has driven the anti-mine movement forward are word-of-mouth exchanges and the sharing of traditional beliefs and stories in impacted local communities. Peni, who wrote the SSD, understands that he and the local people are the spiritual guardians of the land: "I felt that the *N'gego* [the traditional spirit house] leaders held much more power than the formal local governors and councilors or the provincial government assembly ... That's why we went to the *N'gego*" (quoted in Tan 2023). The SSD, reports Falck (2023), builds on traditional governance structures, in which men's houses play a key role.

According to David Lipset, a professor of anthropology at the University of Minnesota, while the government of Papua New Guinea has attempted to be sensitive to customary concepts of person and environment, "when the chips are down, the government bends over backwards to make deals with extractive industry" (quoted in Tan 2023). Thus far, the campaign to stop the mine has been successful and the Papua New Guinean government has not yet issued an environmental permit or mining license.

Embracing a belief in self-determination and the inherent sovereignty of Indigenous peoples to decide what happens on their own territories, various forms of Indigenous resistance to fossil fuel companies and their operations have emerged in recent decades around the world. Indigenous movements seek to exert social and moral authority to protect their traditional home-lands from oil and gas development. Fossil fuel production causes significant environmental health damage in Indigenous communities and frontline Indigenous activists have been effective in fighting what Spiegel (2021) calls "petro-colonialism," namely the ecological destruction, health threats, and moral and legal transgressions perpetrated by fossil fuel companies and state institutions. As explained by Aaron Berstein at Harvard University's Center for Climate, Health and the Global Environment at a 2022 campus con-ference in advance of Earth Day, "Few communities know better the harms of fossil fuel reliance than Indigenous communities" (quoted in Lau 2022). At the conference, this sentiment was given firsthand support by Kandi White, native energy and climate campaign coordinator at the Indigenous Environ-mental Network, who said: "Natural resources in the form of oil, coal and gas, uranium, were discovered on our tribal lands, and we were in the way, and so we became the first and worst impacted," In 2014, a wastewater spill from an oil pipeline turned her community's drinking water a bright, eerie blue. Additionally, Kandi complained about fracking that produces gas flares that "everybody living there—my family, my friends—are breathing in every single day, 24/7 nonstop."

Based on an analysis of 20 fossil fuel projects that have been stopped or delayed through Indigenous action, the Indigenous Environmental Network, an alliance of Indigenous peoples and Oil Change International (Goldtooth and Saldamando 2021) issued a report titled Indigenous Resistance Against Carbon. It describes the impact Indigenous resistance to fossil fuel projects in the United States and Canada has had on GHG emissions over the past ten years. More specifically, the report quantifies the metric tons of carbon dioxide equivalent emissions that have either been stopped or delayed in the past decade by Indigenous land defenders. The report concludes that these frontline struggles have halted gas pollution equivalent to at least one-quarter of the United States' and Canada's annual emissions. From the struggle against Teck's Frontier Oil Sands Mine project to challenging the construction of pipelines crossing critical waterways, Indigenous land defenders have not only successfully stopped fossil fuel projects in their tracks, they have established precedents for organizing successful environmental justice movements.

Some Indigenous-led campaigns have targeted tar sands oil, the world's most climate-damaging oil, producing three times the GHG emissions of conventionally produced oil due to the enormous amount of energy required to extract and process it. Teck's Frontier Oil Sands Mine in Alberta, Canada, was intended to be the largest open-pit tar sands mine ever developed. The mine would have been located on Indigenous Dene and Cree territory in an area with little to no existing industrial development. These lands and waters are home to one of the last free-roaming, disease-free herds of wood bison. The area also lies near many Indigenous settlements, along the migration route for the only wild population of the endangered whooping crane and just 30 kilometers from the boundary of United Nations Educational, Scientific and Cultural Organization World Heritage Site Wood Buffalo National Park. This is an area valued and protected both for its cultural sites and its rich animal biodiversity (55 species of mammals, 250 species of birds, and 59 species of fish, along with many species of reptiles, amphibians, insects and other invertebrates). Following sustained resistance from the Dene Nation, Indigenous Climate Action, and many others, Teck Resources Ltd, the owner, cancelled the US $15.7 billion (C$26.6 billion) project in early 2020. In the weeks prior to Teck's withdrawal, protesters blocked railway lines in Ontario, Quebec, and Alberta in solidarity with the resistance campaign.

The 800,000 barrel per day Keystone XL tar sands pipeline was planned to be a 1,200 mile-long project that crossed from Alberta, Canada, into the United States, first to Steele City, Nebraska, and thence to refineries in Illinois and Texas, as well as to oil tank farms and an oil pipeline distribution center in Cushing, Oklahoma. The massive project was commissioned in 2010 and owned by TransCanada Keystone Pipeline GP Ltd, a Canadian company headquartered in the TC Energy Tower building in Calgary. This company, which operates in Canada, the United States, and Mexico, has three core businesses: natural gas pipelines; liquids pipelines; and energy.

The pipeline was fiercely opposed by Indigenous and other groups in both the United States and Canada, including the Dene, Cree, Metis, Oceti Sakowin, and Ponca tribes and communities. Indigenous leaders helped to lead a coalition that also involved Nebraska landowners and environmentalists in more than a decade-long struggle. On January 24, 2017, President Donald Trump took action to ensure the pipeline's completion. On January 20, 2021, however, President Joe Biden signed an executive order to revoke the permit that had been granted to TC Energy Corporation for the Keystone XL

Pipeline. On June 9, 2021, TC Energy abandoned plans for the Keystone XL Pipeline and the project was officially dead.

These victories were not won without a fight. Across Canada, Indigenous communities have at times been inundated with police using video cameras (intended to intimidate and gather evidence) while they make arrests, particularly when fossil fuel development projects are at stake. The reason: there is a longstanding close relationship between Canada's energy sector and its national security agencies which was strengthened to protect pipeline infrastructure (Spiegel 2021). Similar patterns have been seen around the globe.

Conclusion

Climate change is not just a feature of a transforming physical environment, it is both an expression of social inequality and an ever-growing force contributing to it. The policies and strategies supported by the social and polluting elites accelerate the daily production and discharge of GHGs and the pollutants that foul the air and create black carbon. Their motive: the rich benefits of unregulated production. This pattern, however, constitutes a form of structural violence against the poor. Under current societal arrangements, the poor often are treated as if their bodies were expendable, while their deaths are self-servingly attributed to behavioral causes, such as poor eating and drinking habits, drug use, unruly social behaviors, lack of good family standards, unsupportive adult males, laziness, and inferior values. Oftentimes these falsehoods attributed to the poor are spoken or written about by those who also ignore the overwhelming amount of validated scientific evidence and call climate change a hoax.

While social and economic inequality is inscribed on their bodies by the diseases they suffer and the bodily insults of climate change and environmental pollution, poor and marginalized people are not silent nor are they docile. They develop understandings of changing conditions, find new ways to cope with the impacts of climate change, and band together to politically challenge the perpetrators of their plight in public protests and related activities.

Ultimately, there can be no sustainable ecology without social equity and no social equity without sustainable human lifeways. This "eco-equity" perspective has been embraced by some sectors of the global grassroots effort to build an alternative approach to elite strategies and dominance and to create societies committed to sustainable production and product use, true democracy, and the protection of biodiversity.

References

Adelekan, Ibidun. 2010. Vulnerability of poor urban coastal communities to flooding in Lagos, Nigeria. *Environment and Urbanization*, 22(2), 433–450

Akther, Hasina and Ahmad, Mokbul. 2022. Livelihood in the pluvial flood prone slum communities in Dhaka, Bangladesh. *Progress in Disaster Science* 14: 100227

Banerjee, Neela. 2016. Oil industry group's own report shows early knowledge of climate impacts. *Inside Climate News*. https://insideclimatenews.org/news/0502201 6/oil-industry-report-shows-early-knowledge-climate-change-impact-api-american-petroleum-institute/

Benjamin, Alison. 2007. Stern: Climate change a "market failure." *The Guardian*. https://www.theguardian.com/environment/2007/nov/29/Nclimatechange.carbonemissions

Benjamin, Walter 2005. Communism: A family drama in the epic theater. In *Walter Benjamin: Selected Writings: 1931–1934*, Vol. 2. Cambridge, MA: Harvard University Press

Bermeo, Sarah, Leblang, David, and Alverio, Gabriela Nagle. 2022. *Root Causes of Migration from Guatemala: Analysis of Subnational Trends*. Durham, NC: Duke Center for International Development. https://sanford-dcid-files.cloud.duke.edu/sites/default/files/Migration-Policy-Brief-Guatemala.pdf

Carbon Disclosure Project (CDP). 2017. New report shows just 100 companies are source of over 70% of emissions. https://www.cdp.net/en/articles/media/new-report-shows-just-100-companies-are-source-of-over-70-of-emissions

Center for Climate Integrity. 2021. Chevron CEO Michael K. Wirth. https://climateintegrity.org/news/chevron-ceo-michael-k-wirth

Coleman, Clayton and Dietz, Emma. 2019. Fossil fuel subsidies: A closer look at tax breaks and Societal Costs. Environment and Energy Study Institute. https://www.eesi.org/papers/view/fact-sheet-fossil-fuel-subsidies-a-closer-look-at-tax-breaks-and-societal-costs

Colenbrander, Sarah and Picciariello, Angela. 2020. An exposé of Trump's finances: Ending US public support for the fossil fuel industry. Overseas Development Institute. https://odi.org/en/insights/an-exposé-of-trumps-finances-ending-us-public-support-for-the-fossil-fuel-industry/

Colman, Zack . 2017. Industry to EPA: We want rule "fixed, not just gone." *E&E News*. https://subscriber.politicopro.com/article/eenews/1060058186

DeSmog. 2020. American Petroleum Institute. https://www.desmog.com/american-petroleum-institute/

Falck, Christiane. 2023. The Supreme Sukundimi Declaration: Sacred water, moral ecologies and ontological politics in a mining encounter in Papua New Guinea. *Anthropological Forum*, 1–26. doi:10.1080/00664677.2022.2162847

Goldtooth, Dallas and Saldamando, Alberto. 2021. Indigenous resistance against carbon. Indigenous Environmental Network. https://www.ienearth.org/indigenous-resistance-against-carbon/

Guivarch, Celine, Taconet, Nicolas, and Mejean, Aurelie. 2021. *Linking Climate and Inequality*. New York: International Monetary Fund. https://www.imf.org/en/Publications/fandd/issues/2021/09/climate-change-and-inequality-guivarch-mejean-taconet

Hasan, Mubashar and Macdonald, Geoffrey. 2021. *How Climate Change Deepens Bangladesh's Fragility*. United States Institute of Peace. https://www.usip.org/publications/2021/09/how-climate-change-deepens-bangladeshs-fragility

Heede, Richard. 2019a. Accounting for carbon and methane emissions, top twenty investor-owned and state-owned oil, gas, and coal companies 1965–2017. Climate Accountability Institute, https://climateaccountability.org/carbonmajors.html

Heede, Richard. 2019b. It's time to rein in the fossil fuel giants before their greed chokes the planet. *The Guardian*. https://www.theguardian.com/commentisfree/2019/oct/09/fossil-fuel-giants-greed-carbon-emissions

Kamruzzaman, Palash. 2021. Understanding extreme poverty in the words of the poor: A Bangladesh case study. *Journal of Poverty* 25(3). https://www.tandfonline.com/doi/full/10.1080/10875549.2020.1784352

Keshavarz, Marzieh, Maleksaeidi, Hamideh, and Karami, Ezatollah. 2017. Livelihood vulnerability to drought: A case of rural Iran. *International Journal of Disaster Risk Reduction* 21: 223–230

Khan, M. 2010. *Impact of Climate Change on the Livelihood of the Urban Poor: A Case of Dhaka City.* Master of Public Policy and Governance Program thesis, North South University, Dhaka, Bangladesh

Lau, Jessica. 2022. Fossil fuel extraction is harming Indigenous communities, say experts. Harvard T. H. Chan School of Public Health. https://www.hsph.harvard.edu/news/features/fossil-fuel-extraction-harming-indigenous-communities/

Laville, Sandra. 2019. Top oil firms spending millions lobbying to block climate change policies, says report. *The Guardian.* https://www.theguardian.com/business/2019/mar/22/top-oil-firms-spending-millions-lobbying-to-block-climate-change-policies-says-report

McPherson, I. 2015. Dhaka: The city where climate refugees are already a reality. *The Guardian.* www.theguardian.com/cities/2015/dec/1/dhaka-city-climate-refugees-reality

Open Secrets. 2023. Client Profile: American Petroleum Institute. https://www.opensecrets.org/federal-lobbying/clients/summary?id=D000031493

Pandey, Rajiv, Alatalo, Juha, Thapliyal, Kavita, Chauhan, Sharmela, Archie, Kelli, Gupta, Ajay, Jha, Shashidhar Kumar, and Kumar, Manoj. 2017. Climate change vulnerability in urban slum communities: Investigating household adaptation and decision-making capacity in the Indian Himalaya. *Ecological Indicators* 90: 379–391. doi:10.1016/j.ecolind.2018.03.031

Park, Madison. 2017. 6 Obama climate policies that Trump orders change. CNN. https://www.cnn.com/2017/03/28/politics/climate-change-obama-rules-trump/index.html

Rahman, Atiq. 2008. Climate change and its impact on health in Bangladesh. *Regional Health Forum* 12(1): 16–26

Rana, Ifan and Routray, Jayant. 2018. Integrated methodology for flood risk assessment and application in urban communities of Pakistan. *Natural Hazards: Journal for the Prevention and Mitigation of Natural Hazards* 91(1): 239–266

Sahlins, Marshall. 1968. Notes on the Original Affluent Society. In *Man the Hunter.* Richard B. Lee and Irvin DeVore, eds. pp. 85–89. New York: Aldine Publishing Company

Save the Sepik. 2020. Supreme Sukundimi Declaration. https://savethesepik.org/the-supreme-sukundimi-declaration/

Scott, James. 1985. *Weapons of the Weak: Everyday Forms of Peasant Resistance.* New Haven, CT: Yale University Press

Scott, James. 1987. *Domination and the Arts of Resistance: Hidden.* New Haven, CT: Yale University Press

Spiegel, Samuel. 2021. Fossil fuel violence and visual practices on Indigenous land: Watching, witnessing and resisting settler-colonial injustices. *Energy Research & Social Science* 79:102189

Tan, Jim. 2023. Power of traditional beliefs lies at the heart of an anti-mine campaign in PNG. *Mongabay.* https://news.mongabay.com/2023/04/power-of-traditional-beliefs-lies-at-the-heart-of-an-anti-mine-campaign-in-png/

Taylor, Matthew and Watts, Jonathan. 2019. Revealed: The 20 firms behind a third of all carbon emissions. *The Guardian.* https://www.theguardian.com/environment/2019/oct/09/revealed-20-firms-third-carbon-emissions

Upton, John. 2014. Chevron and BP are pulling out of wind and solar. *Grist*. https://grist.org/business-technology/chevron-and-bp-are-pulling-out-of-wind-and-solar/

Valle, Sabina. 2023. Chevron annual profit hits record but Q4 miss hits shares. Reuters. https://www.reuters.com/business/energy/chevron-annual-profit-doubles-record-365-bln-misses-estimates-2023-01-27/

Volcovici, Valorie. 2018. Big oil eyes U.S. minority groups to build offshore drilling support. Reuters. https://archive.fo/84gDE (archived Aug 1, 2018)

Wolf, Eric E. 1969. *Peasant Wars of the Twentieth Century*. London: Faber

World Bank. 2020. Global action urgently needed to halt historic threats to poverty reduction. https://www.worldbank.org/en/news/feature/2020/10/07/global-action-urgently-needed-to-halt-historic-threats-to-poverty-reduction

Wright, George, Olenick, Liat, and Westervelt, Amy. 2021. The dirty dozen: Meet America's top climate villains. *The Guardian*. https://www.theguardian.com/commentisfree/2021/oct/27/climate-crisis-villains-americas-dirty-dozen

5
PLANETARY HEALTH
A Critical Health Anthropological Perspective

Case Study: Electric Batteries and Environmental and Human Suffering

Until 20 years ago, the substance known as lithium had a few mundane and low-key uses: as a glaze for heatproof cookware, or as a grease to lubricate hot moving motor parts. Today lithium is being touted as offering a way out of our reliance on fossil fuel production. The lightest known metal on the planet, it has been widely adopted for use in electric devices from mobile phones and laptops to cars and aircraft. Lithium-ion batteries are best known for powering electric vehicles, which, with rising gas prices, have become increasingly popular and are on track to account for up to 60% of new car sales by 2030. The Tesla Model S, for example, a full-luxury sedan powered by a lithium-ion battery, uses around 26.4 pounds (12 kilograms) of lithium. The battery pack can weigh more than 1,200 pounds (540 kilograms) and is the heaviest part in the vehicle. Globally, demand for lithium is soaring, as are the prices paid for it.

However, meeting the new demand for lithium does not come without environmental and health costs. Mining the chemical element can be harmful to the environment, and the fiscal costs associated with lithium mining may be especially high. Found naturally in underground deposits in one of the driest regions on Earth—an area where the borders of Chile, Argentina, and Bolivia meet known as the "lithium triangle" and in the Salar de Atacama salt flats in northern Chile—lithium requires a water-intensive mining process that consumes approximately 500,000 gallons of water to extract one ton of the metal. As demand for lithium rises, the mining impacts are increasingly affecting surrounding communities where the harmful extraction process takes place, jeopardizing inhabitants' access to clean water. Further, like fossil fuels, lithium is a non-renewable resource.

DOI: 10.4324/9781003469940-6

In South America, to extract lithium, miners start by drilling a hole in the salt flats and pumping salty, mineral-rich brine to the surface. It is then left to evaporate for months at a time. The drying pools first turn into a mixture of manganese, potassium, borax, and lithium salts. This mixture is then filtered and placed into another evaporation pool, and the process continues. After between 12 and 18 months, the mixture has been filtered sufficiently that lithium carbonate—known as white gold—can be extracted.

Extraction is a relatively cheap and effective process, but it uses a lot of water. In Chile's Salar de Atacama salt flats, mining activities consume 65% of the region's water. This is having a big impact on local farmers—who grow quinoa and herd llamas—in an area where some communities already have to get water driven in from elsewhere. The extraction of lithium has caused water-related conflicts with local communities, such as the village of Toconao in the north of Chile where people have protested against the lithium extractors. Protests reflect the fact that water rights are highly concentrated in the hands of transnational and national corporations which use them to expand their extractivist operations aimed for export to the Global North.

There is also the potential—as occurred in Tibet—for toxic chemicals to leak from the evaporation pools into the water supply. These include chemicals, such as hydrochloric acid, which are used in the processing of lithium into a form that can be sold, as well as those waste products that are filtered out of the brine at each stage. In May 2016, hundreds of upset protestors threw dead fish onto the streets of Tagong, a frontier town of about 8,000 inhabitants (mainly Tibetans) on the eastern edge of the Tibetan plateau in Sichuan, People's Republic of China. They had removed the fish carcasses from the waters of the Liqi River, where a toxic chemical leak from the Ganzizhou Rongda Lithium mine had wreaked havoc on the local ecosystem. Some eyewitnesses reported seeing dead cows and yak floating downstream, poisoned by drinking the contaminated water. It was the third such incident in just seven years in an area that has seen a sharp rise in mining activity, including operations run by BYD, a car manufacturer and the world's biggest supplier of lithium-ion batteries for smartphones and electric cars (including some Tesla models). After the second incident, in 2013, government officials shut down the mine, but it reopened in April 2016 and the fish started dying again.

When people protested, they were surrounded by dozens of baton-wielding riot police. The government has denied that the lithium mine is the culprit, but after the latest poisonings it has issued promises that all the environmental issues will be ameliorated. During the protests, an article in the local *Ganzi Daily* newspaper described plans to make the area "China's lithium capital" (Denver 2016).

In Australia and North America, lithium is mined from rock using more traditional methods, but still requires the use of chemicals in order to extract it in a useable form. Research in Nevada found impacts on fish as far as 150 miles downstream from a lithium processing operation.

In Argentina's Salar de Hombre Muerto, locals similarly claim that lithium operations have contaminated streams used by humans and livestock, and for the irrigation of crops. As in Tibet, in South America there have been clashes between lithium mining companies and local communities. People say that lithium mining is leaving the local landscape filled with mountains of discarded salt and canals filled with contaminated water with an eerie blue hue. According to Guillermo Gonzalez, a lithium battery expert from the University of Chile, "Like any mining process, it is invasive, it scars the landscape, it destroys the water table and it pollutes the earth and the local wells … This isn't a green solution—it's not a solution at all" (quoted in Katwala 2018).

It is, in fact, a solution for companies in the Global North that see great financial rewards in green capitalism. For involved countries in South America, however, it conjures up memories of "the catastrophic trauma of conquest and integration in a subordinate, colonial position in the international system, as a necessary and hidden reverse of modernity" (Alimonda 2011: 21). Consequently, Jerez et al. (2021: 2) argue that lithium mining constitutes "a new imposition of extractive territoriality that reproduces colonial stratifications in the high Andean salt flats of the Southern Cone."

But lithium may not be the most problematic ingredient of modern rechargeable electric batteries. Another key component is cobalt, and its extraction is even more costly in terms of human health and environmental integrity than lithium. The metal is found in huge quantities in the Democratic Republic of Congo (DRC)—a deeply impoverished nation that has been subjected to generations of pillage and ransacking by colonial powers going back to the slave trade— but rarely elsewhere. The price has quadrupled in the last few years. Unlike most metals, which are not toxic when they are mined from the ground as metal ores, according to Gleb Yushin, chief technical officer and founder of the battery materials company Sila Nanotechnologies, cobalt is "uniquely terrible. One of the biggest challenges with cobalt is that it's located in one country" (quoted in Katwala 2018). Adds Siddharth Kara, a fellow at Harvard's T.H. Chan School of Public Health and at the Kennedy School, and author of *Cobalt Red*,

> Cobalt is toxic to touch and breathe—and there are hundreds of thousands of poor Congolese people touching and breathing it day in and day out. Young mothers with babies strapped to their backs, all breathing in this toxic cobalt dust.
>
> *(Quoted in Gross 2023)*

Cobalt can harm the eyes, skin, heart, and lungs. Exposure to cobalt may cause cancer.

In areas where cobalt is found, people can literally just dig up the land and find it. Consequently, says Yushin, "there's a lot of motivation for unsafe and unethical behaviour" (quoted in Katwala 2018). The DRC is home to 'artisanal mines', where cobalt is extracted from the ground by hand, often using child labor, without any protective equipment and for just the equivalent of a few dollars per day. Notes Kara,

> You have to imagine walking around some of these mining areas and dialing back our clock centuries … People are working in subhuman, grinding, degrading conditions. They use pickaxes, shovels, stretches of rebar to hack and scrounge at the earth in trenches and pits and tunnels to gather cobalt and feed it up the formal supply chain.
>
> *(Quoted in Gross 2023)*

Millions of trees around the mines have been cut down, the air is thick with dust and grit, and nearby water sources are contaminated with toxic effluents from the mining processing. Moreover, work in the mines is extremely hazardous. Mining shafts and walls collapse sending a mountain of gravel and stone avalanching down on people causing horrific injuries.

People are displaced because their villages were bulldozed over to make way for large mining operations. So, they are left with no alternative, no other source of income, except to work for meager pay in the cobalt mines. Moreover, the military is used to pressure people to dig, forcing parents to make the painful decision to send their children into the mines to make enough money for food or to send them to school and not eat that day.

To understand how this state of affairs developed it is necessary to examine the modern history of the DRC as it transitioned from a Belgian colony into an independent nation. In 1960, the first democratically elected president of the country, Patrice Lumumba, made a pledge that the land's immense mineral riches would be used to benefit the people who live there. But very quickly, Lumumba was killed, likely by Belgian contractors with the support of Allan Dulles, head of the CIA. Lumumba's body was chopped into pieces and dissolved in acid. He was replaced by a corrupt dictator who would keep the country's minerals flowing to the Global North and the people of the county poor (Young 2015).

Some analysts differentiate between the cobalt that is extracted by the DRC's high-tech industrial mining companies and that dug by women and children with their bare hands. Kara maintains that the two are largely intertwined. Cobalt from both sources combine in the global supply chain as the now precious metal passes from under the ground in the Global South to the electric battery plants of the Global North. Hidden behind the much-publicized benefits of resource extraction for the new technologies is the colonial darkening shadow of "green" electromobility (Jerez et al. 2021).

Why Planetary Health?

The term planetary health has been defined as

> [T]he achievement of the highest attainable standard of health, wellbeing, and equity worldwide through judicious attention to the human systems—political, economic, and social—that shape the future of humanity and the Earth's natural systems that define the safe environmental limits within which humanity can flourish.
>
> *(Whitmee et al. 2015)*

This framework accepts that human impacts on Earth have become so profound that there is a need for a new name for the current geophysical epoch; most popularly it has been dubbed the Anthropocene. The core premise of Anthropocene conceptualization is that at this point in planetary history essentially every natural Earth system, from the deepest oceans to the upper atmosphere, has been significantly modified by human activity. Indeed, as discussed in Chapter 1, the tolerable boundaries of these systems are being violated putting life as we know it at risk. Ultimately these changes may threaten the long-term survival of our species. The natural systems that have benefited humans are beginning to collapse. As a result, we cannot engage in siloed thinking about human health, animal/plant health, and environmental health; we need an interconnected multi-species and animate/inanimate lens on these critical issues. In this chapter, we present the reasons for adopting the planetary health paradigm in light of this objective.

Planetary health explicitly recognizes the fundamental importance of "natural systems in terms of averted cases of disease and potential harm that comes from human-caused perturbations of these systems" (Seltenrich 2018). Further, planetary health envisions a more-than-human approach to public health (Kehr 2020). The shift to planetary health thinking was driven by the growing awareness that not only are all human communities linked together by flows of commodities, ideas, people, and health-related influences from vectors to medicines, but that the health of human communities is intimately linked to the environment, including other species within us (e.g., gut flora) and around us (Singer 2014). Our lives also are shaped by our exposure to and interactions with the land, the water, and the atmosphere.

From a planetary health perspective, the problem today is not humans per se. We are not a cancer on the planet and our species has lived on Earth for a long time without trespassing on planetary boundaries. As discussed in previous chapters, the worst environmental problems we face are "the result of special interests knowingly damaging environmental systems for their own benefit while leaving impoverished systems for everyone else" (Myers and Frumpkin 2020: 13). In Chapter 3, the term Capitalocene was introduced to show the fundamental role that the hierarchal capitalist system and its dual drive for ever-expanded production and ever-mounting release of greenhouse gases has played in the creation of the climate crisis. Some argue that one of the reasons why the Anthropocene fails as a concept is that it inadequately accounts for the unequal power relations within and between nations (Grove 2016).

Davis and Todd (2017), consequently argue that the origin of the current epoch should be explicitly linked to the global processes of colonial dispossession, exploitation, genocide, and ecocide that began approximately 500 years ago rather than accept the beginning of the Industrial Revolution as the start of the current epoch. Unlike many colonized people, who embraced an intimacy with "Mother Nature," it was the economic and political elite of Europe who "developed a distinctively mechanistic view of matter, an oppositional relationship to nature, and an economic system indebted to geographical expansion" (Grove 2016). Similarly, Whyte (2016) describes climate change as "an intensification of environmental change imposed on Indigenous peoples by colonialism." The issue is not dating per se but rather what we understand as the fundamental social event(s) that mark the sharp turn toward the current climate and environmental crisis. Without doubt, the takeoff of the colonial era, involving the Euro-centric domination of Indigenous peoples in the Americas, Africa, Asia, and Oceania, and the rise of capitalism, are deeply entwined. The social and economic reorganization involved in the transition from feudalism to capitalism in Europe was in no small part paid for with the gold, silver, and other treasures ripped from the colonies and hauled back to Europe. Capitalism, in turn, an intrinsically expansionist system that requires perpetual growth and the search for new raw materials and new labor forces, pushed for the further global expansion of colonialism and the subjugation and discounting of the lives, ways of being-in-the-world, and ways of knowing of Indigenous peoples.

While the injustice of the existing world economic system—that is, late capitalism or neoliberalism—is now recognized by mainstream global bodies like the United Nations (UN, as expressed in its Sustainable Development Goals) and the World Bank (in its call to eradicate extreme poverty globally), these entities generally do not advocate for a drastic, fair redistribution of the global wealth and resources—or a return of the wealth extracted from colonized or otherwise dominated lands or peoples—nor do they recognize the need for limits to economic growth, stances that are imperative for addressing the health and social ravages of the ecological and climate crises, particularly (but far from solely) for the poor, other marginalized populations, and Indigenous peoples.

As pointedly expressed by Ta-nehisi Coates (2015: 150):

> Once, the Dream's parameters were caged by technology and by the limits of horsepower and wind. But the Dreamers have improved themselves, and the damming of seas for voltage, the extraction of coal, the transmuting of oil into food, have enabled an expansion in plunder with

no known precedent. And this revolution has freed the Dreamers to plunder not just the bodies of humans but the body of the Earth itself.

These concerns among a growing number of health-, society-, and environment-focused scholars motivated their adoption of the planetary health perspective. To borrow and reapply the words of award-winning author Tiya Miles (2021: 15), planetary health stresses that "there is one earth and one humanity, one social fabric of many folds in dire need of mending."

The planetary health paradigm is not alone in attempting to develop a broader, more-than-human understanding of health. The One Health model is another approach that recognizes that the health of people is closely entwined with the health of animals and our shared physical environment (Atlas 2013). One Health (human and animal) highlights the fact that most people live in close contact with wild and domestic animals, both as pets and livestock. Animals play a significant role in human daily lives, whether as food, a source of products and livelihoods, travel, sport, education, or companionship. Close contact with animals and their environments provides more opportunities for diseases to pass between animals and people (in both directions). Human changes to the climate/environment help to drive interspecies disease movement. Additionally, the movement of people, animals, and animal products has increased through international travel and trade. As a result, diseases can spread quickly across national borders and around the globe. Many animals also share our susceptibility to diseases and environmental hazards. As a result, One Health stresses that disease in animals can sometimes serve as early warning signs of potential human illness. For example, avian die-offs from West Nile virus usually occur before people in the same area are infected and become sick with West Nile disease. Issues of primary concern from a One Health perspective include the human-animal-environment interface, vector-borne diseases, food-borne diseases, antimicrobial-resistant microbes, and water contamination.

An antecedent of One Health can be found in the work of physician/activist Rudolf Virchow, a founder of modern pathology and of social medicine, who argued over 100 years ago that "[b]etween animal and human medicine there can be no dividing line—nor should there be" (quoted in Saunders 2000). The modern origin of One Health traces to an article by a physician named Laura H. Kahn (2006) entitled "Confronting zoonoses, linking human and veterinary medicine." Kahn argued that closer collaboration is needed among veterinarians, physicians, and public health professionals in addressing individual health, population health, and comparative medicine research. Reading this paper prompted Bruce Kaplan, a veterinarian, to contact her and together they started the One Health Initiative team.

While similar in many ways, there are differences between One Health and planetary health. For the years 2020 and 2021, for example, an analysis of the literature found that topics related to infectious diseases were the most prominent in One Health publications (e.g., the COVID-19 pandemic, anti-microbial resistance, and diseases spread from animals to humans). Planetary health publications also frequently addressed the COVID-19 pandemic, but climate change was the dominant topic. Non-communicable diseases and issues related to food systems or physical activity and inactivity were also a concern to planetary health, but not to One Health research during these years (Ruiz de Castañeda et al. 2023). Given its greater focus on climate change, planetary health best integrates with the critical anthropology perspective framing this book.

The Environment and the Spread of Infectious Disease

From ancient times (especially after the emergence of settled populations and the domestication of animals) to the present day, humans have experienced many debilitating and lethal infectious diseases. Caused by an array of pathogenic agents such as bacteria, viruses, fungi, parasites, protozoa, and prions, these conditions have helped to shape human history, behavior, and biology. Despite centuries of trying, we can only claim success in eliminating one pathogen—the smallpox virus. Characterized by their ability to enter the human body, reproduce there, and disrupt cells and body systems, pathogens continue to have devastating consequences across the globe but not equal consequences for all populations. Based on data from 2019, before the deadly COVID-19 pandemic, the World Health Organization (WHO 2020) reported the toll that infectious diseases still take in low-income countries: six of the top ten causes of death in low-income countries are communicable diseases, including malaria (sixth), tuberculosis (eighth) and HIV/AIDS (ninth). In the past 25 years, over 30 new infectious diseases have spread through human populations. Meanwhile, there has been a resurgence of previously well-established infectious diseases, including cholera, malaria, tuberculosis, and diphtheria. Of concern in this section is answering the question: what is the relationship between infectious disease in humans and planetary health?

Pathogens are highly sensitive to changes in environmental conditions, including temperature, soil moisture, precipitation patterns, deforestation, dam building, and irrigation projects. A considerable body of research indicates that ecosystem disruptions and simplifications wrought by human hands and technologies can significantly exacerbate infectious diseases like influenza that have long been a scourge on humankind or cause novel pathogens like COVID-19 to spillover from animal to human populations. Most of the increasingly frequent emergence of what are called zoonotic diseases—because of their origin in other animal species— underscores the importance of a planetary health perspective. This also applies to nonzoonotic infectious diseases that derive from the soil. Infectious outbreaks threaten to become a greater threat as humans continue to encroach on animal habitat and ecosystems are degraded, forcing wildlife and humans into closer proximity with one another.

The range of such impacts was examined in 2002 by a Working Group on Land Use Change and Infectious Disease Emergence (Patz and Confalonieri 2004), made up of several dozen scientists from around the world, who also ranked the top 12 environmental changes in descending order of importance, as follows:

1. agricultural development;
2. urbanization;
3. deforestation;
4. human population movement;
5. introduced species/pathogens;
6. biodiversity loss;
7. habitat fragmentation;
8. water and air pollution;
9. road building;
10. impact of HIV/AIDS;
11. climatic changes (which would be ranked much higher today);
12. hydrological changes, including dams.

On International Day for Biological Diversity in 2020, António Guterres (2020a), the UN Secretary-General, stated, "COVID-19, arising from nature, has highlighted the intimate connection that exists between human health and our relationship with the natural world." To mitigate climatic disturbances, guarantee food and water security as well as to prevent future pandemics, Guterres (2020b) emphasized, it is essential to conserve and sustainably manage biological diversity: "Our solutions lie in nature."

Research shows that air pollution, for example, can be a contributor to infection. Studies reveal that the health consequences of infection with COVID-19, including mortality, may be influenced by the extent to which an individual's respiratory system has been compromised already by breathing polluted air (Pozzer et al. 2020). A study in China found that high levels of particulate matter pollution in the air may increase the susceptibility of people to acute respiratory complications arising from COVID-19 (Chen et al. 2020). In the United States, research indicates that COVID-19-associated death rates are about 15% higher in areas where even a small increase in the local fine-particle pollution level was recorded during the pre-COVID-19 period (Wu et al. 2020). Adherence of the COVID-19 virus on airborne dust and particulate matter from air pollution may contribute to the transport of the virus (Comunian et al. 2020). Increasingly, there is an interest in understanding the role of climate change, habitat destruction, and urban pollution in the appearance and spread of COVID-19. One environmental impact of COVID-19 has been a rise in the demand for plastic. In 2019, the UN declared plastic pollution a global crisis and proposed that 2020 mark a significant shift away from plastic usage to achieve positive environmental changes. The COVID-19 pandemic, however, sidetracked this goal by exerting tremendous pressure on healthcare systems that find plastic to be the most reliable and affordable solution for personal viral protection.

Milder winters, warmer summers, and fewer days of frost make it easier for infectious diseases like dengue, giardiasis, hantavirus, Lyme disease and other tick-borne diseases, plague, rabies, and West Nile disease to expand into new geographic locations and infect more people. Changing weather conditions, for example, are giving disease-carrying ticks more time to reproduce, expand their habitats, and spread diseases. Thus, the geographic ranges where ticks spread Lyme disease, anaplasmosis, ehrlichiosis, and spotted fever rickettsiosis have expanded, and researchers predict that tick-borne diseases will continue to increase (Singer and Bulled 2016). Indeed, vector-borne diseases have been identified as one of the biggest health threats produced by climate change (Randolph 2009). Research by Solveig Jore et al. (2014), for example, found that the disease-carrying *Ixodes ricinus* ticks have spread over an ever wider geographic range and at higher altitudes in mountainous areas because of climate change. Jore and colleagues studied the relationship between the distribution of infected ticks and microclimatic conditions at seven data collection points along the southern coast of Norway. They reported that the level of humidity has a substantial impact on the ability of the tick-borne encephalitis virus to survive and reproduce in the *I. ricinus* tick and then be transmitted to humans in areas where the tick previously was absent. Supporting Jore et al.'s findings, Munderloh and Kurtti (2011: A148) note that "areas that are likely to experience increased or prolonged seasonal tick activity are … those located at the current extremes of the current range of distribution, areas where climate change will be felt most acutely." Especially important, then, is climate warming along the cold edges of a species' distribution, such as the changes now promoting the movement of *Ixodes scapularis*, a Lyme disease tick vector, from the United States into Canada and tick-borne encephalitis to higher altitudes in Slovakia (Kilpatrick and Randolph 2012; Ogden et al. 2019).

Changing weather conditions also are providing disease-carrying mosquitoes with new opportunities to transmit diseases like West Nile. Multiple studies have found that the warmer temperatures associated with climate change facilitate mosquito development, biting rates, and the incubation of the pathogen inside the mosquito. Also, a factor in long-range virus movement is the effect of climate change on the timing of bird migration and breeding patterns. Mild winters and drought have been associated with West Nile virus disease outbreaks, while rainfall can also contribute by creating breeding sites for mosquitoes. (Beard et al. 2016; Hahn et al. 2015). Human settlement patterns that put them in closer proximity to vector species (e.g., birds) also is a factor in the spread of West Nile disease.

As noted previously, climate change also has forced some animal species into new habitats as their natural ecosystems disappear. This movement of animals into new areas increases opportunities for contact with humans and the potential spread of zoonotic diseases. Moreover, rising temperatures have allowed disease-causing fungi to spread into new locations that previously were too cold for them to survive. For example, Valley Fever, which is caused by a fungus that lives in the soil in hot and dry areas, has spread for the first time into the Pacific Northwest. This fungus can cause severe infections and death. As the difference between environmental temperatures and human body temperatures narrows, new fungal diseases may emerge as some fungi species become more adapted to living in humans.

As also noted, climate change increases the risk for natural disasters and flooding, which can heighten the risk for mold (fungus) to grow in people's homes. Certain molds can cause deadly infections of the lungs and brain. Black mold (*stachybotrys*) is one of the most dangerous molds and can cause flu-like symptoms, diarrhea, headaches, memory loss, and severe respiratory damage. Because their lungs are still developing, children are at particular risk for health problems associated with exposure to black mold as are people with pre-existing pulmonary conditions.

Scientists predict that climate change will have devastating effects on freshwater and marine environments. One outcome will be more frequent and more severe outbreaks of harmful algal blooms, which are the rapid growth of algae or cyanobacteria in lakes, rivers, oceans, and bays. Warming temperatures in Lake Erie in the United States, for example, have contributed to extensive toxic blooms that last into the early winter months.

In early December 2022, the Malawi government declared a public health emergency. Malawi, a landlocked country in southeastern Africa, was experiencing a widespread cholera outbreak, with almost 37,000 cases and 1,210 associated deaths reported by March 2022. Cholera is an acute diarrheal disease caused by infection of the small intestine with some strains of the *Vibrio cholerae* bacteria. People can get sick when they swallow food or water that has been contaminated with cholera bacteria. Left untreated cholera can lead to dehydration and electrolyte derangements that can progress to hypovolemic shock (an emergency condition in which severe blood or other fluid loss disables the heart's ability to pump enough blood to the body), and ultimately death. Although cholera has been endemic in Malawi since 1998, the new outbreak was the deadliest in the country's history, with a case fatality rate of 3%. The large geographic spread of the disease and the high number of reported cases in the country stretched medical capacity to respond to the outbreak, sharply increasing the risk of serious public health impact. The outbreak in Malawi occurred against a backdrop of a surge in cholera outbreaks globally, which has constrained the availability of vaccines, tests, and treatments (WHO 2023). According to Inas Haman, a regional spokeswoman for WHO, "[c]holera thrives in poverty and conflict but is now turbocharged by climate change" (quoted in Hamilton 2023).

The outbreak in Malawi started following tropical storm Ana (January 2022) and hurricane/cyclone Gombe (March 2022), which caused flooding leading to the displacement of a population with low pre-existing immunity and a lack of access to safe water, sanitation, and hygiene. The outbreak was mainly limited to flood-affected areas until August 2022 when it spread to the northern and central parts of the country. Since the beginning of the outbreak, those between 21 and 30 years of age have been the most affected (27.7% of cases), but most deaths have been among a particularly vulnerable population, those aged 60 years and above.

While infectious disease has played an outsized role in shaping our past, and with HIV/AIDS and COVID-19 especially, it is having a profound effect in the present time, and given climate change and other environmental disruptions—barring a sharp change in human societies—it promises to cause much suffering, disability, and death in the future.

Environmental Violence

Pollution has been defined as unwanted waste of human origin released into the air, on land, and into water without regard for cost or consequence, a multifaceted process that presents an existential threat to planetary health while jeopardizing the sustainability of modern societies. Pollution references contamination of the air by fine particulate matter (PM2·5), ground-level ozone, oxides of sulphur and nitrogen, chemicalization and toxication of freshwater, defilement of the oceans with mercury, nitrogen, phosphorus, plastic, and petroleum waste among other chemical and metallic substances, and poisoning of the land with lead, mercury, pesticides, industrial chemicals, electronic waste, and radioactive waste. The effects of toxic pollution alone are arguably one of the largest direct threats to global human health.

The 2017 Lancet Commission on pollution and health, a body composed of health experts, using data collected in the Global Burden of Diseases, Injuries, and Risk Factors Study 2015, concluded that in 2015 pollution was responsible for an estimated nine million deaths (16% of all deaths globally) and for economic losses totaling US $4.6 trillion (6.2% of global economic output) (Landrigan et al. 2018). The Commission also noted pollution's deep inequity, 92% of pollution-related deaths, and the greatest burden of pollution's economic losses occur in low-income and middle-income countries. Based on a re-examination of these issues since 2000, the Commission found that the situation has not improved, and that pollution remains a major global threat to health and prosperity, especially in low- and middle-income countries. Since 2000, the steady decline in the number of deaths from the ancient scourges of household air pollution (e.g., indoor wood-burning stoves), unsafe drinking water, and inadequate sanitation have been offset by increasing deaths attributable to industrial forms of pollution. These modern forms of pollution—e.g., ambient air pollution, lead pollution, and chemical pollution—pose severe threats to health and the environment (Fuller et al. 2022).

Human activities are one of the main causes of marine pollution as well and have an impact on the array of marine and coastal ecosystems on which the diet, livelihoods, and health of millions of people on Earth depend (Denchak 2022). All streams flow into rivers and all rivers spill into the sea. As a result, the oceans are the end point for much of the pollution produced on land. From carbon dioxide emissions to leaking oil, there are a vast number of different types of ocean pollution that humans generate. This impact is degrading the health of the sea at an accelerating rate. The damage includes ocean acidification (by carbon dioxide that destroys mussels, clams, coral, and oysters that require calcium carbonate to make their shells); fossil fuel releases (including oil from boats, airplanes, cars, and trucks, as well as leakages and spills from offshore drilling and oil tankers); chemical discharges (from factories, raw sewage overflows from water treatment systems; plastic in a myriad of shapes, sizes, and colors that is dumped in or finds its way to the sea; agricultural runoff of pesticides, herbicides, and fertilizers; and human-generated ocean noise pollution from commercial shipping (involving over 100,000 ocean-going ships that both disrupt communication by marine mammals and fish and their location of food and mating partners).

Marcantonio and Fuentes (2023) propose the use of the term "environmental violence" to characterize the causes and consequences of industrial pollution. An academic literature deploying the term environmental violence has emerged in the past two decades, especially in the social sciences. Radonic (2015) defines environmental violence as "struggles over access to and control of natural resources, and … environmental transformations [that] may disproportionately place subordinate groups at risk." The term environmental violence also has been employed to refer to harm to the environment, which may have indirect effects on people, such as through land degradation and resource extraction. Marcantonio and Fuentes (2023) seek to expand environmental violence to include environmental connections to human health and wellbeing and to use it as a tool to track, measure, and ultimately reduce the impact on human health outcomes. To that end, they propose that environmental violence be redefined as direct and indirect harm experienced by humans due to toxic and non-toxic pollutants put into a local—and concurrently the global—ecosystem through industrial activities and processes. While humans always have and probably always will produce pollution as a by-product of meeting our physical and material needs, elite polluting based on industrial production, in their view, constitutes a violent act because it is the result of intentional, excessive human consumption of resources and energy, and other activities, well beyond what is needed to maximize flourishing. Elite pollution which is prolific in every location worldwide is a form of environmental violence because it leads to mounting environmental hazards that directly and knowingly harm human health.

The persistence and immense inequality in the causes and consequences of environmental violence is facilitated by structural violence — forms of violence institutionalized in society (like racism or sexism) that harms subordinated groups by preventing them from meeting their basic needs. Moreover, environmental violence exacerbates and creates harm and consequential power differentials and hierarchies. This redefinition is consistent with broadly used definitions of violence that describe it as "any violence [that] lowers the real level of needs satisfaction below what is potentially possible" (Kaltung 1964). Focusing on industrial-produced pollution and its harms makes environmental violence tractable, existing at different scales, and transportable without losing its analytical power and potential to contribute to promoting human, and inseparably, planetary health and flourishing.

A person or group's vulnerability mediates their exposure to environmental violence and their ability to resist or cope with it. Vulnerability is defined as the "state of susceptibility to harm from exposure to stresses associated with environmental and social change and from the absence of capacity to adapt" (Adger 2006). In Marcantonio and Fuentes') (2023) assessment, the vital core of human lives "includes the universal and culturally specific, material and non-material elements necessary for people to act on behalf of their interests" (Adger et al. 2014). Contexts of structural violence shape a person's exposure to environmental violence, such as their material ability to protect themselves during an extreme weather event.

The magnitude of an environmental violence occurrence and the vulnerability of a person or group to that hazard interact. This interaction determines the ultimate human impact of the hazard. For example, all people in a region experiencing an extreme heat wave might be similarly exposed to withering temperatures, but they might not all have the wherewithal to leave the region or to access air conditioning. Similarly, people with pre-existing health conditions, such as asthma linked to toxic pollution exposure, generally are more vulnerable to environmental violence.

Is Planetary Health a Colonial Strategy?

Since it emerged, the planetary health paradigm has been subject to criticism by Indigenous and indigenized voices because of its persistent over-reliance on western Eurocentric understandings of health and natural systems. Critics of the paradigm point out its historic and social origins as a much-touted approach to thinking about health. While the term had earlier uses, its primary contemporary use traces back to a collaborative initiative of *The Lancet*, a leading British medical journal, and the Rockefeller Foundation in the United States. The Rockefeller Foundation–Lancet Commission on Planetary Health, which launched the initiative, had its first meeting in Bellagio, Italy, in July 2014.

The Rockefeller Foundation has had a significant international influence on health policies and practices since the 1930s. It served as a model for WHO, the US National Science Foundation, and the US National Institutes of Health. It also has been a source of influence on the UN. According to Vieira-da-Silva (2018), in the early years of the 20th century wealthy capitophilanthropists like John D. Rockefeller anticipated the need for social reform as a way to avoid social unrest that could threaten the capitalist social order. Among other products of this conclusion, the Rockefeller Foundation was formed in 1913 with the initial goal of bringing about the industrialization of the agrarian US South in the service of the capitalist interests of the North. The report of the Rockefeller Foundation–Lancet Commission on Planetary Health maintains that the Foundation is motivated by a deep concern with inequality, the health of the poorest groups, and the environment. Such insistence, however, glosses over a historic colonial desire to preserve the capitalist order in light of increasing environmental deterioration and its perceived negative impacts on profit produced by endangering human health. Additionally, critics point out that in the report of the Rockefeller Foundation–Lancet Commission on Planetary Health, nonhuman beings tend to be of concern only to the degree that they appear to have instrumental value because of their role in human health.

Critiques of planetary health emphasize the problematic incongruity of attempting to solve the global environmental crisis using the same philosophical, ideological, and material frameworks that created the problem in the first place (Redvers 2021). As Audre Lorde (1984) has said, "the master's tools will never dismantle the master's house. They may allow us temporarily to beat him at his own game, but they will never enable us to bring about genuine change." Planetary health was developed within Western health and environmental sciences and has been promoted by Western social science. These disciplines are deemed problematic because their ideological foundations historically separated humans from each other and from the rest of nature. Consider anthropology for example. The problem is effectively summarized by the Statement on Anthropology, Colonialism, and Racism issued by the Department of Anthropology at the University of Pennsylvania (n.d.):

> No form of scholarly enquiry is neutral, and anthropology is no exception. Anthropology began as a colonial science, the product of a settler colonialism uniquely focused on the study of the languages, history, culture, and biology of non-European peoples seen as "primitive," or "ancient" all around the world. Anthropology was, until recently, primarily the study of the exotic "other" in space or time, an orientation that presumes an unmarked normative "self"—white, Euro-American, and often male—positioned as the distanced and "objective" observer. While this conceit has been thoroughly discredited, it helped obscure the field's historical implication in projects of domination, rule, and control.

Coming to terms with our discipline's distant and recent history has involved a painful process of self-critique and reframing, an ongoing endeavor that has not been easy or without debate, nor is it close to being completed. It has entailed recognizing and directly confronting anthropology's enduring colonial legacies that contributed to the marginalization, exploitation, and erasure of Indigenous peoples, woman, ethnic minorities, sexual minorities, and the poor and working classes along with their knowledge and understandings of the world (Allen and Jobson 2016; Bolles 2013). In a campus talk entitled "Tibet, Ferguson, Gaza: On political crisis and anthropological responsibility" presented at American University, Carole McGranahan (2014) reflected on political changes in our discipline over the last decade, including the need not only to address anthropology's colonial past, but also our imperial present. In this process, how we do research, how we frame our findings, who we include as partners in the research and writing-up process, and the words and concepts we use all matter. Similarly, Fay Harrison, a leader in the decolonizing movement in anthropology, stresses the need for really

> [T]aking the subaltern, the Indigenous, the indigen*ized* and the minor-it*ized* seriously. To the extent that we truly recognize people's full humanity, that of course means we recognize their wisdom, their intelligence, their capacity to produce forms of knowledge that include potentially powerful interpretations and explanatory accounts of the world, which give us the clues to then create strategies to change the world.
> *(Quoted in McGranahan and Rizvi 2016; emphasis in the original)*

As discussed in Chapter 2, the critical anthropology of climate change has its roots in political ecology, an approach that has "long worked to expose the uneven power dynamics involved in knowing and managing nature, including the knowledge politics that privilege Western science and marginalize local knowledge and land-use practices" (Goldman et al. 2018). Analysis from this perspective exposes the persistently uneven power dynamics that shape what is accepted as valid knowledge, including climate knowledge. Failure to pay attention to these issues leads to blaming the victims of climate change, such as accusations that the Fulani of the Sahel in Africa south of the Sahara retain "bad cultural practices" that spread the desert (e.g., We Are Water Foundation 2019). Similarly, Maasai herders in Kenya and northern Tanzania have been described as having limited knowledge of novel changes in their habitat and as holding on to ancient, outdated cultural practices that result in overgrazing and desertification of woodlands and savannahs (e.g., Fuchs 2017). For these reasons, political ecologists of all disciplines, including the critical anthropologists of climate change, call for the recognition of other ways of knowing about and experiencing "nature," while learning from the negative results of development or conservation initiatives that do not consider local understandings and knowledge.

All these issues arise in the Indigenous assessment of the planetary health paradigm. One expression of this assessment has been offered by a collaborative team composed of Indigenous scholars, practitioners, land and water defenders, respected elders, and knowledge-holders from around the world who came together as a consensus panel to define the determinants of planetary health from an Indigenous perspective (Redvers 2021).

Participants began with the premise that Indigenous peoples globally have the sacred mandate and right to give voice to Mother Nature. Although this right is often not respected by the legal statues of common law, Indigenous peoples have always seen the need to stand up to protect their human and nonhuman relatives. Instead, past and current Eurocentric political and economic narratives deprive the land, water, and air of being in the world as equal rights-holders. Redvers (2021) points out that this "denial of the right of being is a direct product of ongoing capitalist and colonial mandates, which will continue to exacerbate the environmental crisis." The author proceeds to lay out their consensus perspective, which, in part, asserts the following points.

First, they call attention to gender issues, noting that Indigenous worldviews commonly recognize Mother Earth's creative power as the primordial First Mother. Planetary health, by contrast, emerged from a patriarchal competitive world of colonization, militarism, racism, social exclusion, biomedicine, and poverty-inducing economic and so-called development policies.

Second, since the rise of colonialism, people have lost their identity as organisms within a larger natural system and with it they have lost awareness of how to live sustainably with Mother Earth. The rapidly advancing ecological demise now occurring brings to light an impaired human relationship with our inner self (i.e., we are Nature and not separate from it).

Third, Indigenous societies are predominantly collective by nature. Individuals in collective-based societies learn while growing up that interdependence with others and with place (i.e., with the area where they live) helps to maintain wellness and balance. This collective focus contrasts with that of individualist societies like the Eurocentric world that created the planetary health paradigm.

Finally, Western science is a paradigm that uses the scientific method to theorize, hypothesize, find variables, measure, and describe a relationship, usually framed in mathematical, economic, or even political terms. However, this paradigm is limited in its attempt to explain complex, multidimensional relationships over time, because it tends to be linear, reductionistic, and mechanistic. There is, too, an overarching interest of Western science in finding ways to influence and even control the natural systems and processes of the world, including human behavior. In the early 1990s, Per Fugelli, a nationally celebrated Norwegian general medical practitioner and professor of social medicine, expressed the medico-centric mindset from which the planetary health paradigm emerged: "The patient Earth is sick. Global environmental disruptions can have serious consequences for human health. It's time for doctors to give a world diagnosis and advice on treatment" (quoted in Casassus 2017).

With the rise of "Indigenous science," the pendulum in the 21st century is swinging towards the need for a community-centered, systems-oriented, eco-logical-based networking approach to knowing and being. Consensus panel participants conclude by stating:

> As equitable and inclusive societies, institutions, and fields are built, embracing diverse knowledges will get us closer to a well and just planet for all. Indigenous voices are a powerful and beneficial solutions-orien-tated force for Mother Earth's wellbeing and for all living beings that inhabit her. We therefore call for an inclusion of wisdom that is not mere knowledge or information but is an insight that comes from the heart—from the heart of Mother Earth
>
> *(Redvers 2021)*

In light of these criticisms, a number of advocates of the planetary health approach have concluded that a new conception of planetary health is needed. In the view of Rhys Jones, a Māori academic, Papaarangi Reid, also a Māori academic, and Alexandra Macmillan, a scholar-activist living in Aotearoa New Zealand (2022), from a reoriented decolonial conception of planetary health any benefit that accrues to some through the exploitation of others cannot be considered to be health-promoting, and such treatment is antithetical to good health. These researchers, who believe that planetary health has an important role to play in guiding humanity away from injustice and a dystopic future, use as an example lithium mining and the building of electric vehicles: "Planetary health action cannot simply be about replacing one form of extractive capitalism with another, protecting elite lifestyles and nationalising notions of ecological sustainability while accepting racist vio-lence against some populations as a regrettable, but unavoidable, externality" (Jones et al. 2022). Transportation and other infrastructure, they argue, should be shaped by the landscape, local relationships, and histories of the land, rather than the now dominant approach of subjugating the environment to accommodate the transport needs of governments, corporations, and car owners based on Western values of efficiency and time saving. Similarly, con-temporary health campaigns to reduce dietary consumption of meat incor-porates a colonial disregard of customary and culturally meaningful food sources among Indigenous peoples forcing them to rely on commercial food systems that perpetuate exploitative dynamics and carbon-intensive supply chains. Further, initiatives undertaken to improve planetary health must avoid acts of injustice against Indigenous peoples. Such breaches of Indigenou rights are occurring already, including the forced removal of Indigenous communities to make way for forestry projects (e.g., forcibly moving the Maasai in the name of conservation) and efforts to expand biofuel production, which also significantly increases the price of food and exacerbates food insecurity for Indigenous communities.

This understanding of meaningful change involves dismantling established systems of power and the fundamental reorientation of governance at all levels. A change of this magnitude requires a transformational shift in dominant values and ideologies. Such a world-reshaping process, they believe, should begin with a commitment to "epistemic disobedience" (Mignolo 2009), which involves the interrogation of the "naturalness" and "superiority" of the Western, objective, and individualistic approach to knowing and being-in-the-world. This project consists of far more than reductive goals such as avoiding 1.5 °C (2.7 °F) of global warming. It must include a disinvestment from business-as-usual activities but does not preclude various actions to mitigate the adverse impacts of existing systems. While planetary health scholar-activists may need to continue working in Western academic institutions, they must focus on creating alternative systems that are grounded in principles of relational wellbeing and engagement in community climate change and social justice actions such as legislation rewriting Aotearoa New Zealand's Resource Management Act to including reference to the concept of *Te Oranga o te Taiao* (the health and wellbeing of the natural environment). This example, although nascent, envisions a planetary health that is grounded in relationality and kinship with the natural world. At the same time, all systems of global governance (e.g., the Intergovernmental Panel on Climate Change) need to be built on a commitment to Indigenous rights, providing a forum that brings together sovereign Indigenous peoples and reinforces their claims to self-determination.

The potential for the kind of collaboration needed to build a decolonized planetary heath is perhaps, by way of analogy, illustrated by the conservation effort in New Zealand to save the kākāpō—a slow, moss-colored flightless parrot—from extinction. Kākāpō once lived throughout Aotearoa, the Indigenous Māori name for New Zealand. Found nowhere else in the world, these birds have become a national icon on the island nation. On the brink of extinction caused by imported predators (including rats, mice, stoats, and Australian possums), the few wild kākāpō were evacuated to three tiny islets around New Zealand to live free from predators. There, they also have the close protection of conservation rangers, who operate under the watchful eye of the Ngāi Tahu, the Māori *iwi*, or tribe, that calls New Zealand's South Island home. This collaboration has produced great success: the kākāpō population has quadrupled in number. To solve the overcrowding that resulted, ten birds were airlifted in 2023 to Sanctuary Mountain, a mainland ecological area in the North Island of New Zealand that is surrounded by one of the world's longest pest-proof fences. The kākāpō story and its revelations about relations among people, highlights the successes achieved by marrying Western and Indigenous conservation approaches to revive an endangered species, and reintroduce it to its native land.

Sealers and whalers, who arrived around 1810, were the first Europeans to land in Aotearoa, foreshadowing the wave of colonial settlers to come. Ultimately, these colonial settlers took over. In 1840 Britain formally annexed the islands. That year, 540 Māori *rangatira* (chiefs) were forced to sign the Treaty of Waitangi by which they accepted British sovereignty in exchange for guaranteed possession of their land. But armed assaults continued against the Māori, who fought back as best they could. By 1870, the Māori population was greatly diminished. In 1986, the Wildlife Service (later the Department of Conservation) turned Sanctuary Mountain into a nature reserve and barred Māori from entering without permission.

Eventually, however, New Zealand's government began to address the country's iwi over the colonial theft of their land. As a first step, in 2023, Andrew Little, New Zealand's Minister of Health announced,

> The Crown acknowledges that Ngāti Maru's relationship with the Crown has been one characterized by loss of land, of identity, and of autonomy. For Ngāti Maru, this loss has left a legacy of dislocation and dispossession ... For those actions which rendered your iwi almost completely landless, severed your connection to your *whenua* [land], and inflicted economic hardship and suffering on generations of your people, the Crown sincerely apologizes.
>
> *(Quoted in Cineas 2023)*

As a second step, the *iwi* was awarded financial redress of US $30 million plus interest, the return of 16 sites of cultural significance, funding of $1,023,454 to aid with cultural revitalization, and Ngāti Maru's right to purchase the Te Wera Crown Forest. This followed the Ngāi Tahu signing a settlement in 1996 giving them a special role in the management of Whenua Hou (Codfish Island), the most important of the three kākāpō islands, and, in recognition of the tribe's centuries-long relationship with the bird, in the conservation of the kākāpō.

Based on this settlement, a Kākāpō Recovery Group, composed of representatives from Ngāi Tahu, the scientific community, and the Department of Conservation, seeks to shepherd the species' survival. It took years, however, to begin reconciling the Department of Conservation's largely Western attitudes towards Indigenous values. One of the difficult moments in that process occurred in 2008 when rangers proposed that they should artificially inseminate female birds with manually gathered sperm. The Ngāi Tahu at first rejected the idea saying that the procedure was unnatural and would harm the kākāpō. The rangers explained that the artificial insemination would help to avoid the harm of inbreeding. This connected for the Ngāti Maru with the familiar Indigenous concept of *whakapapa*, meaning a line of descent from the ancestors down to the present day. *Whakapapa* links people to all other living things, and to Earth and the sky, and it traces the universe back to its origins. With this understanding, the insemination program was approved. Over time and working together collaborative, the *iwi* and the rangers have deepened their shared understandings and found ways to advance the cause of saving the kākāpō.

In short, despite its origin, planetary health (and anthropology too) can transition passed coloniality and become a truly emancipatory and just approach to thinking and acting in a world beset with mounting climate/environment threats, threats to all people, Indigenous or, like most people today, the descendants of migrants.

Conclusion

Planetary health faces three vexing challenges:

1. The challenge of attribution: It is often very difficult to definitively prove that particular changes in the climate/environment cause specific health outcomes in humans or other species.
2. The challenge of Indigenous critiques: The long ignored or even actively silenced voices of Indigenous Nations have taken planetary health to task for not hearing and responding to their traditional understandings of relations among humans, plants and animals, and "Mother Earth."
3. The challenge of competing models. One Health and other approaches are also attempting to broaden Western understanding of health.

Like all approaches to environment-based health, planetary health seeks through ever-improving and more powerful research technologies and a broadening knowledge base to identify the specific linkages between changes in the climate/environment and health. Additionally, efforts have been initiated to include Indigenous ideas and people in conceptualizing and operationalizing planetary health, although there is still a long way to go. Finally, rather than stressing competition, there is movement towards collaboration of planetary health and One Health. As climate change and environmental disturbances and stressors advance, planetary health offers one framework for grassroots/professional/Indigenous collaboration in building a sustainable world.

References

Adger, W. 2006. Vulnerability. *Global Environmental Change* 16: 268–281

Adger, W.N., Pulhin, J.M., Barnett, J., Dabelko, G.D., Hovelsrud, G.K., Levy, M., Oswald Spring, Ú., and Vogel, C.H. 2014. Human security. In *Climate Change 2014: Impacts, Adaptation, and Vulnerability. Part A: Global and Sectoral Aspects. Contribution of Working Group II to the Fifth Assessment Report of the Intergovernmental Panel on Climate Change.* C.B. Field, V.R. Barros, D.J. Dokken, K.J. Mach, M.D. Mastrandrea, T.E. Bilir, M. Chatterjee, K.L. Ebi, Y.O. Estrada, R.C. Genova, B. Girma, E.S. Kissel, A.N. Levy, S. MacCracken, P.R. Mastrandrea, and L.L. White, eds. pp. 755–791. Cambridge and New York: Cambridge University Press

Alimonda, H. (2011). La colonialidad de la naturaleza. Una aproximaci on a la Ecología Política Latinoamericana. In *La naturaleza Colonizada: Ecología Política y Minería en America Latina*. H. Alimonda, ed. Buenos Aires: CLACSO

Allen, Jafari Sinclair and Jobson, RyanCecil. 2016. The decolonizing generation: (Race and) theory in anthropology since the eighties. *Current Anthropology* 57(2): 129–148

Atlas, Ronald. 2013. One Health: Its origins and future. *Current Topics in Microbiology and Immunology* 365: 1–13

Baquero, Oswaldo, Fernández, Mario Nestor Benavidez, and Aguilar, Myriam Acero. 2021. From modern planetary health to decolonial promotion of One Health of peripheries. *Frontiers in Public Health* 9

Beard, C., Eisen, R., Barker, C., Garofalo, J., Hahn, M., Hayden, M., Monaghan, A., Ogden, N., and Schramm, P. 2016. Vector-borne diseases. In *The Impacts of Climate Change on Human Health in the United States: A Scientific Assessment*. Washington, DC: US Global Change Research Program. https://health2016.globalchange.gov

Bolles, A. Lynn. 2013. Telling the story straight: Black feminist intellectual thought in anthropology. *Transforming Anthropology* 21(1): 57–71

Casassus, Barbara. 2017. Per Fugelli. *The Lancet* 390: 2032

Chen K., Wang M., Huang C., Kinney P.L., and Anastas P.T. 2020. Air pollution reduction and mortality benefit during the COVID-19 outbreak in China. *The Lancet Planetary Health* 4(6): e210–e212

Cineas, Fabiola. 2023. New Zealand's Māori fought for reparations—and won. *Vox*. https://www.vox.com/the-highlight/23518642/new-zealand-reparations-maori-settlements

Coates, Ta-nehisi. 2015. *Between the World and Me*. New York: Spiegel and Grau

Comunian S., Dongo D., Milani C., and Palestini P. 2020. Air pollution and Covid-19: The role of particulate matter in the spread and increase of Covid-19's morbidity and mortality. *International Journal of Environmental Research and Public Health* 17(12): 4487

Davis, Heather and Todd, Zoe. 2017. On the importance of a date, or, decolonizing the Anthropocene. *ACME: An International Journal for Critical Geographies* 16(4): 761–780

Denchak, Melisa. 2022. Ocean pollution: The dirty facts. Natural Resources Defense Council. https://www.nrdc.org/stories/ocean-pollution-dirty-facts

Denver, Simon. 2016. Tibetans in anguish as Chinese mines pollute their sacred grasslands. *The Washington Post*. https://www.washingtonpost.com/world/asia_pa cific/tibetans-in-anguish-as-chinese-mines-pollute-their-sacred-grasslands/2016/12/ 25/bb6aad06-63bc-11e6-b4d8-33e931b5a26d_story.html

Department of Anthropology at the University of Pennsylvania n.d. Statement on Anthropology, Colonialism, and Racism. https://anthropology.sas.upenn.edu/news/ 2021/04/28/statement-anthropology-colonialism-and-racism

Fuchs, S. 2017. Opinion: Are Maasai cattle to blame for overgrazing in Tanzania? *Africa Geographic*. https://africageographic.com/stories/maasai-cattle-blame-overgra zing-tanzania/

Fuller, Richard, Landrigan, Philip, Balakishnan, Kalpana, Bathan, Gynda, Bose-O'Reilly, Stephen, and Brauer, Michael. 2022. Pollution and health: A progress update. *The Lancet Planetary Health*. https://www.thelancet.com/journals/lanplh/article/PIIS2542-51 96(22)00090–0/fulltext#bib1

Galtung, J. 1964. A structural theory of aggression. *Journal of Peace Research* 1: 95–119

Goldman, M., Turner, M., and Daly, M. 2018. A critical political ecology of human dimensions of climate change: Epistemology, ontology, and ethics. *WIREs Clim Change* 9: e526

Gross, Terry. 2023. How "modern-day slavery" in the Congo powers the rechargeable battery economy. Goats and Soda, NPR. https://www.npr.org/sections/goatsandsoda/2023/02/01/1152893248/red-cobalt-congo-drc-mining-siddharth-kara

Grove, Jairus. 2016. Response to Jedediah Purdy. Forum: The new nature. *Boston Review.* https://www.bostonreview.net/forum_response/jairus-grove-response-nature-anthropocene/

Guterres, António. 2020a. *Secretary-General's remarks to United Nations Biodiversity Summit.* New York: United Nations. https://www.youtube.com/watch?v=5WoQ0JjMzP8&t=485s

Guterres, António. 2020b. *Inequality Defines Our Time. UN Chief Delivers Hard-Hitting Mandela Day.* New York: United Nations. https://news.un.org/en/story/2020/07/1068611

Hahn, M., Monaghan, M. Hayden, R. Eisen, M. Delorey, N. Lindsey, R. Nasci, and Fischer, M. 2015. Meteorological conditions associated with increased incidence of West Nile virus disease in the United States, 2004–2012. *American Journal of Tropical Medicine and Hygiene* 92(5): 1013–1022

Hamilton, M. 2023. Turbocharged by climate change: Malawi's Cholera outbreak is worsened by Covid-19 misinformation. Rockefeller Foundation. https://www.rockefellerfoundation.org/case-study/turbocharged-by-climate-change-malawis-cholera-outbreak-is-worsened-by-covid-19-misinformation/#:~:text=Climate%20Change%20Triggers%20and%20Worsens%20Cholera%20Outbreak&text=Malawi%27s%20outbreak%2C%20worsened%20by%20climate,for%20cholera%20outbreaks%20in%20Africa

Horton, Richard and Lo, Selina. 2015. Planetary health: A new science for exceptional action. *The Lancet* 386. https://www.thelancet.com/pdfs/journals/lancet/PIIS0140-6736(15)61038-61038.pdf

Jerez, Barbara, Garces, Ingrid, and Torres, Robinson. 2021. Lithium extractivism and water injustices in the Salar de Atacama, Chile: The colonial shadow of green electromobility. *Political Geography* 87: 102–382

Jones, Rhys, Papaarangi Reid, and Alexandra Macmillan. 2022. Navigating fundamental tensions towards a decolonial relational vision of planetary health. *The Lancet* 6(10). doi:10.1016/S2542-5196(22)00197-8

Jore, Solveig, Vanwambeke, Sophie, Viljugrein, Hildegunn, Isaken, Keteil, Kristoffersen, Anja, Woldehiwet, Zerai, Johansen, Bernt, Brun, Edgar, Brunn-Hansen, Hege, Westerman, Sebastian, Larsen, Inger-Lise, Ytrehus, Bjornar, and Hofshagen, Metate. 2014. Climate and environmental change drives *Ixodes ricinus* geographical expansion at the northern range margin. *Parasites & Vectors* 7:11. https://parasitesandvectors.biomedcentral.com/articles/10.1186/1756-3305-7-11#auth-Anja_B-Kristoffersen-Aff1-Aff5

Kahn, Laura. 2006. Confronting zoonoses, linking human and veterinary medicine. *Emerging Infectious Diseases* 12(4). https://wwwnc.cdc.gov/eid/article/12/4/05-0956_article

Kaltung, Johan. 1964. A structural theory of aggression. *Journal of Peace Research* 1(2): 95–119

Katwala, Amit. 2018. The spiralling environmental cost of our lithium battery addiction. *Wired.* https://www.wired.co.uk/article/lithium-batteries-environment-impact

Kehr, Janina. 2020. "For a more-than-human public health." *BioSocieties* 15: 650–663. doi:10.1057/s41292-020-00210-8

Kilpatrick, M. and Randolph, S. 2012. Drivers, dynamics, and control of emerging vector-borne zoonotic diseases. *Lancet* 380: 1946–1955

Landrigan, Philip, Fuller, Richard, Acosta *et al.*2018. The Lancet Commission on pollution and health. *The Lancet* 391: 462–512

Lorde, Audry. 1984. The Master's tools will never dismantle the Master's house. In *Sister Outsider: Essays and Speeches*. Audrey Lorde, ed. pp. 110–114. Trumansburg, NY: Crossing Press

Marcantonio, Rice and Fuentes, Augustin. 2023. Environmental violence. *The Lancet Planetary Health*. https://doi.org/10.1016/S2542-5196(23)00190-0

McGranahan, Carole. 2014. Tibet, Ferguson, Gaza: On political crisis and anthropological responsibility. American University. https://www.youtube.com/watch?v=Pf2HX0IbGhw

McGranahan, Carole and Rizvi, Uzma Z. 2016. *Decolonizing Anthropology: A Conversation with Faye V. Harrison, Part 1*. Decolonizing Anthropology. Savage Minds. https://savageminds.org/2016/05/02/decolonizing-anthropology-a-conversation-with-faye-v-harrison-part-i/

McKenzie, Pete. 2023. How New Zealand saved a flightless parrot from extinction. *National Geographic*. https://www.nationalgeographic.com/animals/article/kakapo-release-new-zealand-maoriconservation#:~:text=On%20the%20brink%20of%20extinction,pests%20like%20cats%20and%20stoats

Mignolo, W. 2009. Epistemic disobedience, independent thought and decolonial freedom. *Theory, Culture and Society* 26(7–8): 159–181

Miles, T. 2021. *All That She Carried*. New York: Random House

Munderloh, U. and Kurtti, T. 2011. Emerging and re-emergent tick-borne diseases: New challenges at the interface of human and animal health. In *Critical Needs and Gaps in Understanding Prevention, Amelioration and Resolution of Lyme Disease and Other Tick-Borne Diseases: The Short-term and Long-term Outcomes*. pp. A142–A166. Washington, DC: The National Academies Press

Myers, S. and Frumpkin, H. 2020. *Planetary Health: Protecting Nature to Protect Ourselves*. Washington, DC: Island Press

Ogden, N., Bouchard, C., Badcock, J., Drebot, E., Elias, S., Hatchette, T., Koffi, J., Leighton, P., Lindsay, L., Lubelczyk, C., Peregrone, A., Smith, R., and Webster, D. 2019. What is the real number of Lyme disease cases in Canada? *BMC Public Health* 19, 849. https://doi.org/10.1186/s12889-019-7219-x

Pachauri, A, Sevilla, N.P.M., Kedia, S., Pathak, D., Mittal, K., and Magdalene, A.P. 2021. COVID-19: A wake-up call to protect planetary health. *Environmental Resilience and Transformation in Times of COVID-19*. pp. 3–16. Amsterdam: Elsevier. doi:10.1016/B978-0-323-85512-9.00017-6

Patz, Jonathon and Confalonieri, Ulisses. 2004. Human health: Infectious and parasitic diseases. In *Millennium Ecosystem Assessment: Conditions and Trends*. Washington, DC: Island Press

Pozzer A., Dominici, F., Haines, A., Witt, C., Münzel, T., and Lieveld, J. 2020. Regional and global contributions of air pollution to risk of death from COVID-19. *Cardiovascular Research* 116(14): 2247–2253

Radonic, L. 2015. Environmental violence, water rights, and (un) due process in northwestern Mexico. *Latin American Perspectives* 42: 27–47

Randolph, S. 2009. Tick-borne disease systems emerge from the shadows: The beauty lies in molecular detail, the message in epidemiology. *Parasitology* 136: 1403–1413

Redvers, Nicole. 2021. The determinants of planetary health. *The Lancet* 5(3): E111–E112. https://www.thelancet.com/journals/lanplh/article/PIIS2542-5196(21)00354-003 55/fulltext?trk=public_post_main-feed-card_reshare_feed-article-content

Ruiz de Castañeda, Rafael, Villers, Jennifer, Guzmán, Carlos, Eslanloo, Turan, de Paula, Nicole, Machalaba, Catherine, Zinsstag, Jakob, Utiziner, Jürg, Flahault, Antione, and

Bolon, Isabelle. 2023. One Health and planetary health research: Leveraging differences to grow together. *The Lancet* 7(2): E109–E111

Saunders, L. 2000. Virchow's contributions to veterinary medicine: Celebrated then, forgotten now. *Veterinary Pathology* 37(3):199–207. doi:doi:10.1354/vp.37-3-199.

Seltenrich, Nate. 2018. Down to earth: The emerging field of planetary health. *Environmental Health Perspectives* 126(7): 1–7

Singer, Merrill. 2014. Climate change and planetary health. *Somatosphere.* September 8. http://somatosphere.net/2014/09/climate-change-and-planetary-health.html.it

Singer, Merrill and Bulled, Nicola. 2016. Ectoparasitic syndemics: Polymicrobial tick-borne disease interactions in a changing anthropogenic landscape. *Medical Anthropology Quarterly* 30(4): 442–461

Vieira-da-Silva, L.M. 2018. *O Campo da Saúde Coletiva: Gênese, Transformações e Articulações Coma Reforma Sanitária.* Salvador: Editora da UFBA, Editora Fiocruz

We Are Water Foundation. 2019. The Sahel, desertification beyond drought. https://www.wearewater.org/en/the-sahel-desertification-beyond-drought_318262

Whitmee, S., Haines, A., Beyrer, C., Boltz, F., Capon, A.G., de Souza Dias, B.F., Ezeh, A., Frumkin, H., Gong, P., Head, P., Horton, R., Mace, G.M., Marten, R., Myers, S. S., Nishtar, S., Osofsky, S.A., Pattanayak, S.K., Pongsiri, M.J., Romanelli, C., Soucat, A., Vega, J., and Yach, D. 2010. Safeguarding human health in the Anthropocene epoch: Report of The Rockefeller Foundation-Lancet Commission on planetary health. *The Lancet* 386: 1973–2028. doi:10.1016/S0140-6736(15)60901-1

Whyte, Kyle. 2016. Is it déjà vu? Indigenous peoples and climate injustice. In *Humanities or the Environment: Integrating Knowledges, Forging New Constellations of Practice.* Joni Adamson, Michael Davis, and Huang Hsinya, eds. pp. 88–104. London: Earthscan Publications

World Health Organization (WHO). 2020. WHO reveals leading causes of death and disability worldwide: 2000–2019. https://www.who.int/news/item/09-12-2020-who-reveals-leading-causes-of-death-and-disability-worldwide-2000-2019

World Health Organization (WHO). 2023. Cholera: Malawi. https://www.who.int/emergencies/disease-outbreak-news/item/2022-DON435

Wu, X., Nethery, R.C., Sabath, M.B., Braun, D., and Dominici, F. 2020. Air pollution and COVID-19 mortality in the United States: Strengths and limitations of an ecological regression analysis. *Sci. Adv.* 6(45): eabd4049

Young, Crawford. 2015. *Politics in Congo: Decolonization and Independence.* Princeton, NJ: Princeton University Press

6

TOWARDS A CRITICAL ANTHROPOLOGY OF CLIMATE REFUGEES

Case Study: The Floods of Starvation

Record flooding has inundated the country of South Sudan—a landlocked country in eastern Central Africa—for several years. One of the least developed and poorest countries in the world, South Sudan only gained independence from Sudan in 2011 after a grisly civil war. The intense rains have severely upset the lives of people in South Sudan. An estimated 1.7 million people have already been displaced from their homes and migration has increased each year, with people reporting being forced to pack up their few belongings and flee to higher ground multiple times. Others have given up entirely on internal migration and crossed the border into Sudan. One of the climate migrants in South Sudan is Nyathak. When the floods covered everything she owned in her hometown of Niahldiu, she took her children through the water to the shore, hoping to be rescued. Flooding during the rainy season is not uncommon in South Sudan, but now the water is no longer receding and has turned into a muddy swamp. Like many others, Nyathak decided to leave Niahldiu with her family in search of dry land. It took the family six days to reach a drier place almost 30 miles away in Bentiu, the seat of Unity State, and now Bentiu is basically an island. Since 2021, tens of thousands of fellow migrants have come to Bentiu seeking dry shelter, medicine, and food, as the flood water inundates more communities and cuts off many others from needed supplies. More than one million people now face severe food insecurity. Because of the flooding, one in five children in the country is severely malnourished. Sanitation is also a major issue here, as well as diarrheal diseases (quoted in Castillejo 2023).

The stressful experience is echoed in the words of Nyadiag Gak, another refugee:

DOI: 10.4324/9781003469940-7

The flood destroyed everything people had ... But we couldn't plant here either, because it's also flooded ... All the people who came here tried planting maize, but all of it was destroyed by the floods ... Before we migrated here, we had already lost our house. When we arrived, we settled in a small space on this island. But even here, we don't have any food to eat ... When there were no floods, we used to plant sorghum and maize, but now we don't have any seeds in our hands, so we make do with what we have.

(Quoted in ReliefWeb 2021)

Without seeds to plant and or somewhere to plant them, there was no food to save for the future. Most of the local cattle—which provide milk—died after grazing in disease-infested floodwaters. Life has become hand-to-mouth and every day people struggle to survive.

Women, desperate to find food for their families, have turned to harvesting the water lilies that grow in the floodwaters as a last resort. Mothers, grandmothers, and even pregnant women travel by canoe or walk long distances through the muddy water in order to dive in and out for hours or the entire day searching for the lily bulbs which grow there. Another refugee woman, Bol Kek, who has seven children, explains: "We are not used to collecting water lilies, but the flood water forces us to collect them ... We women are very strong in our hearts, and we bring food to our families" (quoted in ReliefWeb 2021). After she has collected several dozen bulbs, she returns home and mixes the bulb hearts with edible weeds that she gathers to make a soup for her family. It takes at least two bunches of bulbs to make a small meal for a child. The nutritional value of the lilies is low, but it is all there is apart from the occasional small fish that the migrants have to prevent empty stomachs. Bol Kek adds, "My children depend on what I get. I cook for them, and they wait for me while I'm away collecting water lilies ... Life is so hard for us, but we keep strong" (quoted in ReliefWeb 2021).

There are health consequences for the bulb collectors. According to Nyadiang,

We go in the water when the sun is high and we return when the sun sets ... You get cold in the water, especially if it is deep, this is the reason why some of us have developed a cough ... I feel pain in my chest ... the reason might be the work. When we bring the water lilies from the river, we must also grind them and, when you breathe in that dust, it makes you cough.

(Quoted in Castillejo 2023)

While the climate factors, places, languages, clothing, and faces of the people differ, the story of South Sudan exemplifies the growing crisis of climate refugees.

Anthropological Study of Refugees of War, Conflict, and Oppression

The modern history of refugees, encapsulating the 20th and 21st centuries, began with fall of the old multiethnic empires of Europe (e.g., the Russian Empire, the Habsburg Empire, the Ottoman Empire, and the German Empire) and their replacement by smaller nation states (e.g., the French Republic, the Republic of Italy, the United Kingdom of Great Britain and Northern Ireland, etc.). In this wrenching transition, hundreds of thousands of people were forced to flee their homes because they did not match the ethnic/cultural identity embraced by the nation in which they were living.

Gatrell (2015) comments that these refugees were portrayed by the media and by sections of the public to be unhealthy and undisciplined and to pose a potential threat to the social order. By contrast, the League of Nations, which was established after the First World War with the mission of averting future armed combat in Europe, labeled these displaced persons "refugees" and began developing international laws to protect them. Refugees become the focus of various international agreements and assistance programs, which continued after the Second World War with the founding of the United Nations and its extensive refugee agency, the UN High Commissioner for Refugees (UNHCR).

In establishing this humanitarian regime, refugees were defined as people who had been forced to leave their countries in order to escape war, violence, conflict, or persecution. As this definition indicates, people were only considered to be refugees by UNHCR if they had crossed an international border in order to seek safety in another country. The 1951 Refugee Convention, the centerpiece of international refugee protection, explicitly defines a refugee as "someone who is unable or unwilling to return to their country of origin owing to a well-founded fear of being persecuted for reasons of race, religion, nationality, membership of a particular social group, or political opinion" (UNHCR 2023a). By the end of 2017, there were 25.4 million people across the world who met this description. UNHCR does not officially recognize climate/environment displaced persons as refugees and entitled to refugee aid because they are not included in the 1951 Refugee Convention.

In light of the growing numbers of men, women, and children now fleeting climate-related and other environmental disasters, there has been debate between those who want to retain the old definition of refugees (and hence focus aid on smaller populations in need) and those who accept that people who flee climate/environment crises are no less displaced and no less in need than those fleeing conflict or political oppression. As it stands, those displaced from their homes by environment and climate factors fall through the cracks of international refugee policies. The Environmental Justice Foundation (2014) refers to this deficit as a "protection gap" stressing that "the experience of involuntarily leaving one's home due to persecution … is an inherent feature of globally unequal distribution of responsibility for climate change, which has systematically marginalized the world's most vulnerable communities." Certainly, climate change has the well-demonstrated ability to destroy peoples' homes, rob them of all their possessions, steal their livelihoods, cause injury, shatter social ties, and sometimes render them stateless in a culturally foreign environment. To say that they should not be called refugees is a semantic and legalistic splitting of hairs.

As Castañeda et al. (2016) observe, anthropologists and others have long wrangled with "the ambiguous and contested nature of the category refugee." Central to this discussion is the issue of "deservingness," not only of the label refugee but also in terms of access to the assistance benefits that may accrue to documented refugees (Willen 2012; Yarris and Castañeda 2015). Terrio (2015) adds that these immigrants have been overlooked because they are not judged to be "good victims" or are seen as being "questionable symbols of vulnerability." While some prefer the term "climate migrant," former US President Barrack Obama accepted climate scientists' findings and the designation of "climate refugee" (Miller 2019: 64), a usage we follow in this book. However, it should be noted that some people displaced by climate change
reject the refugee label as "undignified," since they do not wish to move, and they do not want to be treated like political refugees.

Recognizing a change in thinking in many circles about climate-displaced people, UNHCR (2023b) has now defined climate action as a central element of its mission to address displaced persons and provide protection to vulnerable populations, although it still does not define them as refugees unless climate change helps to drive conflict. In 2023, it sought US $845.1 million for activities supporting climate action. As a result of this emerging perspective, in 2023, UNHCR's Assistant High Commissioner for Operations, Raouf Mazou, traveled to Kenya's Dadaab refugee complex together with Ambassador Majid Al Suwaidi of the United Arab Emirates, a Special Representative of the UN Climate Change Conference. In Dadaab, they witnessed the impact on both refugees and host communities of climate change. During the visit, Al Suwaidi stated:

> The Dadaab refugee complex, which is one of the largest refugee settlements in the world, is a striking example of the urgent and interconnected challenges we face … Unpredictable weather patterns and the devastating Horn of Africa droughts have disrupted communities and livelihoods, pushing them to the brink of survival … The stories I have heard from refugees and host communities in Dadaab are a stark reminder of why we must continue to drive equitable and just climate action that leaves no one behind.
>
> *(Quoted in UNHCR 2023b)*

Beginning especially during the 1980s, the study of forced migration has steadily gained scholarly attention, although for many years this remained a fairly limited endeavor. It is noteworthy in this regard that the anthropology of human migration and its effects dates at least the 1930s. In both ethnographic inquiry and theorizing, the movement of peoples across the planet has captured the attention of anthropologists, especially archeologists and anthropological linguists. As Baba (2013) notes, "[i]t is tensions inherent in the dynamics of globalization, transnationalism, and the nation state that figure prominently in contemporary anthropological literature on migration." The issue of refugees and other forced migrants has today become a major area of study in anthropology although the road to embracing this focus has been rocky.

One of the first anthropological ethnographies on refugees was Peter Loizos's (1981) *The Heart Grown Bitter: A Chronicle of Cypriot War Refugees*, followed a few years later by B.E. Harrell-Bond's (1986) *Imposing Aid: Emergency Assistance to Refugees*. These ethnographies described the troubled experiences of refugees and internally displaced people living in camps, spontaneous settlements, and countries of asylum. Conquergood (1988) described the refugee camp where he did his research as a "liminal zone" in which refugees try out new identities and new strategies to cope with the challenges they face.

Despite these early efforts, for many years refugee studies in anthropology remained few and far between. The primary reason for this neglect appears to be an early unwillingness to focus on change because of a desire to find and describe enduring cultural patterns. This blinkered view of ongoing social life was frequently accompanied by inattention to violence, oppression, and suffering in the places where anthropologists worked. As a new generation of anthropologists began to challenge the resistance to studying change, there emerged an acceptance of displacement as a legitimate anthropological domain of knowledge and an accompanying rise in studies on refugees, refugee policies, the relationship of migration to social suffering and the governmentality of immigration, issues that now have gained considerable anthropological interest (Harrell-Bond and Voutira 1992).

More recent studies with an anthropological lens have tended to focus on understanding culture as an inherently political issue, including conceptualizing displacement as a product of global capitalism and neoliberal restructuring (Ramsay 2020). Therefore, anthropology has taken an interest in power dynamics and structures, which sometimes put anthropologists in a position to contribute meaningfully to national and international policy on refugees and migration. Anthropologists also have documented the experience of people repatriated to their home countries, undertaken an examination of the institutions created to deal with massive population displacements, studied how host populations are affected by the arrival of a large number of refugees or other displaced people, and reported on the emergence of new international diasporas.

The anthropology of refugees is conducted by researchers who try to depict and capture the realities faced by refugees, and to understand the world from their perspective. The field also addresses the historical contexts that drives displacement, as well as how refugees interact with their cultural, social, religious, and economic environments, and how doing so compels cultural change for them and their hosts. One avenue for anthropology to play a greater role in this arena is by helping to bring migrant experiences and migrant voices to the decision-making table. This stems from the understanding that it is fundamental to have refugees participate directly in the policy discussions that impact their lives. Most recently, anthropologists have turned their attention to climate change refugees (Baer and Singer 2021; Oliver-Smith 2009).

The Climate Migration Nexus

Human migration in response to ecological change, including climate/environment-linked change, has been occurring since the origin of our species. But the widely dispersed push that climate change currently is exerting on human migration is relatively new and is gradually and consequentially intensifying. While a range of economic, political, and social factors trigger the in-country or cross-border migration of populations, climate change is proving to be a significant "threat multiplier" that magnifies the effects of other drivers of human relocation (Schwerdtle et al. 2018) while compelling people to seek refuge away from their homelands. Amar Rahman, Global Head of Climate Resilience at Zurich Resilience Solutions, explains that climate risks are interconnected and that their interaction can cause a domino effect:

> When temperatures rise in a country, for instance, it can reduce water availability and water quality. This may increase the spread of disease and raise the likelihood of drought leading to crop failures that will reduce incomes and food supplies. All this can potentially lead to social disruption and political instability.
>
> *(Quoted in McAllister 2023)*

The term "environmental refugees" was first used in 1985 by UN Environment Programme expert Essam El-Hinnawi (1985) who defined them as people who are forced to flee their traditional place of residence due to local disruption such as an avalanche or earthquake; those who migrate because environmental degradation has undermined their livelihood or poses unacceptable risks to health; and those who resettle because land degradation has resulted in desertification or because of other permanent and untenable changes in their habitat. Jenny Stoutenberg (2011) subsequently called attention to the fact that "[c]limate change-induced migration now needs to be distinguished not only from the social, economic, and political factors compelling human movement, but also from the 'background noise' of general environmental change that might cause people to seek a livelihood elsewhere." This statement was made in a review of the groundbreaking volume entitled *Climate Change and Displacement: Multidisciplinary Perspectives* and edited by Jane McAdams (2010).

Even at this early stage, researchers recognized that identifying climate migrants is often a challenge, especially in places already burdened by the weight of crime, poverty, violence, and conflicts. Because worsening weather conditions often exacerbate other existing problems, climate change commonly is overlooked as a contributing factor to migration. In El Salvador, for example, hundreds of people leave their home villages each year because of crop failure owing to drought or flooding, and end up in cities where they become victims of gang violence and may ultimately flee into neighboring or even distant countries because of those attacks. As Elizabeth Ferris, a research professor at the Institute for the Study of International Migration at Georgetown University in the United States observes,

> It's hard to say that someone moves just because of climate change. Is everyone who leaves Honduras after a hurricane a climate migrant? And then there are non-climate related environmental hazards—people flee earthquakes, volcanic eruptions and tsunamis—should they be treated differently than those displaced by weather-related phenomena?
>
> *(Quoted in Watson 2022)*

The definitional issue is not as trivial as it first seems as it impacts the development of social policies and the allocation of aid. The questions that must be asked are who does this policy apply to and who is eligible for assistance?

According to UNHCR (2016), between 2008 and 2016 an average of 2.15 million people were displaced annually by climate-related events, including floods, storms, wildfires, and extreme temperatures. Thousands more flee their homes in the face of slow-onset hazards, such as droughts or coastal erosion linked to sea level rise.

The experiences of climate refugees are expressed in the following two statements made by displaced people in Greece (Mailes 2023). An interviewee from Somalia explains:

> Most people when they lost their animals, and they find there is no rain, they try to escape. And they try to reach the cities so they can get food, and once they get there, they think they have no way to go back, because all the animals died—cows, goats, camels—so they stay there in the cities. And now there are many people in the capital, and there are no more opportunities for jobs. You can find every street, many people sitting there, and the main reason is climate change. And now a lot of the people start to come to the cities, especially the capital. And when you walk you see other people packing. They said we are in famine. There is a line in

the city, father, children, grandfather, grandmother … they suffer and they pack, and they walk to camps … because of climate change.

Similar sentiments were expressed by an interviewee from the Gambia:

> Climate change really has affected my country. The rain really dropped … it's not raining anymore as it was before. And then also so many countries on my journey coming to Greece here which are also facing the same kind of dry, especially the countries close to the Sahara, like the desert—Mali, Niger—they are very dry, and climate change is affecting so many of those countries as well, places relying on agriculture. 'Cause if the climate is not suitable for them, the agriculture system goes down. They don't have enough rain to grow the crops … If you're farming, if your livelihood is on the farm, in the end you see your family suffering all the time, like famine. They would rather go out and look for greener pasture. It affects so many people.

Climate change-induced migration is expected to surge in coming decades, as outlined by the Institute for Economics & Peace (IEP, 2020), an international think tank with offices in six countries, which predicts that 1.2 billion people could be displaced globally by 2050 by climate change and "natural" disasters. According to the IEP (2020):

- Sub-Saharan Africa, South Asia, the Middle East and North Africa are climate "hotspots" that are facing the largest number of ecological threats.
- Many of the countries at greatest risk from ecological threats are also predicted to experience significant population increases, such as Nigeria, Angola, Burkina Faso, and Uganda. These countries, which are already struggling to address ecological problems, suffer from resource scarcity, low levels of peacefulness, and high rates of poverty.
- By 2040, an estimated 5.4 billion people—more than half of the world's projected population—will live in 59 Global South countries that increasingly are experiencing high or extreme water stress, including India and the People's Republic of China.
- The lack of resilience in poorer countries will lead to worsening food insecurity and competition over resources, an increase in civil unrest, and mass displacement.
- Regions that have high resilience, such as Europe and North America, will not be immune from the wider impact of ecological threats. The refugee crisis in the wake of wars in the Syrian Arab Republic and Iraq in 2015, for example, saw two million people flee to Europe. This pattern may be repeated as the planet warms.

Most climate migrants move within the borders of their home countries, usually from rural areas to cities after losing their homes or livelihoods (or both, as well as their belongings and perhaps family members) because of drought, rising seas or flooding, as seen in Chapter 4 in the case of Parul Akter and other rural-to-urban migrants in Dhaka, Bangladesh. Because cities also are facing their own climate-related crises, including soaring, intensive, enduring heat waves and water scarcity, and coastal inundation from rising seas, people are increasingly being forced to flee across international borders to seek some form of refuge. The reception they receive varies but is sometimes lukewarm or even quite hostile.

As this discussion suggests, life for most climate/environment refugees is perilous. Nonetheless, their numbers continue to grow as failure to make the kinds of fundamental changes needed to slow climate change are resisted by elite polluters.

Climate Migration and Health

While there is a growing body of increasingly nuanced research on human migration as a response to climate change, the full range and extent of impacts on human health are thus far under-studied. Nonetheless, many public health experts see climate change as the greatest threat to health facing the world in the 21st century. It is estimated that 250,000 deaths per year will occur in the next few decades as a result of climate change. This threat to health is rising because climate change affects all aspects of our lives: the food we eat; the air we breathe; the water we drink; the pathogens around us; the places and structures that provide us with shelter; even our relations with other people.

Environmental changes linked to rising greenhouse gas (GHG) concentrations is one of the major causes of human migration through droughts, crop failures, clean water scarcity, and a rise in sea levels that makes living in coastal areas untenable. Climate migration negatively impacts people's physical and mental health. Health challenges that are frequent among climate refugees are a result of factors such as lack of adequate shelter, poor hygiene, injuries, prolonged heat exposure, a shortage of food and undernutrition both during and following migration, and exposure to violence. Rural populations forced by climate change to migrate to cities have been found to suffer from a jump in the prevalence of non-communicable diseases such as obesity, cardiovascular disease, and type 2 diabetes. Also important are various climate-sensitive infectious diseases, including malaria, dengue, leishmaniasis, and diarrheal diseases, especially among children. Climate migrants also face health risks from food and water-borne diseases such as cholera, typhoid, and hepatitis because of inadequate sanitation and polluted drinking water sources. Skin diseases are among the most commonly observed health conditions observed in migrant populations. The warmer cities that climate migrants flee to are seeing a rise in respiratory diseases due to the high level of pollution.

The myriad number of diseases suffered by climate refugees can be seen in a study carried out with 426 participants in Khulna city, the third largest metropolitan area in Bangladesh (Rahaman et al. 2018). Khulna is known as the gateway to the Sundarbans, the world's largest mangrove forest and home to the Bengal tiger. It is also the most vulnerable climatic region in the country. Khulna is increasingly experiencing climate change-induced urban problems. These include a growing influx of impoverished climate migrants, drainage congestion, water logging, and diminishing freshwater availability. In the last decade, the population of Khulna has increased by more than 20% because of in-migration from nearby climate-vulnerable rural districts.

The study found that many respondents were suffering from climate-sensitive diseases as well as a lack of access to adequate health facilities. Respondents suffered from notable levels of diarrhea, cholera, dysentery, skin diseases, asthma, hypertension, malnutrition, malaria, fever, cough, reproductive disorders, jaundice, recurrent pregnancy loss, early or delayed menarche, and urinary tract infection. Diarrhea was the most common disease among respondents (60% of participants), with malaria being the second most common disease (36%).

Mental health among climate migrants is undercut by the loss of supportive social relationships, in addition to the direct effects of trauma (Torres and Casey 2017). Climate change poses serious risks to mental health from emotional distress, anxiety, depression, grief, existential dread (known as "eco-anxiety"), and suicidal behavior. For example, individuals who were displaced from their homes by Hurricane Katrina in Louisiana and Alabama in the United States and who were forced to move to places with a low level of social cohesion had significantly higher odds of past-month depression compared to those who experienced only one of these factors (Lê et al. 2013). Hurricane Katrina was the cause of the largest forced relocation in the United States since the Dust Bowl of the 1930s. A study that examined attitudes about climate change among 10,000 children and young adults from across the world found that about 62% said that they were anxious about climate change and about 67% reported that they were sad and afraid. Others felt powerless, helpless, and guilty (Hickman et al. 2021).

The existence and quality of social ties have been linked to population health, as they can serve as protective mechanisms of resource sharing and emotional support, but they can also become stress-inducing mechanisms of social and financial burden. Migrating populations may face stressors related to the disruption of their social networks, for example, through separation from family and community. Mobile populations may also face substantial social isolation and marginalization in their new settlements. Once separated, refugees may face substantial administrative barriers to reunifying even with immediate family members in other countries. For example, documents that serve as evidence of biological family connections and dependency (e.g., birth certificates) may be lost or destroyed during migration. Even if immediate family members can migrate together or reunite after separation, extended or emotionally close non-biological family and community members may be geographically dispersed as a result of administrative assignment to different settlement camps or other locations.

Importantly, all these health risks can potentially interact synergistically with disastrous consequences for climate refugees. Heffernan (2013), using a term coined by Singer (2009b), refers to these interactions as climate-change syndemics. Conceptualizing climate-change syndemics begins with the recognition that climate change impacts a wide range of physical and mental health conditions, and frames understanding at the population health level. Syndemics are complex biosocial events involving the interaction of two or more diseases or other health conditions and social (e.g., injustice, discrimination, poverty) or anthropogenic environmental factors, the latter of which are termed ecosyndemics (Singer 2009b). With reference to the latter, unlike traditional environmental models, syndemics theory does not conceive of "nature" as natural, in the sense of being a place separate from and independent of human action, an untouched reserve of nonhuman things. Rather, there is a strong concern from the syndemic perspective with identifying the historical ways in which the "natural" environment, no less than the explicitly human-built environment (e.g., cities), has been shaped and influenced by human action, intentional and otherwise. Of special concern are the ways in which affected environments reflect social inequalities within and across societies, inequalities that are clearly evident in the lives of climate refugees. Inequality is deemed important for three primary reasons. First, syndemics are found disproportionately among subordinated groups, precisely because of their oppressed social status. Second, the exercise of power in society routinely shapes the human imprint on the physical environment (e.g., fossil fuel companies). Third, climate change causes or contributes to a myriad number of diseases among refugees that have the potential to interact in ways that increase the disease burden of affected populations.

From an ecosyndemics perspective, an issue of great concern is what happens to diseases as they are spread by climate change to new environments, including diseases brought with them by climate refugees, those they acquire while in migration, and those they first encounter in host contexts. Do these diseases interact and with what consequences for health?

Island Nations at Growing Risk

Several Pacific Island countries, including Fiji, Vanuatu, the Solomon Islands, and Papua New Guinea, have initiated planning for and implemented plans for relocation in response to continuing sea level rise. The independent state of Papua New Guinea, with a population of almost 12 million (National Statistics Office 2021), has seen the lives and livelihoods of the communities on the small atoll being affected by both slow onset and sudden climatic changes (IPCC 2014). The 2,700 residents of the Carteret Islands in Papua New Guinea, a chain of seven small raised coral atolls in a circular shape, face multiple challenges including sea level rise, inundation, soil salinization, and land loss. Despite the construction of sea walls and the planting of mangroves, more than 50% of their island has eroded since 1994. One of the original islands, Huene, was split into two parts, a partitioning that islanders believe was caused by rising sea levels. Today, an almost 100-foot-long (30–40 meters) channel separates the two halves of Huene Island.

Atolls like the Carteret Islands have a high ratio of coastline to surface land area, which makes them particularly vulnerable to slow onset sea level rise and rapid onset storm surges, as well as to devastating "king tides," which are perennial high tide events that pose a significant threat to coastal communities. There is virtually no arable agriculture on any of the Carteret Islands, making the residents almost exclusively dependent on marine resources for their livelihoods and food. Among the main sources of economic activity and sustenance have been, respectively, the exportation of farmed seaweed to Asian markets and the capture of sea cucumber, an important dietary staple. The island chain historically had a notable level of marine biodiversity and natural resources, but this ecological wealth has rapidly declined over the past few decades.

In 2006, the Council of Elders of the Carteret Islands formed an organization called Tulele Peisa (Sailing the Waves on Our Own). Tulele Peisa developed the Carteret Integrated Relocation Project, which was one of the first community-driven climate change refugee relocations in the region. Its mission initially was to coordinate the voluntary movement of 1,700 Carteret islanders to Bougainville, the largest island in the Solomon Islands archipelago, 62 miles (100 kilometers) to the north-east. The location of the site was critical to ensure sufficient land for the Carteret families to be economically self-sufficient. The plan prioritized food security by ensuring access to fishing grounds, which is important for nutritional and cultural reasons. The relocation plan also included the construction of housing and infrastructure, and envisaged the development of income generation projects, food security measures, and sustainable land use management strategies. Additionally, the Carteret Integrated Relocation Project attempted to ensure that host communities would benefit from the relocation through an upgrading of health facilities and schools (Bronen 2014). Nevertheless, despite the apparent opportunity for livelihoods, food security, and access to health services, many islanders have rejected relocation. In 2006, only three families expressed a wish to resettle. Two years later that number had increased to 38 families. Staff at Tulele Peisa now realize that the relocation process will take considerably longer than initially predicted. The current plan is to move 1,350 people, or 50% of the total population, with the remaining 50% staying on the islands. There Tulele Peisa is working on protecting the remaining biodiversity and planting mangroves as a natural buffer against storm surges.

In a time of rising oceans, a particularly fragile environment is the very low-lying South Pacific island nation of Tuvalu. Tuvalu's population of 12,000 is spread across nine islands. Around 26% of Tuvalu's population lives below the established national poverty level, and a majority of the country's limited land area is devoted to subsistence agriculture. Climate variability and limited terrestrial resources result in Tuvalu having low food and clean water security. Critical subsistence food crops like coconuts and taro are failing because of salinization. These problems produce low levels of resilience to "natural" hazards and a heavy reliance on international aid during disasters. Fresh food is limited, forcing the population to be reliant on imported products, which are expensive and have lower levels of nutritional value. On average, at least one tropical hurricane/cyclone passes close to Tuvalu's islands every year and multiple hurricane/cyclones have hit Tuvalu in quick succession. The intense climate events expose Tuvalu to high wind speeds, extreme rainfall, and storm surges, all of which cause significant economic and social damage to the islands and their social infrastructure. Cyclone Pam, for example, which struck Tuvalu in 2015, caused damage that amounted to more than 25% of the nation's gross domestic product in that year (World Bank 2021).

The sea level in Tuvalu is nearly 6 inches (0.15 meters) higher than it was 30 years ago (Brennan 2023). This level of ocean rise is 1.5 times higher than the global average and is expected to increase by as much as 20–40 inches (0.5–1 meter) by the end of the century. As projected, much of Tuvalu's critical infrastructure (e.g., government buildings, healthcare facilities, airport), will be below the average high tide by the year 2050. At current rates of sea level rise, some estimates suggest that half the land area of Funafuti, the island nation's capital city, will be flooded by tidal waters by that year. Comments Lily Teafa, who works with a youth-led organization that undertakes climate change resilience projects: "It's the worst feeling ever; worse than being afraid of heights, afraid of the dark. Now we're afraid of the future" (quoted in Fainu 2023). Residents of the islands report that erosion is evident along the coastlines as are large amounts of washed-up debris. The remnants of infrastructure and abandoned homes dot the edges of the shore. Even cemeteries are being worn away forcing people to build small tombs next to their homes.

About one-fifth of Tuvalu's population has already relocated, many to New Zealand under a ballot measure that allows up to 150 people to be granted residence in New Zealand every year. Kelesoma Saloa, who moved from Tuvalu to New Zealand in 2013, reports still feeling a strong sense of dislocation:

> Coming from a self-sufficient society to a very commercialised society is so, so difficult … If you have no money here, you can't survive. Not like in the islands, if you have no money, you have your family, your small land, your fish.
>
> *(Quoted in Fainu 2023)*

Saloa now works as a guide and educator at the Auckland War Memorial museum. He comments: "I feel sometimes that I betrayed my people, that I walked away from my people. But I have the chance to talk about the plight of my people here."

Operations have been initiated by the Tuvalu Coastal Adaptation Project to use dredging to reclaim land. Launched in 2017 with monies from the global Green Climate Fund and in partnership with the UN Development Programme, the aim of the Te Lafiga o Tuvalu (Tuvalu's Refuge) Project is to reduce exposure to coastal hazards and provide a longer-term adaptation strategy for the country. Continually rising sea levels and more powerful hurricanes/cyclones make such adaptation a difficult challenge.

Climate Change and High-Altitude Communities: The Andean Altiplano

The Altiplano (high plain), also known as the Andean Plateau, is the largest high-altitude plateau (an area of relatively level high ground) outside of Tibet. The plateau lies primarily in Bolivia. The Northern Altiplano in particular is very vulnerable to climate change which threatens numerous local communities and ecosystems. The region is already experiencing rising temperatures as well as changes in rainfall patterns and the water cycle. The most visible impact of climate change in high mountain areas like the Altiplano is glacier retreat. The Intergovernmental Panel on Climate Change reports that Andean glaciers are moving towards complete disappearance in coming decades, a process that severely threatens water availability. The Indigenous Aymara people of the Altiplano have adapted to climate variability in the past and have developed an extensive knowledge about the local climate and the environment. They have domesticated various subsistence crops and animals that are well suited to the harsh terrain. But now they are facing more rapid and impactful changes. Even though they have generations of experience to draw on in developing ways of coping, contemporary climate change is threatening their low climate/environmental impact way of life.

Rural people in the Altiplano rely on natural resource-based ways of sustaining their families and communities. In farming villages around Lake Titicaca, people also engage in small-scale fishing. Titicaca is one of the highest navigable lakes in the world at an elevation of 12,507 feet (3,812 meters). Local families store a part of their production for their own consumption and sell the remainder to buy things they cannot produce themselves. However, poverty sometimes forces small farmers to sell their produce, a dangerous decision because it threatens their food security. Seeking nonfarming employment and migration are the two main risk management strategies people resort to in hard times. Migration, pushed by the effects of climate change, has been rising over the past few decades, with some people moving in anticipation of or in order to adapt to crop-threatening environmental and climate impacts, while others are displaced by extreme weather events. There is now a high level of internal migration from rural to urban areas and even international migration, especially among men and youth. Rural areas are now mostly populated by women and older adults who face growing problems in terms of access to land, reduced productivity, fragmentation of land tenure, and the expanding effects of climate change. Although some level of migration is not a new phenomenon in the area, climate change appears likely to become a major force prompting future population movements, probably mostly through internal migration, but also to some extent through migration across national borders. Currently people feel that farming practices may also be forced to change as a result of reductions in water availability due to less precipitation and melting glaciers. While traditional weather and climate forecasting practices are still used by Andean communities to make decisions about the management of their farming strategies, the intensity, frequency, and extent of climate impacts are challenging this traditional knowledge base (Flores-Palacios 2021).

Fleeing Drought and Rain in Kenya

Kenya is confronting multiple expressions of climate change impacts, including floods, drought, landslides, rising lake waters, and locust infestations. Drylands occupy as much as 90% of Kenya, and the impact of drought on arid and semi-arid lands is especially acute. Climate change has caused displacement, as well as cultural and economic loss, poorer health, and social disruption. The number of people in the Horn of Africa region forced by drought to migrate as climate refugees is estimated to be as high as 285,000. Kenya is also one of the leading refugee-hosting countries in Africa and is heavily impacted by both sides of the displacement crisis.

In Kenya, it is estimated that almost 40 million people have been affected by dire food insecurity and near famine as a result of climate change. In recent years, herders have lost nine million livestock, destroying their pastoralist livelihoods and further contributing to malnutrition among children (Climate Refugees 2020). Exemplary of the crisis is the Kiwaja Ndege Internally Displaced Persons Camp in Marigat, Baringo County. About 1,000 ethnically marginalized Indigenous Ilchamus people living in 150 households reside in this refugee camp that has limited access to humanitarian services and protection programming. The Ilchamus are one of the smallest ethnic groups in Kenya. Resident homes were submerged in 2020 when Lake Baringo waters swelled past human habitability due to climate change. Water levels in the lake rose sharply by over 109% and the size of the lake increased from 128 square kilometers to 228 square kilometers. The rising levels in Lake Baringo aggravated human–wildlife conflicts and exposed thousands to water-borne diseases. According to a Kenya government report:

Some households have rebuilt their homes up to five times since this began. Further, the rising water levels have disrupted communities' livelihoods by destroying and sweeping away the little available possessions and restricting access to natural resources and even markets.

(Chebet 2022)

Super Flooding in Pakistan

From June to October 2022, Pakistan was hit by a "monster monsoon" that caused "super floods" and killed over 1,700 people, one-third of them children. The sweeping floodwaters caused US $14.9 billion in damage and $15.2 billion in other economic losses. Over 2.1 million people were left homeless. In all, over one million homes were destroyed, two million acres of vital crops were flooded, bridges and other infrastructure was washed away, dams were breached, and 310,685 miles (500,000 kilometers) of roads were damaged. When the flooding was at its height, more than one-third of the country was underwater.

The immediate cause of the floods was a disastrous combination of torrential monsoon rain and melting glaciers that followed a severe heat wave, the worst since 1901. Both of these dramatic changes are linked to climate change. The flooding has been described as the worst in the country's history. It was also recorded as one of the costliest disasters in world history.

Two of the country's providences, Sindh and Balochistan, were hit the worst. They received far more rainfall than the August average, an eye-watering increase of 784% and 500%, respectively. The Indian Ocean, where the storms ultimately originated, is one of the fastest warming oceans in the world, with an average temperature of 1 degree Celsius (1.8 °F) compared to a global ocean average of around 0.7 °C (1.3 °F) (Tunio 2022). A rise in sea surface temperature is believed to increase monsoon rainfall. In addition to the rise in sea surface temperatures, the frequency and intensity of extreme warming events, called marine heat waves, in the Indian Ocean have increased significantly in recent years. Moreover, on land, southern Pakistan experienced back-to-back heat waves in May and June 2022. This deadly heating was record-setting itself and was made more likely by ongoing climate change. The heat waves created a strong thermal low that triggered heavier rains than usual. The heat waves also caused glacial flooding in Gilgit-Baltistan, which is the northern portion of the larger Kashmir region. With more than 7,000 glaciers, Gilgit-Baltistan has been called "the land of glaciers" but this is changing. Over 3,000 lakes have formed in the area due to melting glaciers.

Pakistan contributes less than 0.3% of global GHG emissions but is one of the places most vulnerable to climate change. Research by an international team of climate scientists reported that "the 5-day maximum rainfall over the provinces Sindh and Balochistan is now about 75% more intense than it would have been had the climate not warmed" (quoted in Cleetus 2023).

The science is also clear that climate change will continue to make these kinds of disasters more likely and/or frequent, highlighting the growing risk to Pakistan and surrounding countries, including the creation of millions of destitute climate refugees. According to the World Bank (Knippenberg et al. 2023), the flooding in 2022 forced as many as 9.1 million more people in Pakistan into poverty, increasing the nation's poverty rate by 4% above the 2018–2019 poverty rate. Many of the hardest-hit areas already had high rates of poverty and many children are suffering from malnutrition. A report by UNICEF (2023) estimated that "20.6 million people, including 9.6 million children, need humanitarian assistance." Meanwhile, 1.8 million people were still living near stagnant floodwaters eight months after the floods. According to the report, "[t]he prolonged lack of safe drinking water and toilets, along with the continued proximity of vulnerable families to bodies of stagnant water, are contributing to the widespread outbreaks of waterborne diseases such as cholera, diarrhoea, dengue, and malaria." The World Health Organization (WHO 2023) adds that such flooding constitutes "the perfect storm for malaria" as mosquitoes arrive en masse.

According to Allan Schapira, a WHO consultant who had recent conducted a review of Pakistan's national malaria program,

> It largely hit the districts that had the most difficult malaria situation already, and it had the biggest impacts on the poorest people—those who normally live hand to mouth, in subsistence agriculture, or very small shops, and so on.
>
> *(Quoted in WHO 2023)*

Ali and Hamid (2022) note that in addition to creating opportunities for exposure, "[c]hild morbidity and mortality may further increase as routine vaccination has been affected during the floods, mainly due to two main reasons: shifting of resources at the government level and the dislocation of children due to flooding." Flooding can make it very difficult for people to get to healthcare facilities, cause the suspending of routine vaccination drives, and limit the ability of mobile vaccination teams to reach their targeted populations. It can also create challenges in maintaining the "cold chain," which refers to the temperature-controlled environments needed to store, manage, and transport vaccines. Failure to maintain the cold chain can render a vaccine ineffective. Another problem is the overcrowding of people in flood rescue camps, which increases the risk of the spread of infections as people's immune systems are weakened owing to inadequate food distribution, unhygienic living conditions, and polluted water.

The making of flood-related climate refugees has causes that began during the British colonization of the region. Much of Pakistan's rural population faces poverty and the enduring pressure to migrate to the city owing to the long-term and slow dismantling of historic rural resilience. This occurred during a time of debt-inducing infrastructure and policy interventions. This story of rural vulnerability dates back 150 years to the British and their canal colonization and deforestation drives, the World Bank-funded rapid expansion of dam and canal building, the Green Revolution, and the indebtedness of small farmers caused by taxation and landgrabs by global investors and the Pakistani military for industrial farming. Further, floods are linked to the building of the Indus Basin Irrigation System whose designers assumed stable climates and forcibly limited the historic seasonal mobility of pastoralists, fishers, and traditional irrigators who often welcomed the monsoon floods and moved around during drought years. With limited flexibility, people have no choice in times of crisis, like the flooding of 2022, but to consider permanent migration to urban centers, or ideally, from their perspective, internationally. Preliminary research on the people affected by the 2022 floods confirms that most migrate to large cities and urban centers, but issues of water scarcity and precarity in cities create the desire to return to their farmlands. While the first experience of floods may allow for this return, ecological destruction by government-funded projects and the recurring nature of floods and systemic water insecurity means that return may no longer be an option for many climate refugees in Pakistan (Kamal 2023).

Conclusion

Over the next few decades, given the likelihood that GHG emissions will not decrease anytime soon, the climate refugee crisis will continue to intensify. Humanity is at a critical existential juncture as it faces numerous interrelated threats manifested as climate chaos. Many of these crises ultimately are related to contradictions of global capitalism, a political economy oriented toward profit-making, and a continual economic growth model of production that retains (its denials aside) a heavy reliance on fossil fuels which emit an array of GHGs. Within the context of the capitalist world system, individual nation states, as well as the UN, operate as border-making institutions that legitimize the exclusion of millions of people from land and resources essential to their livelihoods and enforce these exclusions through legally sanctioned violence when needed. In contrast to poor people throughout the world, very wealthy people operate in an essentially borderless world that allows them to manage their overseas trade networks, spend their money on luxury consumer goods and services, and jet around the globe as tourists visiting sites that have not yet been despoiled by climate/environment change. Indeed, the 2015 UN Paris Agreement on climate change, a global treaty aimed at limiting planetary warming to 1.5–2 °C (2.7–3.6 °F), proved to be unrealistic given its embeddedness within the parameters of the growth paradigm, and the exemptions of airplanes and shipping from its stipulations. Many climate scientists and social scientists maintain that humanity faces a dystopic scenario marked by ongoing social inequality and catastrophic climate change resulting in a growing number of economic, political, and climate refugees, many of whom will seek to cross existing national borders (Lustgarden 2020).

Of course, people already are beginning to flee the ravages of climate chaos. In Southeast Asia, where increasingly unpredictable monsoon rainfall and drought have made subsistence farming far more difficult, eight million people have, despite great obstacles, fled to the Middle East, Europe, and North America. In Africa, millions of rural people have fled to the coasts and the cities in the face of drought and widespread crop failures. Trying to escape the relentless confluence of drought, flood, bankruptcy, and starvation, thousands have fled Central America to the United States in recent years. The pattern is clear: while most GHG production and its disastrous release occurs in the Global North, the worst impacts—those that force internal and external survival migration—are experienced in the Global South. At the same time, ironically, we are witnessing the creation of "privatized green enclaves" and opulent enclaves of luxury by rich environmental elites within those countries most threatened by climate change (e.g., Nigeria's Eko Atlantic on Victoria Island adjacent to the capital of Lagos advertised by its developers as a gateway to the emerging markets of Africa). This has been described by some as an emerging "climate apartheid" (Brisman et al. 2018). It is "a world in which the rich and powerful exploit the global ecological crisis to widen and entrench extreme inequalities and seal themselves off from its impacts" (Lukacs 2014). As this example affirms, climate change and all the transformations it ushers in is very much a social issue—an alarming environmental reflection of grossly unequal and unjust social structures.

References

Ali, I. and Hamid, S. 2022. Implications of COVID-19 and "super floods" for routine vaccination in Pakistan: The reemergence of vaccine preventable-diseases such as polio and measles. *Human Vaccines and Immunotherapeutics*, 2154009. https://www.tandfonline.com/doi/full/10.1080/21645515.2022.2154099

Baba, M. 2013. Anthropology and transnational migration: A focus on policy. *International Migration* 51(2). doi:10.1111/imig.12083

Baer, H.A. and Singer, M. 2021. The growing climate-driven refugee crisis. *Polish Migration Review* 6:10–23

Brennan, P. 2023. NASA-UN partnership gauges sea level threat to Tuvalu. NASA. https://sealevel.nasa.gov/news/265/nasa-un-partnership-gauges-sea-level-threat-to-tuvalu/#:~:text=This%20will%20pose%20challenges%20for,and%20severity%20of%20periodic%20flooding

Brisman, A.South, N., and Walters, R. 2018. *Climate apartheid and environmental refugees*. In *The Palgrave Handbook of Criminology and the Global South*. L. Carryington, R. Hogg, J. Scott, and M. Sozzo, eds. Cham: Palgrave Macmillan

Bronen, R. 2014. *Choice and Necessity: Relocations in the Arctic and South Pacific*. http://unfccc.int/files/adaptation/groups_committees/loss_and_damage_executive_committee/application/pdf/bronen_choice_and_necessity_relocations_in_the_arctic_and__south_pacific_2014.pdf

Canadian Association for Refugee and Forced Migration Studies. 2023. Environmental refugee. In *Online Research and Teaching Tools.* http://rfmsot.apps01.yorku.ca/glossary-of-terms/environmental-refugee/#:~:text=those%20people%20who%20have%20been,of%20their%20life%20%20

Castañeda, H., Holmes, S., Kallius, A., and Monterescu, D. 2016. Anthropology and human displacement: Mobilities, ex/inclusions and activism. *American Ethnologist.* https://anthrosource.onlinelibrary.wiley.com/hub/journal/15481425/refugees-and-immigrants

Castillejo, E. 2023. In South Sudan, a new front line of climate change after historic flooding. ABC News. https://abcnews.go.com/International/south-sudan-new-front-line-climate-change-historicflooding/story?id=98636490#:~:text=One%20in%20five%20children%20here,to%20keep%20their%20bellies%20full

Chebet, C. 2022. Report reveals nightmare of swelling lakes in Rift Valley. *The Sunday Standard.* https://www.standardmedia.co.ke/rift-valley/article/2001427360/report-reveals-nightmare-of-swelling-lakes-in-rift-valley#

Cleetus, R. 2023. A year after the deadly Pakistan floods began, hard lessons about climate loss and damage. Union of Concerned Scientists. https://blog.ucsusa.org/rachel-cleetus/a-year-after-the-deadly-pakistan-floods-began-hard-lessons-about-climate-loss-and-damage/#:~:text=Last%20summer%2C%20from%20June%20through,long%20flooding%20across%20the%20country

Climate Refugees. 2020. *Submission to the UN Special Rapporteur on the Promotion and Protection of Human Rights in the Context of Climate Change.* https://www.climate-refugees.org

Colson, E. 2003. Forced migration and the anthropological response. *Journal of Refugee Studies* 16(1)

Conquergood, D. 1988. Health theatre in a Hmong refugee camp: Performance, Communication and culture. *Journal of Performance Studies* 32(3)

El-Hinnawi, E. 1985. *Environmental Refugees.* UNEP(02)_E52-E – PDF

Environmental Justice Foundation. 2014. *Falling through the Cracks.* https://ejfoundation.org/resources/downloads/EJF-Falling-Through-the-Cracks-briefing.pdf

Fainu, K. 2023. Facing extinction, Tuvalu considers the digital clone of a country. *The Guardian.* https://www.theguardian.com/world/2023/jun/27/tuvalu-climate-crisis-rising-sea-levels-pacific-island-nation-country-digital-clone

Flores-Palacios, X. 2021. *Climate Hazards in the Northern Bolivian Altiplano.* https://blogs.ucl.ac.uk/irdr/2021/07/14/climate-bolivia/#:~:text=The%20Northern%20Altiplano%20is%20particularly,patterns%20and%20the%20water%20cycle

Gatrell, P. 2015. *The Making of the Modern Refugee.* Oxford: Oxford University Press

Harrell-Bond, B. and Voutira, E. 1992. Anthropology and the study of refugees. *Anthropology Today* 8(4): 6–10

Heffernan, C. 2013. The climate change-infectious disease nexus: Is it time for climate change syndemic. *Animal Health Research Reviews* 14(2): 151–154

Hickman, C., Marks, E.P., Panu, C.S., Lewandowski, E.R., and Mayall, E. 2021. Climate anxiety in children and young people and their beliefs about government responses to climate change: A global survey. *The Lancet Planetary Health.* https://www.thelancet.com/journals/lanplh/article/PIIS2542-5196(21)00278–00273/fulltext

Institute for Economics & Peace. 2020. Over one billion people at threat of being displaced by 2050 due to environmental change, conflict and civil unrest. https://www.economicsandpeace.org/wp-content/uploads/2020/09/Ecological-Threat-Register-Press-Release-27.08-FINAL.pdf

Intergovernmental Panel on Climate Change (IPCC). 2014. *Climate Change 2014: Impacts, Adaptation, and Vulnerability. Part A: Global and Sectoral Aspects. Contribution of Working Group II to the Fifth Assessment Report of the Intergovernmental Panel on Climate Change.* C. Field, V. Barros, D. Dokken, K. Mach, M. Mastrandrea, T. Bilir, M. Chatterjee, K. Ebi, Y. Estrada, and R. Genova, eds. Cambridge: Cambridge University Press

Kamal, A. 2023. Climate, floods, and migration in Pakistan. *International Migration.* https://onlinelibrary.wiley.com/doi/10.1111/imig.13170#:~:text=Preliminary%20research%20on%20the%20people,desire%20to%20return%20to%20farmlands

Knippenberg, E., Amarion, M., Javaid, N. and Meyer, M. 2023. *Quantifying the poverty impact of the 2022 floods in Pakistan.* https://blogs.worldbank.org/developmenttalk/quantifying-poverty-impact-2022-floods-pakistan

Lê, F., Tracy, M., Norris, F., and Galea, S. 2013. Displacement, county social cohesion, and depression after a large-scale traumatic event. *Social Psychiatry and Psychiatric Epidemiology* 48(11): 1729–1741

Loizos, P. 1981. *The Heart Grown Bitter: A Chronicle of Cypriot War Refugee.* Cambridge: Cambridge University Press

Lukacs, M. 2014. New, privatized African city heralds climate apartheid. *The Guardian.* https://www.theguardian.com/environment/true-north/2014/jan/21/new-privatized-african-city-heralds-climate-apartheid.

Lustgarden, A. 2020. Where will everyone go? *Propublica.* https://features.propublica.org/climate-migration/model-how-climate-refugees-move-across-continents/

Mailes, J. 2023. Climate migration: The value of refugee perspectives. *Latitude.* https://latitude.plos.org/2023/04/climate-migration-the-value-of-refugee-perspectives/

McAdams, J. 2010. *Climate Change and Displacement: Multidisciplinary Perspectives.* Oxford: Hart Publishing

McAllister, S. 2023. There could be 1.2 billion climate refugees by 2050. Here's what you need to know. *Zurich.* https://www.zurich.com/en/media/magazine/2022/there-could-be-1-2-billion-climate-refugees-by-2050-here-s-what-you-need-to-know

Miller, T. 2019. *Emperor of Borders: The Expansion of the U.S. Border around the World.* London: Verso

National Statistics Office. 2021. *PNG, population.* https://www.nso.gov.pg

Oliver-Smith, A. 2009. Climate change and population displacement: Disasters and diasporas in the twenty-first century. In *Anthropology and Climate Change.* S. Crate and M. Nuttall, eds. London: Routledge

Rahaman, M., Rahaman, M., Bahauddin, K., Khan, S., and Hassan, S. 2018. Health disorder of climate migrants in Khulna City: An urban slum perspective. *International Migration* 56(5). https://www.researchgate.net/profile/Muhammad-Rahaman-5/publication/324778185_Health_Disorder_of_Climate_Migrants_in_Khulna_City_An_Urban__Slum_Perspective/links/5d0a76fc299bf1f539d16a7d/Health-Disorder-of-Climate-Migrants-in-Khulna-City-An-Urban-Slum-Perspective.pdf

Ramsay, G. 2020. Time and the other in crisis: How anthropology makes its displaced object. *Anthropological Theory* 20(4). https://journals.sagepub.com/doi/abs/10.1177/1463499619840464

ReliefWeb. 2021. Desperate for food, mothers turn to water lilies. https://reliefweb.int/report/south-sudan/desperate-food-mothers-turn-water-lilies

Schwerdtle, P., Bowen, K., and McMichael, C. 2018. The health impacts of climate-related migration. *BMC Medicine* 16(1): 1

Singer, M. 2009a. Global warming and the coming plagues of the 21st Century. In *Plagues and Epidemics.* D.A. Herring and A. Swedlund, eds. London: Routledge

Singer, M. 2009b. *Introduction to Syndemics: A Critical Systems Approach to Public and Community Health.* San Francisco, CA: Jossey-Bass

Stoutenberg, J. 2011. Review of climate change and displacement: Multidisciplinary perspectives. *European Journal of International Law* 22(4): 1196–1200

Terrio, S. 2015. *Whose Child Am I? Unaccompanied, Undocumented Children in U.S. Immigration Custody.* Berkeley: University of California Press

Torres, J. and Casey, J. 2017. The centrality of social ties to climate migration and mental health. *BMC Public Health* 17(1): 600

Tunio, Z. 2022. After unprecedent heatwaves, monsoon rains and the worst floods in over a century devastate South Asia. *Inside Climate News.* https://insideclimatenews. org/news/02082022/flooding-flooding-monsoons-india-pakistan-bangladesh/

United Nations High Commissioner for Refugees (UNHCR). 2016. Frequently asked questions on climate change and disaster displacement. https://www.unhcr.org/uk/ news/stories/frequently-asked-questions-climate-change-and-disaster-displacement

United Nations High Commissioner for Refugees (UNHCR). 2023a. Convention and protocol relating to the status of refugees. https://www.unhcr.org/us/media/con vention-and-protocol-relating-status-refugees

United Nations High Commissioner for Refugees (UNHCR). 2023b. UNHCR commits to climate action in Africa to protect displaced populations and foster resilience. https://www.unhcr.org/us/news/press-releases/unhcr-commits-climate-action-a frica-protect-displaced-populations-and-foster#:~:text=UNHCR%20is%20taking% 20action%20to,affected%20populations%2C"%20said%20Mazou

UNICEF. 2023. *Pakistan: Humanitarian crisis: Report 11.* https://www.unicef.org/m edia/138756/file/Pakistan-Floods-SitRep-March-2023.pdf

Watson, J. 2022. Climate change is already fueling global migration. The world isn't ready to meet people's changing needs, experts say. PBS. https://www.pbs.org/news hour/world/climate-change-is-already-fueling-global-migration-the-world-isnt-ready-to-meet-peoples-needs-experts-say

Willen, S. 2012. How is health-related "deservingness" reckoned? Perspectives from unauthorized im/migrants in Tel Aviv. *Social Science & Medicine* 74(6): 812–821

World Bank. 2021. *Climate Change Knowledge Portal: Tuvalu.* https://climateknowl edgeportal.worldbank.org/country/tuvalu/vulnerability

World Health Organization (WHO). 2023. "It was just the perfect storm for malaria": Pakistan responds to surge in cases following the 2022 floods. https://www.who.int/ news-room/feature-stories/detail/It-was-just-the-perfect-storm-for-malaria-pakistan-responds-to-surge-in-cases-following-the-2022-floods

Yarris, K. and Castañeda, H. 2015. Discourses of displacement and deservingness: Interrogating distinctions between "economic" and "forced" migration. *International Migration.* https://doi.org/10.1111/imig.12170

7

CAN ECOLOGICAL MODERNIZATION CONTAIN CLIMATE CHANGE?

How the Rich and Powerful Seek to Address the Ecological Crisis

Case Study

In a detailed ethnography called *Who Owns the Wind?*, David McDermott Hughes (2021) discusses anti-industrial, anti-corporate resistance through an account of the impact of constructing windfarms as an alternative non-carbon-based energy source in a small village situated near the Atlantic coast of Andalusia, Spain, near the Straits of Gibraltar. Villagers initially resisted the new construction because they found the wind turbines surrounding their land unsightly and noisy, but they slowly warmed to their presence. The only local person, however, to economically benefit from the turbines was a large land-owner who received handsome compensation for the turbines installed on his farm. Although Hughes (2021: 207) argues that wind power has the potential to reduce greenhouse gas (GHG) emissions, thus playing a role in mitigating climate change, in fact no net benefit is achieved within a green capitalist framework. While the grid density of wind power in Spain rose from 1% to 10%, national carbon emissions soared from 242 megatons to 338 megatons, a 40% jump. In effect, wind power did not so much replace fossil fuel power as supplement it. Spaniards increased their electricity consumption as they operated their newly acquired household appliances, ranging from microwaves to smartphones, thereby creating a classic illustration of the Jevons paradox, or rebound effect, which arises when technological progress or government policy increases the efficiency with which a resource is used, but the falling cost thereof then leads to rising consumption. Hughes (2021: 29) therefore advocates a "socialism of the wind" as an "alternative to this atmospheric phase of capitalism under which governments would socialise or nationalise wind energy," a step which we view as potentially being part and parcel of eco-socialism. Hughes shows that a climate-stabilizing energy revolution must socialize renewables so that wind power

DOI: 10.4324/9781003469940-8

comes to be equated with social justice as opposed to private gain. Left out of the bargain, with no skin in the game, but subject to living in a turbine-filled environment, Hughes says, people will insist that local governments veto new installations, which could stall the wind energy revolution before it really takes off.

The Technical Fix: A Full or Partial Solution?

Ecological modernization is the process of seeking environmental sustainability and climate change mitigation and adaptation through the adoption of more efficient, environmentally sustainable, and low-carbon energy sources and manufacturing processes. It has become a virtually hegemonic stance, particularly in the European Union (EU) but also in North America, Australasia, and the People's Republic of China (Machin 2019). Dieter Helm (2017: 245) exhibits the hubris of ecological modernization in his bold assertion that climate change is a "solvable problem only with the march of new technologies." Over the past few decades, an international networked movement for legal action intended to prevent catastrophic climate change has promoted the replacement of fossil fuels with purportedly renewable ones, particularly solar and wind power but, more recently, hydrogen and lithium-based batteries for electric vehicles too. Members of this movement include former US Vice-President Al Gore, various conservation and wilderness societies, and many environmentalists including within the climate movement.

However, while renewable energy sources, including solar, wind, and geothermal, have the potential to be part of the process of mitigating climate change, they are not a panacea. These sources of energy also have the problem of what renewable-energy engineers call *Dunkelflaute*, which means the "dark doldrums." At night, solar panels do not capture any energy, while wind speed is generally lower. Moreover, renewable energy generation requires infrastructure that is produced by manufacturing processes that depend on fossil fuels, entail much embedded energy and land space, and require rare earth metals, the mining of which is environmentally destructive. Last but not least, ecological modernization is very weak in addressing social justice issues, such as who has access to the benefits of new technologies and who does not (Singer 2019), and who most suffers the consequences of their development.

Mainstream Ecological Modernization

Carter (2007: 228) laid out the key components of ecological modernization thus:

> Ecological criteria must be built into the production process. On the supply side, costs can be reduced by improving efficiency in ways that have environmental benefits. Savings can be made by straightforward

technological fixes to reduce waste, and hence pollution, through a more fundamental thinking of manufacturing processes so that large-scale systems such as "smoke-stack" industries, that can never be made ecologically sound, are gradually phased out.

Lester Brown (2009), the author of *Plan B* and director of the Earth Policy Institute, is a staunch proponent of ecological modernization. *Plan B* advocates hope to replace nonrenewable energy sources with renewable ones. The "sunrise industries" that are touting energy conservation, efficiency, and renewable energy resources have "been working with some environmental groups to make the case for tough targets to stimulate markets for their products (for example groups like E7 and the European Wind Energy Association)" (Newell and Paterson 2010: 42). Renewable energy generators will require equipment and buildings that will have to be produced by manufacturing processes that require fossil fuels and mineral resources.

While Al Gore in his film and associated book *An Inconvenient Truth* (2006) popularized the findings of climate science to millions of people around the world, his solutions for mitigating climate change are framed very much within the parameters of green capitalism and ecological modernization by advocating carbon trading, green consumerism, tree plantations, and techno-fixes as sufficient climate change mitigation strategies. He proposes implementation of a Global Marshall Plan, which would entail the following elements: (1) stabilization of the world population; (2) the development and sharing of "appropriate technologies"; and (3) the development of a "new global economics" (Gore 2006: 307–337). Gore (2006: 346) argues that the definition of gross national product should be changed to include environmental costs and benefits, or treated as what mainstream environmental economists term an *externality.*

Gore (2009) lays out his views on ecological modernization, which include an overall endorsement of energy efficiency, retrofitted buildings, hybrid cars, greater reliance on public transport, and renewable sources of energy (solar, wind, and geothermal) as climate change mitigation strategies. Overall, he is ambivalent about the viability of carbon capture and sequestration (Gore 2009: 134–149), at least for the immediate future, noting that

> [m]ost experts who have studied the CCS [carbon capture and storage] option have concluded that it is probably impracticable for many years to come, because the technology for capturing CO_2 [carbon dioxide] would either require a dramatic increase in the use of coal and gas for the same amount of electricity, or sharply reduce the amount of electricity obtained from burning the same amount of fuel as of present—and because every one of the potential geological repositories presents a unique and extremely difficult challenge in characterizing its geology deep underground and estimating both storage capacity and the safety of storing CO_2 [carbon dioxide] there.

In keeping with the premises of ecological modernization, John Urry (2011: 16) advocates a post-carbon sociology that "emphasizes how modernity has consisted of an essentially carbonized modern world." He defines a post-carbon society as one that has transcended the current reliance on fossil fuels and fossil fuel driven technologies, such as cars dependent on the internal combustion engine. Urry (2011: 100) forcefully contends: "Climate change politics involves campaigning not for abundance or growing abundance now, so as to ensure reasonable abundance in the long term and in other parts of the globe." While much of what he writes is critical of existing capitalism, he does not call for the transcendence of the capitalist world system per se. Instead, Urry (2011: 118) advocates "resource capitalism" which relies upon ecological modernization and recognizes that "there is a limited capability to supply resources and to absorb pollution."

Ecomodernism

Ecomodernism constitutes a more recent genre of ecological modernization. It has taken on different guises, some of them politically right-of-center and others left-of-center. In its conservative form, US ecomodernism tends to be more critical of environmentalism than does the European ecological modernization which mainstream environmentalists and climate activists generally advocate. Its character and influence owe much to the establishment of the Breakthrough Institute, a US-based think tank set up by Michael Shellenberger and Ted Nordhaus in 2003. The Breakthrough Institute (n.d.) boldly and idealistically proclaims on its website:

> We believe that ecological vibrancy results from human prosperity, not the other way around. Meeting people's needs is both an ethical imperative and a pre-condition for societal concern about nature. Technological innovation, particularly in energy and agriculture, can enable us to meet both human needs and reduce our reliance on natural resources. And clean energy technologies are key to creating a high energy planet without overheating the climate.

The Breakthrough Institute favors "'clean energy," which includes not only solar and wind energy but also nuclear energy, over carbon pricing mechanisms. It also promotes industrial agriculture, including genetically modified foods. It adopted a bipartisan approach in a report entitled *Partisan Power*, published by the liberal Brookings Institute and the conservative American Enterprise Institute. However, much to the annoyance of traditional conservatives, the report advocates some government support for its technological proposals.

A more politically centrist version of ecomodernism is implied by the work of Jonathan Symons (2019) who regards ecomodernism as a social democratic response to global ecological problems. He argues that "ecomodernists generally advocate reforming, rather than overthrowing, capitalism" and "seek to invest wealth generated by capitalism in low-carbon innovation" (Symons 2019: 61). Symons (2019: 68) also asserts that ecomodernists advocate "greater social equality." He offers six summary propositions for his version of ecomodernism: first, climate change constitutes a serious threat; second, it should be a public policy priority; third, aggressive climate change mitigation would be politically impossible with existing technologies, including renewable energy sources such as those that currently exist; fourth, there is therefore a need for more additional low-carbon technologies; fifth, the state has played a major role in implementing technological innovations: and sixth, climate activists should therefore push for greater state involvement in low-carbon technologies (Symons 2019: 113–114).

Further to the left are some self-described socialists who have also adopted ecomodernist thinking. One of them is Leigh Phillips (2015), a Marxist and a staunch advocate of economic growth and a wide array of technological innovations, such as new materials to replace steel and concrete, improved battery and energy storage technologies, and electric cars. He confidently exclaims: "We must push through the Anthropocene, indeed accelerate our modernity, and accept our species' dominion over the Earth" (Phillips 2015: 186). The US radical magazine *Jacobin* devoted articles in the second half of its summer 2017 special issue entitled "Earth, Wind, and Fire" to ecomodernism, prompting a critical assessment by Foster and Clark (2020: 211) in which they argue:

> The ecological crisis brought by capitalism is used here to justify the setting aside of all genuine ecological values. The issue's contributors endorse a "Good Anthropocene," or a renewed conquest of nature, as a means of perpetuating the basic contours of present-day commodity society, including most disastrously, it's imperative for unlimited exponential growth. Capitalism is a global economic system that, in its drive for profits, requires ongoing accumulation and expansion.

As Harvey (2014: 222) observes: "Capital is always about growth and it necessarily grows at a compound rate." Global capitalism fosters a treadmill of production and consumption primarily for the purpose of generating profits for the few and, in the process, because they are of lesser importance relative to profitmaking, sacrifices both human needs and environmental sustainability.

Rich Proponents of Ecological Modernization

In this section, we discuss the views of various wealthy proponents of ecological modernization as an overarching climate change mitigation and adaptation strategy. The first is Bill Gates, an enigmatic figure who has functioned as a software developer, entrepreneur, and philanthropist, who several times has been the richest person in the world and who at the time of writing is ranked as the fourth richest person in the world. The second is John Doerr, a venture capitalist, who sought to build a cleantech network by bringing together the likes of Steven Chu (former US President Barack Obama's Secretary of Energy), Al Gore, and the writer and physicist Amory Lovins.

Bill Gates

Bill Gates as a multi-billionaire capitalist has fully embraced ecological modernization. In his own words, Gates asserts that "the world needs to provide more energy so the poorest can thrive, but we need to provide energy without releasing any more greenhouse gases." Despite the emissions generated directly by his own activities and the companies in which he has investments—and these emissions are considerable, Gates (2021: 8) boldly asserts that humanity needs to take three steps to avoid a global climate catastrophe: achieve net zero emissions; deploy existing tools, such as solar and wind energy, at a faster and smarter pace; and create and roll out "breakthrough technologies that can take us the rest of the way." To walk the talk, Gates has invested in various clean energy companies, putting US $100 million into a start-up company to design a next-generation nuclear power plant that would produce clean electricity and very little nuclear waste. In 2019, he divested all his direct holdings in oil and natural and gas companies—as did the trust that manages the Bill and Melinda Gates Foundation's endowment—noting that he has not had investments in coal companies for several years (Gates 2021: 10). Shortly before the 2015 UN Climate Change conference in Paris, France, Gates joined the Breakthrough Energy Coalition, now simply called Breakthrough Energy, which has invested in zero-carbon technologies and supports early-stage clean energy research. In this context, it should be re-emphasized that subalterns around the world are increasingly having their land and labor expropriated by mining companies, including ones that are providing resources for renewable energy operations and supposedly green technologies, such as electric cars and autonomous vehicles (Arboleda 2020).

To his credit, he admits that "building infrastructure such as wind turbines, solar panels, nuclear plants, electricity storage facilities, and so on will involve releasing more greenhouse gases" (Gates 2021: 40). Indeed, he is now the largest farm owner in the United States (Schwab 2023: 294). Although Gates reports that the richest 16% of the world's population accounts for 40% of its emissions and acknowledges the need to lift the poorest people out of extreme poverty so that they can survive, he fails to call for global redistribution of wealth. Instead of advocating for redistribution, Gates (2021: 42) argues that

[W]e need to make it possible for low-income people to climb the ladder without making climate change worse. We need to get to zero—producing even more energy than we do today, but without adding any carbon to the atmosphere—as soon as possible.

Gates (2021: 59) observes that most zero-carbon solutions at the present time are more expensive than their fossil-fuel counterparts, entailing what he terms Green Premiums that need to fall in price to zero or preferably less than zero. He sees the path to zero emissions in manufacturing as entailing the following steps, all of which will require considerable innovation: (1) electrification of every process possible; (2) transmitting that electricity from a decarbonized power grid; (3) utilizing carbon capture to sequester the remaining emissions; and (4) utilizing materials more efficiently (Gates 2021: 111).

In order to reduce emissions generated by meat production, Gates (2021: 119) recommends the production of artificial meat products, although there is evidence that this also creates emissions. While he is a fan of synthetic fertilizers for their contribution to food production and the Green Revolution, Gates (2021: 117) acknowledges that nitrogen runs into surface waters causing pollution or escapes as nitrous oxide, a powerful GHG. While reporting that transportation results in large amounts of emissions, he is staunch advocate of electric vehicles but admits that battery-powered airplanes on a large scale are not yet technologically viable. Gates has not commented whether solar-powered train travel could be an alternative to air travel. Regarding the viability of biofuels, Gates (2021: 138) expresses guarded optimism, but admits "it's a touchy field."

While Gates focuses upon techno-fixes for mitigating GHG emissions, he devotes some attention to climate change adaptation. He recommends three pathways for climate adaptation: (1) reducing the risks of climate change by climate-proofing buildings and other infrastructure, protecting wetlands as a bulwark against flooding, and encouraging people to relocate permanently from areas that are no longer habitable; (2) preparing for and responding to climate emergencies; and (3) recovery plans, including services for displaced people and insurance that helps people at all income levels rebuild (Gates 2021: 170). While acknowledging that geoengineering, studies of which he has been funding, is a highly contested topic, he views it as a back-up strategy, one that constitutes the "only known way we could hope to lower earth's temperature within a year or even decades without crippling the economy (Gates 2021: 178).

Despite his anti-trust battles with the US government, Gates is not a market fundamentalist and believes that governments can play a significant role in investing in reach and development projects in developing clean-energy technologies. His listing of potential clean energy technologies includes hydrogen produced without emitting carbon, grid-scale electricity storage that can last a full season, electro-fuels, advanced biofuels, zero-carbon cement, plant- and cell-based meat and dairy, zero-carbon fertilizer, next-generation nuclear fission, nuclear fusion, carbon capture, underground electricity transmissions, zero-carbon plastics, geothermal energy, pumped hydro, thermal storage, drought- and flood-tolerant food crops, zero-carbon alternatives to palm oil, and coolants that don't contain fluorinated gases (Gates 2021: 200). Gates also advocates government environmental regulations and carbon pricing, either through carbon taxes or emission trading schemes.

Curiously, Gates has said that he does think that using a private jet and campaigning on the issue of climate change represents a contradiction open to allegations of hypocrisy, saying "I spend billions of dollars on ... climate innovation. So, you know, should I stay at home and not come to Kenya and learn about farming and malaria?" (quoted in Frangoul 2023). Consequently, Gates sees himself as part of the solution not part of the problem. In his book, *The Bill Gates Problem*, journalist Tim Schwab (2023a) rejects this idea, writing that Gates is exactly who he was when at Microsoft: a bully and monopolist, convinced of his own righteousness and committed to imposing his ideas, his solutions, and his leadership on the world. At his core, Gates is not a selfless philanthropist or the good billionaire he claims to be, says Schwab; rather, he is a dedicated power broker who uses his extreme wealth to acquire immense political influence. A question that rarely gets asked is why we are not questioning the US $184 billion in wealth—a figure that continues to grow—that Gates has not given away? Says Gates, an impoverished child's life can be saved with $1,000. Using Gates's own figures, his $184 billion could save 184 million lives. To do so, however, he would have to give it away, which he has not done (Schwab 2023b).

Other Wealthy Proponents of Ecological Modernization

In a similar vein to Bill Gates's proposals to solve the climate crisis is the work of John Doerr (2021), an American investor and venture capitalist with a net worth of almost US $10 billion. A staunch advocate of OKRs (objectives and key results), in collaboration with a team consisting of some of the world's leading experts on climate change and cleantech, he delineates a plan by which "we can drive greenhouse emissions to net zero by 2050" (Doerr, 2021: xiii). Doerr's (2021: xxvi) plan includes the following strategies:

- Electrifying transportation by "switching from gasoline and diesel engines to fleets of plug-in electric bikes, cars, trucks, and buses":
- Decarbonizing the grid by "substituting fossil fuels with solar, wind, and other zero-emissions fuel sources";
- Fixing food by "restoring our lost carbon-rich topsoil, adopting less polluting fertilization practices, motivating consumers to eat more lower-emissions proteins and less beef, and reducing food waste";

- Protecting nature by introducing proven "interventions and protections for forests, soil, and oceans";
- Removal of carbon dioxide from the atmosphere and storing it long term, using "both natural and engineered solutions."

These objectives are to be achieved through the implementation of vital public policies, transforming the efforts of various actors (such as voters, government, business, educational institutions) to adopt climate action, developing and implementing new technologies, and deploying capital on a grand scale. Doerr's book, *A Global Action Plan for Solving Our Climate Crisis Now* (2021) includes conversations with Jeff Bezos (Amazon and Bezos Earth Fund), Margot Brown (Environmental Defense Fund), Christiana Figueres (Global Optimism), Bill Gates (Breakthrough Energy), Al Gore (Climate Reality Project), John Kerry (US State Department), Amory Lovins (Rocky Mountain Institute), Eric Trusiewicz (Breakthrough Energy), and Tenise Whelan (Rainbow Alliance).

HRH The Prince of Wales is another wealthy proponent of ecological modernization and the principal sponsor of the Earthshot project. He observes that science "tells us that if we do not act to restore our planet by 2030 damage will be irreversible" (William 2021: ix). Earthshot draws inspiration from the Moonshot project announced by US President John F. Kennedy stating the ambition of the United States to land a man on the moon. After speaking to activists, business leaders, prime ministers, conservationists, and filmmakers around the world, the Earthshot team delineated five overarching goals, namely (1) protecting and restoring nature; (2) cleaning the planet's air; (3) reviving the planet's oceans; (4) building a waste-free world; and (5) fixing the climate. If met, these changes will offer us the greatest chance of a stable and thriving future (William 2021: xi). The project seeks to support 50 solutions, five per year (one in each of the categories listed above) over the course of a decade (termed "the Decade of Change"). Winners in 2023, the third year of the awards, include Acción Andina, which uses ancient Inca principles to bring together tens of thousands of people in local and Indigenous communities to protect and restore depleted high Andean forests and ecosystems. Earthshot provides critical resources such as salary support, project and financial management, and technical training for local conservation leaders, organizations and communities. Another 2023 winner was WildAid, a global conservation non-profit that focuses on effective enforcement of laws established for Marine Protected Areas. WildAid's approach builds law enforcement capacity by ensuring that people have the tools, technology, and resources needed to stop illegal fishing, allow wildlife to recover, and improve the livelihoods of coastal communities.

While it is too soon to tell if Earthshot projects are making a difference in the world, some barriers to their success have been identified. One of the 2022 Earthshot winners was the London-based start-up Notpla, which received US

$1.2 million to develop a biodegradable plastic alternative to seaweed. The same day as the award was made, the efforts of the Intergovernmental Negotiating Committee established by the UN Environment Assembly to develop a binding agreement to reduce plastic pollution ended in failure. While a number of countries, including EU member states and various nations in the Global South, advocated instituting mandatory global restrictions on plastics production, the United States and other leading plastic- and petrochemical-producing countries, fought for voluntary cutbacks on a country-by-country basis. Increasing plastic production, in fact, is one way the petrochemical industry has planned to stay profitable as renewable energy becomes cheaper.

Ecological Modernization "Down Under": Greenwashing a Developed Country

Can developed countries use capitalist ecological modernization to redevelop themselves as carbon-free green havens? The case of Australis is instructive.

Ross Garnaut, a high-profile economist based at the University of Melbourne, exemplifies the hegemony of ecological modernization as a climate change mitigation strategy par excellence. He maintains that Australia has the potential of becoming an economic and energy superpower in a future post-carbon world due to its ready access to renewable energy sources, particularly solar, wind, and geothermal. Garnaut (2019: 11) boldly asserts: "The full emergence of Australia as an energy superpower of the low-carbon world economy would encompass large-scale early-stage processing of Australian iron, aluminium and other minerals."

He maintains that Australia could provide not only for its own energy needs with renewables but could prosper economically by exporting renewable energy to its long-time trading partners Japan, South Korea, the United Kingdom, and the EU, along with its newer trading partners, particularly China, as well as Indonesia, India, and other Southeast Asian and South Asian countries. As a neoclassical economist, Garnaut frames his version of ecological modernization within a capitalist framework: He argues: "Climate change will not be stopped by ending development. The challenge is to change the relationship between economic growth and emissions of the greenhouse gases that cause climate change" (Garnaut 2019: 16).

As a techno-optimist, Garnaut (2019: 17) firmly believes that the transition from fossil fuels to alternative forms of energy is "possible without sacrificing living standards in currently wealthy countries or disappointing hopes for improving conditions in the developing world." In his view, battery and hydrogen electric-powered vehicles will soon outcompete and thus replace motor vehicles powered by the international combustion engine (Garnaut 2019: 131).

Hydrogen power, however, potentially has its downsides. Hydrogen production methods can result in GHG emissions. Although hydrogen does not emit carbon dioxide when it is used in fuel cells, at which point the principal by-products are water vapor and small amounts of nitrogen oxide, some hydrogen production methods utilize fossil fuels, such as the production of hydrogen through coal gasification and steam methane reforming. Kaitsu et al. (2019: 8) report that the Latrobe Valley project in Australia

> [P]lans to use brown coal to produce hydrogen through coal gasification. Hydrogen made via this method is known as "brown hydrogen." This method of producing hydrogen is highly inefficient and polluting. The pilot project in the Latrobe Valley estimates that it will produce over thirty times more carbon dioxide than hydrogen in weight.

Steam methane reforming utilizes natural gas or methane and produces "blue hydrogen," releasing methane in the process. Conversely hydrogen production through electrolysis is more expensive than the other two production processes, but results in zero-carbon hydrogen or "green hydrogen," with oxygen being the only by-product of the production process.

In his book *Reset: Restoring Australia after the Pandemic Recession* (2021) Garnaut shows his enduring faith in techno-fixes and market mechanisms to shift Australia out of the economic doldrums created by the pandemic and the dangers of the climate crisis. More recently, Garnaut edited *The Superpower Transformation* (2022), a sequel to his highly touted *Superpower*, which was celebrated in government and mass media circles. As a long-time economic advisor to various Australian Labor Party governments, in his introduction in *The Superpower Transformation* Garnaut (2022a: ix) reports: "The new prime minister, Anthony Albanese, said in his acceptance speech as Prime Minister that his government would make Australia a renewable energy Superpower."

In his lead essay, Garnaut (2022b: 5) argues that the anthology is "mainly about how the new government and its successors can capture Australia's opportunities as well as avoiding the traps on the path to building the Superpower." He cites three reasons that Australia needs to contribute to holding the global temperature increase below 1.5 degrees Centigrade (2.7 °F), an aspiration of the 2015 UN Paris Agreement: (1) to mitigate potential damage of global warming to Australia, the developed country that would be most severely impacted by climate change; (2) decarbonization with "zero-carbon goods and services" would be good for Australia's prosperity; and (3) "positive approaches to climate change mitigation are important to Australia's security and international relations" (Garnaut 2022b: 40).

Garnaut (2022b: 40) argues that the "superpower" opportunity is embedded in the "five endowments that Australia has in abundance more than any other country":

- Renewable energy resources;
- Land suitable for growing biomass for zero-emissions industry and sequestering carbon;

- The world's largest access to minerals requiring large amounts of energy in processing;
- Energy transition minerals which will be scarce and valuable over the next several decades;
- Institutions and infrastructure from the old resource economy that can be transferred to the new economy.

He believes that Australian zero-carbon energy exports can adequately facilitate decarbonization in Germany and other densely populated European countries along with China, Japan, and South Korea in northeast Asia (Garnaut 2022b: 39).

Garnaut predicts that Australian hydrogen derived from renewable electricity, namely green hydrogen, will become the primary process for converting iron ore into steel. Green hydrogen can be liquefied or turned into ammonia and other carriers and exported around the world for recovery of hydrogen at points of import or use. Given that high-grade silicon is a crucial element in the development of global solar power, he believes that Australia possesses all the raw materials needed to produce the silicon metal required in manufacturing solar photovoltaic systems and computers, hopefully positioning it to capture the European market after the end of the Russian war in Ukraine. He also maintains that bunker ammonia (ammonia gas that has been turned into a saturated liquid) could eventually replace bunker oil as an emissions-intensive international shipping fuel. Garnaut (2022b: 53) argues that Australia's "superpower" economy "would entail a huge local demand for high-voltage transmissions cables for transporting renewable energy" around the world. Investment in "superpower" export industries would contribute to Australia's prosperity in the form of full employment coupled with rising wages.

As was the case with the mining boom driven largely by China's need for Australia's resources beginning around 2002, Garnaut (2022b: 59–61) maintains that private capital can drive the "superpower" economy, one in which he proposes Indigenous business can play a part in that lands under Indigenous title have some of the world's highest solar radiance. In keeping with his long-time neoliberal approach, Garnaut (2022b: 81) believes that "investment decisions and export are best left to private corporations that comply with Australian law and regulations." In his *Garnaut Reviews* (2008, 2011) he called for Australia to embrace some form of carbon pricing, which occurred briefly in 2012–2015 in the form of the Carbon Pricing Mechanism under Australian Labor Party governments in collaboration with the Greens, until the coalition government comprising the Liberal and the National parties abolished it in 2015. Garnaut believes that Australia will eventually adopt a carbon price but not any time soon. Instead, in keeping with his "superpower" vision, he agrees with "Minister for Energy and Climate, Chris Bowen

[who] said on 24 June 2022 that the best way to reduce electricity prices is to accelerate growth of renewable energy supply" (Garnaut 2022b: 66).

In order to ensure the stability of his "superpower" vision, Garnaut recommends two mechanisms. The first would be the Energy Reserve, a "Commonwealth entity, with the task of achieving high degrees of reliability in electricity supply at the lowest possible cost" which "would be charged with ensuring the supply of power would match demand in all regions under almost all conditions" (Garnaut 2022b: 93). The second would be the Sungrid, namely a "multi-terminal complex of electricity superhighways" as a complement to existing electricity grids in Australia (Garnaut 2022b: 98). Garnaut (2022b: 102) believes that nuclear power would not play a role in an "economically efficient Australian electricity system." Overall, he maintains that his suggestions would put Australia on a pathway to meet its

> [C]ommitment to the UN to reduce global emissions by 43% in 2020 [and] would be seen to be on a path to reducing emissions by 75% from 2005 levels by 2035 and net zero emissions in the 2040s.
>
> *(Garnaut 2022b: 90)*

In *The Big Switch*, Saul Griffith (2022), an inventor, entrepreneur, and engineer, outlines his vision for mitigating climate change by fully electrifying Australia, particularly by utilizing solar energy and batteries, while creating jobs and saving costs for consumers. Relying on renewable energy, Griffith would like to electrify virtually everything, including vehicles, industry, space heating, water heating, and dwelling units.

In terms of climate change mitigation, he, like Garnaut and his collaborators, is an advocate of "creating enormous profitable export industries" (Griffith 2022: 3). In contrast to Garnaut and at least some of his collaborators, he rejects hydrogen as a viable mitigation tool, arguing that "hydrogen doubles or triples the amount of clean energy we have to produce" (Griffith 2022: 7).

Griffith (2022: 46) also maintains that hydroelectricity "can be environmentally destructive to the local area that is flooded to build the reservoir it requires, but it is otherwise very reliable and quite environmentally benign." He views wave and tidal energy as limited in their supply, nuclear energy as highly problematic because of its perceived dangers and high water demands in cooling in reactors, and biofuels, such as ethanol derived from sugar cane, as problematic because they compete with food for land. Griffith (2022: 3) argues that some geothermal systems require a lot of water and most of Australia's best geothermal resources are situated in remote areas and would require new transmission lines. He rejects carbon capture and storage as being too expensive and views ammonia as a long-shot option.

Griffith (2022: 127) maintains that the "big switch" requires good government policy and that the federal government needs to lead with a national electrification plan. Conversely, he argues that most of the "rewiring of

Australia can be carried out by households and investors" (Griffith 2022: 132), although governments can help by stopping fossil fuel industry subsidies and subsidizing electrification pilot projects. Expanding upon their development of reasonable climate action plans, in contrast to the federal government, Griffith maintains that state governments should create renewable energy zones in regional Australia that will host numerous solar, wind, and battery projects. He admits that there are other environmental problems besides climate change, such as plastic waste, the nitrogen problem and algal blooms emanating from modern agriculture, and the overfishing of the oceans.

Alan Finkel, another entrepreneur and engineer who served as the Chief Scientist under the Morrison coalition government, is yet another Australian exemplar of ecological modernization. He maintains that the world needs to shift from *petrostates* to *electrostates*, in his view the "toughest task undertaken by humanity since the taming of fire" (Finkel 2023: 23). Finkel (2023: 76) outlines the following steps in which the "clean energy transition can proceed at the fastest possible pace with reduced mining and refining requirements":

- Recycling and reuse;
- Innovation to reduce demand;
- Stockpiling to minimize the impact of temporary or sudden shortages in supply chains such as occurred at the peak of the COVID-19 pandemic;
- Innovation in mining to reduce GHG emissions and avoid the peaking of resources.

While being a staunch advocate of renewable energy in the form of solar power, wind power, and hydropower, he believes that nuclear power will face formidable challenges, such as having difficulty integrating with an electricity grid dominated by solar and wind, lack of broad public approval in the Australian context, and lack of private investor interest due to the protracted time in approving and constructing nuclear power plants (Finkel 2023: 104).

Given that Finkel (2023: 186) does not see Australians wanting to forego their relish for domestic and international travel, he proposes the development of sustainable aviation fuels, which he maintains can be made from various biomass feedstocks, such as harvest waste, forestry residues, plant oils, and municipal waste. In reality such residue sources would probably not be sufficient to provide a viable alternative for conventional jet kerosene fuel given present and future demands for air travel. Bridger (2013: 3) maintains that "enormous amounts of energy to grow and process crops means that biofuel greenhouse gas emissions are higher than from fossil fuels." Like Garnaut and Griffith, Finkel is a strong proponent of electric vehicles. Even if powered by solar energy and batteries, electric vehicles require a massive amount of embedded energy and infrastructure in the form of roads, bridges, and tunnels.

Finkel (2023: 223) rejects calls for de-globalization, a shift in which countries implement "economic nationalism through import tariffs, localized trading blocs and rejection of international bodies." In recognizing that the Albanese government has opted to permit the export of coal to various countries around the world, he argues:

> Developed countries—especially those that are fossil-fuel exporters—should help developing countries build renewable energy microgrids and attract developed-world finance into large-scale renewable projects. Just as we expect companies to do, as nations we should be investing in building alternatives that will eliminate alternatives that will eliminate the demand for thermal coal and eventually other fossil fuels.
>
> *(Finkel 2023: 226)*

In at a nutshell, Finkel (2023: 258–259) maintains that with the right policy settings Australia has the potential to evolve into a full-fledged electrostate that would entail the following dimensions:

- Mining and refining energy transition resources such as nickel, lithium, cobalt, copper, and rare earth minerals;
- Exporting "sunshine" in the forms of hydrogen, ammonia and synthetic fuels;
- Manufacturing decarbonized products, such as green iron, green aluminum, and green fertilizer;
- Developing "direct air capture" and storage, more commonly referred to as carbon capture and storage by removing carbon dioxide from the air and burying it.

Garnaut, Griffith, and Finkel operate on the premise that both Australia and the global economy require not only a lot of energy but growing amounts of energy, in essence operating under the parameters of what has commonly come to be termed green capitalism. Unfortunately, even green capitalism fails adequately to address the treadmill of production and consumption that contributes to the depletion of natural resources and environmental degradation, including anthropogenic climate change. Just as capitalism operated on other forms of energy prior to the fossil fuel revolution, capitalism could, in theory, operate entirely on renewable sources of energy, which will require enormous resources to develop and maintain, thus leading to new resource curses, particularly in the Global South. Rosewarne (2022: 412) warns that the construction of massive solar and wind farms may result in "restricting traditional custodians' access to country" and constitute a form of "colonisation that is being abetted by federal and State governments in their determination to find a solution to the climate crisis that does not compromise the pace of capital accumulation."

Griffith (2022: 3) also seeks to address the social equity issue arguing that at least in the Australian context there is a need to "make sure that households at all income levels can afford to be part of this [clean energy] transition and benefit from the new abundant Australia." In contrast to Garnaut and Finkel, Griffith (2022: 163) acknowledges that there are limits to the electrification vision and indirectly acknowledges the limits to economic growth, observing:

> We could get to zero emissions but still bury ourselves in things we don't need …The planet would be more verdant, and quieter, with more electric bicycles than electric cars. More trams than electric pick-up trucks. Certainly freeing up wildlands for wild animals by changing our diet would be great.
>
> *(Griffith 2022: 164)*

Finkel (2023: 240) claims that the clean energy transition will facilitate energy equity, namely the provision of energy at a reasonable cost both in developed and developing societies so that "all members of society can meet their basic needs for cooking, heating and transportation." That capitalism has never worked equitably seems to have escaped his notice.

Critique of Ecological Modernization

Over a decade ago, Shaw (2010) queried the 2 degree Celsius (3.6 °F) limit that numerous governments and corporations, the EU, and even many environmental nongovernmental organizations have adopted as a safe temperature rise. Episodic and ongoing instances of regional catastrophic climatic events in recent temperatures that containing global warming at 2 °C (3.6 °F), even 1.5 °C (2.7 °F), belies the notion of a "safe climate" at the present time or in the immediate foreseeable future. Shaw argues that this arbitrarily designated limit "makes climate change a problem for the future which allows humanity to continue on with 'business as usual' whilst the search for a techno-fix continues," an argument that aligns with our critique of ecological modernization. In recognizing the limitations of ecological modernisation, Eriksen and Mendes (2022: 23) observe:

> The production of "clean energy" is in itself a never-ending puzzle. Hydroelectricity dams are big sources of methane and CO_2 [carbon dioxide]; wind turbines use sulphur hexafluoride (SF_6), a potent greenhouse gas; solar energy relies heavily on mining and metallurgical industries and produces large amounts of toxic waste (mainly tetrachloride).

In recent years, various governments, including those of the United States, the United Kingdom, and even Australia, along with selected multinational corporations, have adopted the aspiration of achieving net-zero emissions by 2050. Referring to the US case, Aronoff (2021: 129) argues that this aspiration "doesn't tend to include the millions of barrels of oil and gas the US exports to be burned abroad each day" nor "does it account for the emissions they hope to 'offset' with land grabs to mass produce bioenergy halfway around the world, and speculative new technology." Notably, President Biden entered the White House in 2021 with a vow that fighting climate change would be a driving priority for his administration. But his climate agenda has succumbed to legal, legislative, and political opposition within his party and complete rejection from the Republicans and the energy industry, as well as inflation and rapidly rising gas prices, and the Russian Federation's invasion of Ukraine. The result has been an administrative call for more drilling and less emphasis on fighting climate change. President Biden announced that he would release one million barrels of oil per day from the Strategic Petroleum Reserve over a period of 180 days in an effort to bring down global oil prices, and, without doubt, to attempt to win votes for the Democrats in the 2022 election. The scale and duration of this oil release was unprecedented. The United States also plans to increase natural gas exports to Europe to help them to wean themselves off Russian oil.

While some components of ecological modernization have the potential to serve as important mitigation and adaptation strategies, unfortunately all forms of ecological modernization tend to be oblivious to social justice issues or, at best, pay them lip service. At the same time, its solutions fail to address the features of capitalism that created our current socioecological dilemma. As Foster et al. (2010: 5) assert, under capitalism "energy savings are used to promote new capital formations and the proliferation of commodities, demanding even more resources." Instead, what is needed is to make technological innovations that are more environmentally sustainable and energy-efficient as part of a shift to a steady-state or net zero-growth global economy. Only then are they able to circumvent the Jevons paradox or rebound effect (see above). Ecological modernization as a dimension of green capitalism has become a central component thereof.

As Wall (2010: 11) observes, "Even a renewable energy capitalism would still tend to degrade the environment through commodification of nature." Ted Trainer (2007: 2) acknowledges the superiority of renewable energy sources over fossil fuels but maintains that the "very high levels of production and consumption and therefore of energy use that we have in today's consumer-capitalist society cannot be contained by renewable sources of energy." Wind and solar electricity generation requires very large parcels of land, of which Australia for example has plenty, much of it consisting of land over which Indigenous peoples have native title rights. In other parts of the world, such as much of Europe and Japan, a transition to 100% renewable energy would be difficult because of the land requirements. However, offshore wind is increasingly being used to offset land constraints. It is also important to observe that renewable energy and the digital economy are heavily reliant in their present forms upon rare earth metals (Pitron 2020). These are found in terrestrial rocks in infinitesimal amounts and are a subset of some 30 raw materials often associated with more abundant metals, such as iron, copper, zinc, bauxite, and lead. Extracting rare earth metals also necessitates huge amounts of water in the purification process. As discussed in Chapter 5, the extraction of cobalt to make lithium-ion batteries in electric cars requires heavy physical labor using picks and shovels and has resulted in the pollution of nearby waterways when implemented in the Democratic Republic of Congo. The manufacture of laptop computers and mobile phones accounts for approximately19% of the global production of rare metals such as palladium and 23% of cobalt (Pitron 2020: 42). Furthermore, the extraction of rare metals may result in an array of negative health consequences for humans, ranging from children not developing teeth to getting cancer (Pitron 2020: 27–28). Pitron argues that the scarcity of rare metals is a monumental issue that could spark geopolitical conflict, particularly given that China currently dominates their production.

That ecological modernization constitutes the overarching agenda of climate capitalism is illustrated in by a wind power development project in the Isthmus of Tehuantepec, by Leppert and Barrios (2022) who analyze what they see as a case of green neoliberalism involving a public-private partnership between a renewable energy company and the Mexican state. Ecolica del Sur, a multinational renewable energy corporation supported by Japanese and Mexican investors, approached the community leaders of Binniza (pseudonym), a small Zapotoc Indigenous town. While the project has the potential to reduce GHG emissions, it exacerbated existing class and ethnic inequities in Binniza. Leppert and Barrio (2022: 325) report:

> The installation of wind turbines was to take place on agricultural lands, which required the company to pay a tax to the town's local government for the change of soil use from agricultural to industrial. The municipal president at the time negotiated an amount that landowners considered dismal (3.5 million pesos). The project was deployed over two thousand hectares of land, but the president and company agreed that the tax would only be paid for each eight-square-meter area covered by the base of the turbines ... [T]he landowners attempted to amicably resolve this matter with the state government and the company through talks, they were once again ignored, pushing them to carry out yet another blockade and halt the construction project. In this particular instance, their efforts proved futile, and the tax amount was not changed.

The installation of turbines entailed the construction of raised roads that bisected agricultural fields and adversely impacted water drainage and irrigation patterns, resulting in the flooding of some fields and lack of sufficient water for others. Some landowners also complained that the company had understated the noise that would be generated by the wind turbines.

Conclusion

In contrast to the past when many corporations were drivers of climate change denialism, particularly in the form of the now defunct Global Climate Coalition, as the evidence for climate change has become more apparent and acknowledged in the more progressive sectors of the mass media and on the part of selected conservative politicians, in order to maintain a favorable public face, more and more corporations have adopted what may be termed corporate environmentalism, in some ways a more recent variant of the notion of corporate social responsibility. Like some of the corporate voices in Doerr's book, *A Global Action Plan for Solving Our Climate Crisis Now* (2021), they have come to assert that they are striving to achieve environmental sustainability and reduce GHG emissions in their businesses. Wright and Nyberg (2015) maintain that corporate environmentalism tends to build on the notion of ecological modernization that stresses the ability to come up with technological innovations that are environmentally friendly.

Although proponents of ecological modernization concede that many environmental problems are by-products of a market economy or global capitalism, they generally reject transcending the capitalist mode of production. Proponents of ecological modernization maintain that capitalism can be made more "environmentally friendly" through environmental regulations and technological changes managed by ecologically sensitive governments, or *green states*, that function in concert with corporations. Ecological modernization, which started in northern European countries, has quickly been adopted by corporate elites and politicians in various countries, including the United States and Australia, and even among mainstream environmental groups in both developed and developing societies.

While we firmly believe that renewable energy sources, including solar, wind, and geothermal, have the potential to be part of the process of mitigating climate change, in and of themselves they are not a panacea. As Wall (2010:11) observes, "Even a renewable energy-fueled capitalism would still tend to degrade the environment through commodification of nature." While adopting more environmentally sustainable technologies and achieving energy efficiency are in and of themselves commendable objectives, they will not lead to a "decoupling" from economic growth as mainstream environmental economists maintain because, following the Jevons paradox, in a capitalist economic system, "energy savings are used to promote new capital formation and the proliferation of commodities, demanding ever greater resources" (Foster et al. 2010: 5). Salleh (2010: 196) notes that ecological modernization "will consume vast amounts of front-end fuels—in welding turbines and grids, road making, water supply, component manufacture for housing, [and] air conditioning for shopping malls."

References

Arboleda, M. 2020. *Planetary Mine: Territories of Extraction Under Late Capitalism.* London: Verso

Aronoff, Kate. 2021. *Overheated: How Capitalism Broke the Planet—and How We Fight Back.* New York: Bold Type Books

Beck, Ulrich. 2007. *Environment and Social Theory* (2nd edn). London: Routledge

Beck, Ulrich. 2010. Climate for change, or how to create a modern modernity. *Theory, Culture and Society* 27: 254–266

Breakthrough Institute. (n.d.). About Us. http://www.thebreakthrough.org/about

Bridger, Rose. 2013. *Plane Truth: Aviation's Real Impact on People and the Environment.* London: Pluto Press

Brown, Lester R. 2009. *Plan B 4.0: Mobilizing to Save Civilization.* New York: W.W. Norton

Carlson, C.J., Colwell, R., Hossain, M.S., Rahman, M.M., Robock, A., Ryan, S.J., Shafiul, M.S., and Trisos, C.H. 2022. Solar geoengineering could redistribute malaria risk in developing countries. *Nature Communications* 13, Article 2150

Carter, Neil. 2007. *Politics of the Environment: Ideas, Actions, Policy* (2nd edn). Cambridge: Cambridge University Press

Chomsky, Noam and Robert Pollin. 2020. *Climate Crisis and the Green New Deal.* London: Verso

Doerr, John. 2021. *A Global Action Plan for Solving Our Climate Crisis Now.* London: Penguin Random House

Eriksen, Thomas Hylland and Pablo, Mendes. 2022. *Introduction: Scaling down in order to cool down.* In *Cooling Down: Local Responses to Global Climate Change.* Susanna M. Hoffman, Thomas Hylland Eriksen, and Pablo Mendes, eds. Pp. 1–24. New York: Berghahn Books

Frangoul, A. 2023. I spend billions of dollars on … climate innovation. So, you know, should I stay at home and not come to Kenya and learn about farming and malaria? CNBC. https://www.cnbc.com/2023/02/07/private-jet-use-and-climate-campaigning-not-hypocritical-bill-gates-.html#:~:text=During%20a%20recent%20interview%20with,as%20fast%20as%20they%20can."&text=Bill%20Gates%20does%20not%20agree,open%20to%20allegations%20of%20hypocrisy

Finkel, Alan. 2023. *Powering Up: Unleashing the Clean Energy Supply Chain.* Melbourne: Black, Inc.

Foster, John Bellamy. 2000. *Marx's Ecology: Materialism and Nature.* New York: Monthly Review Press

Foster, John Bellamy and Clark, Brett. 2020. *The Robbery of Nature: Capitalism and the Ecological Rift.* New York: Monthly Review Press

Foster, John Bellamy, Clark, Brett, and York, Richard. 2010. Capitalism and the curse of energy efficiency: The return of the Jevons Paradox. *Monthly Review*, November, pp. 1–12

Gates, Bill. 2021. *How to Avoid a Climate Disaster: The Solutions We Have and the Breakthroughs We Need.* London: Allen Lane

Garnaut, Ross. 2008. *The Garnaut Climate Change Review: Final Report.* Melbourne: Cambridge University Press

Garnaut, Ross. 2011. *The Garnaut Review 2011.* Melbourne: Cambridge University Press

Garnaut, Ross. 2019. *Superpower: Australia's Low-Carbon Opportunity.* Melbourne: La Trobe University Press

Garnaut, Ross. 2021. *Reset: Restoring Australia After the Pandemic Recession.* Melbourne: La Trobe University Press

Garnaut, Ross. 2022a. Introduction. In *The Superpower Transformation: Making Australia's Zero-Carbon Future.* Ross Garnaut, ed. pp. ix–xv. Melbourne: La Trobe University Press

Garnaut, Ross. 2022b. The bridge to the superpower. In *The Superpower Transformation: Making Australia's Zero-Carbon Future.* Ross Garnaut, ed. pp. 1–95. Melbourne: La Trobe University Press

Gore, Al. 2006. *An Inconvenient Truth: The Planetary Emergency of Global Warming and What We Can Do About It.* Emmaus, PA: Rodale

Gore, Al. 2009. *Our Choice: A Plan to Solve the Climate Crisis.* London: Bloomsbury

Greenpeace. 2010. *Koch Industries: Secretly Funding the Climate Denial Machine.* Washington, DC: Greenpeace

Griffith, Saul. 2021. *Electrify: An Optimist's Roadmap to Our Clean Energy Future.* Cambridge, MA: MIT Press

Griffith, Saul. 2022. *The Big Switch: Australia's Electric Future.* Melbourne: Black, Inc.

Helm, Dieter. 2017. *Burn Out: The Endgame for Fossil Fuels.* New Haven, CT: Yale University Press

Harvey, David. 2014. *Seventeen Contradictions and the End of Capitalism.* London: Profile Books

Horn, H. 2022. The depressing reality behind Prince William's Earthshot Prize. *New Republic.* https://newrepublic.com/post/169400/depressing-reality-behind-prince-williams-earthshot-prize

Huber, Martin T. 2022. *Climate Change as Class War: Building Socialism on a Warming Planet.* London: Verso

Hughes, David M. 2021. *Who Owns the Wind? Climate Crisis and the Hope of Renewable Energy.* London: Verso

Jacobs, M. 2021. System change, not climate change. *Inside Story,* 9 November. https://insidestory.org.au/system-change-not-climate-change/

Leppert, Amanda and Barrios, Roberto E. 2022. Emitting inequity: The sociopolitical life of anthropogenic climate change in Oaxaca, Mexico. In *Cooling Down: Local Responses to Global Climate Change.* Susanna M. Hoffman, Thomas Hylland Eriksen, and Paul Mendes, eds. Pp. 313–338. New York: Berghahn Books

Machin, A. 2019. Changing the story? The discourse of ecological modernization. *Environmental Politics* 26: 438–458

Nordhaus, Ted and Schellenger, Michael. 2007. *Break Through: From the Death of Environmentalism to the Politics of Possibility.* Boston, MA: Houghton Mifflin

Phillips, Leigh. 2015. *Austerity Ecology and the Collapse-Porn Addicts: A Defence of Growth, Progress, Industry and Stuff.* Winchester: Zero Books

Pitron, G. 2020. *The Rare Metals War: The Dark Side of Clean Energy and Digital Technologies.* Melbourne: Scribe

Rosewarne, Stuart. 2022. *Contest Energy Futures: Capturing the Renewal Energy Surge in Australia.* Singapore: Palgrave Macmillan

Salleh, Ariel. 2010. Climate strategy: Making the choice between ecological modernisation or living well. *Journal of Australian Political Economy* 66: 118–143

Singer, Merrill. 2019. *Climate Change and Social Inequality: The Health and Social Costs of Global Warming.* London: Routledge

Singer, Merrill. 2021. *Ecosystem Crises Interactions: Human Health and the Changing Environment.* Malden, MA: Wiley

Schwab, Tim. 2023a. *The Bill Gates Problem: Reckoning with the Myth of the Good Billionaire.* New York: Macmillan

Schwab, Tim. 2023b. Why Bill Gates's philanthropy is a problem. *The Nation.* https://www.thenation.com/article/society/bill-gates-philanthropy-misanthropy/

Shaw, Christopher. 2010. Dangerous limits: Climate change and modernity. In *History at the End of the World: History, Climate Change and the Possibility of Closure.* Mark Levene, Rob Johnson, and Penny Roberts, eds. pp. 95–111. Westport, CT: Praeger

Symons, Jonathan. 2019. *Ecomodernism: Technology, Politics and the Climate Crisis.* London: Polity

Trainer, Ted. 2007. *Renewable Energy Cannot Sustain a Consumer Society.* New York: Springer

Urry, John. 2011. *Climate Change and Society.* London: Polity Press

Wall, Derek. 2010. *The No-Nonsense Guide to International Politics.* Oxford: New Internationalist

Wijkman, Andreas and Rockstroem, Johannes. 2011. *Bankrupting Nature: Denying Our Planetary Boundaries.* London: Earthscan

William, HRH The Prince of Wales. 2021. Introduction. In *Earthshot: How to Save Our Planet.* Colin Butfield and Jonnie Hughes. pp. vii–xiv. London: John Murray

Wright, Christopher and Nyberg, Daniel. 2015. *Climate Change, Capitalism, and Corporations: Processes of Creative Self-Destruction.* Cambridge: Cambridge University Press

8

THE SCHOLARLY ELEPHANT IN THE SKY

How Can Anthropologists and Other Scholars Grapple with their Heavy Reliance on Flying in the Era of Climate Crisis?

Case Study

In 2023, Gianluca Grimalda, a climate researcher at the Kiel Institute for the World Economy in Germany, was fired from his job for refusing to take an airplane back from his research site in Bougainville, Papua New Guinea. His return trip had been delayed owing to various factors including "natural" disasters and his employers wanted him home quickly. He refused, explaining that:

> [a] trip by plane from Papua New Guinea to Germany produces, in 32 hours, 5.3 tonnes of CO_2 [carbon dioxide] per passenger. Slow travel [by boat and train] produces approximately 12 times less (420 kg). In the current state of climate emergency, wasting 4.9 tonnes of CO_2—about how much the average person in the world emits in one year—to expedite my return to Europe is not morally acceptable to me.
>
> *(Grimalda 2023)*

Through his actions—giving up what he considers a dream job—he hopes to put small crack into the wall of "selfishness, greed, and apathy," that, in the words of climate lawyer Gus Speth, is the main barrier to stopping runaway climate change. Moreover, Grimalda made a promise to the 1,800 participants in his Bougainville research that he would return home by low-carbon transport. He wanted to keep his promise. White men (of whom he is one, a fact that he is frequently reminded of in Papua New Guinea) are often referred to as *giaman*—liars or fraudsters—probably with good reason given the country's tumultuous colonial history.

DOI: 10.4324/9781003469940-9

Staying Grounded

Within the broader umbrella of global capitalism, there are many drivers of anthropogenic climate change, ranging from fossil fuel use, agricultural and forestry practices, manufacturing, steel and cement production, the construction, heating and cooling of buildings and residences, transport, the "cloud" or telecommunications, etc. The list seems endless. In all of this, one driver that tends to be downplayed is the growing number of airplane flights around the world, even in instances where people could travel long distances by train, coach, or even car with four or five passengers. While global capitalism is often the "elephant" in the room" when it comes to identifying the drivers of anthropogenic climate change, airplanes constitute the "elephant in the sky" as one of the multiplicity of climate change drivers, even among academics, including climate scientists and anthropologists. Yet the environmental consequences of airplanes have been known for some time. Over three decades ago, Stuart McBurney (1990: 35) delineated the following environmental, economic, and social consequences of jet airplane flights.

- Noise pollution;
- Atmospheric and potential climatic disturbances;
- Psychological disruption of family units and communities;
- Profits for some, resulting from airliner manufacture and operation; loss for others, resulting from an imbalanced concentration in the use of the world's resources;
- Air pollution;
- Heat pollution of the globe's atmosphere as a result of excessive carbon dioxide released by jet engines.

The number of airplane flights worldwide has been growing, at least prior to the outbreak of the COVID-19 pandemic, which forced reluctant governments to restrict temporarily the number of flights. Air travel, along with the cruise ship industry, played a key role in turning a localized epidemic in Wuhan, the People's Republic of China, into a global pandemic. Although air travel contributes only about 2.5% of carbon dioxide emissions globally, airplanes also emit nitrous oxide and other contrail or exhaust fumes, meaning that a "factor between two or three is normally applied to the CO_2 emissions from aviation to account for the additional warming impact" (Tickell 2008: 41). Thus, air flights may have been contributing in the order of 5 to 6% of emissions prior to the COVID-19 pandemic; perhaps even more, according to some sources.

Despite repeated claims by airline companies that they were gradually turning to more fuel-efficient and aerodynamic aircraft, these technological innovations were offset by an increase in air travel of roughly 5% per annum (in keeping with the Jevons paradox, or rebound effect, which arises when technological progress or government policy increases the efficiency with which a resource is used, but the falling cost thereof then leads to rising consumption). This rise was even higher among affluent people in China, India, and other developing countries, who started to emulate the habits of their counterparts in developed countries. Airplanes of many sorts (commercial, military, and private) have become sources of tremendous profit and integral components of modernity and the capitalist world system. Furthermore, aviation companies are an excellent example of how corporate profit-making is subsidized by public funds. Airplanes serve to transport both human actors and commodities to keep the world system operating and overheating. However, they do so with dire environmental, climatic, and health consequences (Baer 2020).

The human actors who rely on air travel include businesspeople, politicians, celebrities, the super-rich who own multiple homes in far-flung locations, sports teams, foreign correspondents, tourists, people going home for holidays, academics, international university students, other students studying abroad for short-term stints, and even United Nations climate change conference delegates and observers, environmentalists, and climate activists. The list seems almost endless but, with some exceptions such as low-paid migrant workers, refugees, and rank-and-file military personnel, it consists of relatively affluent people.

Despite a period of government-imposed lockdowns, which reduced the number of flights, and despite the persistence of the pandemic, economic pressures—including on the part of airlines and even universities—air travel began to increase in early 2022 and was predicted by the International Air Transport Association (2022) to return to pre-pandemic levels by 2024.

Increase in Airplane Flights

There is much variability among these passengers: some may be taking a once in a lifetime flight and others, such as businesspeople, politicians, celebrities and the super-rich, may be frequent fliers, even daily. Indeed, the practice of taking brief trips on luxury aircraft appears to be common among the rich and famous. A review of the Celebrity Jets tracking account shows that in just one calendar month, the rapper Drake took an 18-minute flight from Hamilton to Toronto, Canada; the country music singer Kenny Chesney traveled for just 20 minutes between Akron, Ohio, and Pittsburgh, Pennsylvania, USA (a single flight that produced the equivalent of the annual average global carbon dioxide emissions of one person); while the actor Mark Wahlberg was recorded to have taken a 23-minute flight from Dublin to County Clare, Ireland. There are several rationales for these environmentally damaging and unnecessary short flights. Sometimes they are taken to park a personally owned aircraft at a convenient or less expensive location, or they are part of a longer, two-leg journey; however, many appear to have no clear rationale, such as the decision by boxing champion Floyd Mayweather to fly for 14 minutes from Las Vegas to nearby Henderson, Nevada, before taking a 10-minute flight back again (Millman 2022).

Projections for an increase in flights vary: Airbus anticipates that passenger flights will rise by 4.8% per year between 2005 and 2025, with the global airline fleet doubling during this period; Boeing anticipates that passenger flights will increase by 4.5% per year and cargo flights by 6.1% per year between 2006 and 2026 (Bows 2009: 19). While there has been a small decrease in the percentage of flights emanating from airlines based in countries in the Global South, this has been offset by a boom in the number of flights made by airlines such as Cathay Pacific, Emirates, and Qatar, which are based in countries in the Global South.

Flying has for some time been reportedly the fastest-growing single source of greenhouse gas (GHG) emissions (Bridger 2013: 2), although it may now have been superseded by emissions generated by the cloud (Pitron 2023). The Intergovernmental Panel on Climate Change (IPCC, 1999) published a report in which it observed that aircraft released more than 600 million tons of carbon dioxide into the atmosphere per annum and were responsible for about 3.5% of anthropogenic global warming. More recently, the IPCC (2007: 14) projected that even with ongoing efficiency gains, cumulative GHG emissions would rise from 489.29 million tons in 2002 to 1,247.02 million tons in 2030, an increase of more than 250%.

As airplanes emit nitrous oxide and other exhaust fumes, a factor of between two and three is normally applied to the carbon dioxide impact. Clark (2009: 14) reports that a flight between London and Edinburgh produces 140 kilograms of carbon dioxide equivalent per passenger, whereas a single passenger trip over the same distance in a Ford Mondeo 2.0 results in 120 kilograms of carbon dioxide equivalent in emissions, a trip in a Toyota Prius with four passengers results in 16 kilograms carbon dioxide equivalent per passenger, a trip on an ordinary train results in 15 kilograms of carbon dioxide equivalent per passenger and a trip on a coach results in 18 kilograms of carbon dioxide equivalent per passenger. A return flight from London to Hong Kong results in 3.4 tons of carbon dioxide equivalent per passenger in economy class, a whopping 13.5 tons of carbon dioxide equivalent per passenger in first class, and 4.6 tons of carbon dioxide equivalent per passenger on average (Berners-Lee 2010: 135). Flying first class or business class is more environmentally damaging than flying economy because the former requires more space and more amenities, such as higher quality food and beverages, than is the case in the latter (Bridger 2013: 18).

Both the airlines and airline organizations claim that GHG emissions from flights will progressively decline, even to the point that eventually "zero emissions flights" will be achieved. Most airlines now have an environmental policy of some sort. The aviation industry has promised emissions reductions based upon future technological improvements, such as lighter airframes and more energy-efficient forms of propulsion, including solar flight, electric flight, and reliance on alternative fuels, particularly biofuels and even hydrogen (Bridger 2013: 21–22). Fuel efficiency does increase with the size of the airplane, meaning that flights in small aircraft, particularly private jets, are especially energy-intensive in terms of GHG emissions (Bridger 2013: 13).

In terms of making aircraft design more energy efficient, the aircraft industry tends to focus on reducing aircraft weight, reducing aerodynamic drag, and improving engine performance. While there have been significant improvements in energy efficiency over the past few decades, these have been offset by the number of airplane flights in this period. While the newer aircraft are more energy efficient, a shift in the global fleet will at best—given the expense of new airplanes—require enormous embedded energy, resulting in additional GHG emissions. This constitutes yet another example of the Jevons paradox, or rebound effect.

Academic Air Travel

As in many areas of a stratified world system, the affluent contribute much more overall to GHG emissions from flights than working-class people and particularly the poor around the globe. In certain social circles, air travel has become ubiquitous. The documentary series *City in the Sky* (2016) asserted that:

> every day, 100,000 flights crisscross the globe with more than 1 million people in the air at any time. In essence, 'hypermobility' is a "process driven by a relatively small part of society, increasingly comprising new societal groups with new mobility motives.
>
> *(Goessling et al. 2009: 146)*

Air travel, both domestic and international, is much more common among people in developed countries than in developing countries, although its use is growing among affluent sectors in the latter. Watson (2014: 16) asserts:

> Flying is an elite activity: only 5 per cent of people alive today have ever flown and, of those, very few are frequent flyers. It may be that just 1 per cent of humanity is responsible for 80 per cent of the world's flights.

Zygmunt Bauman (1998) maintains that the increasing number of professional trips made by various types of experts contributes to the growing division between cultural elites and marginalized groups. While academics are not generally ranked among the global elites, many academics in full-time positions—including anthropologists—and particularly those at elite institutions, fall into the ranks of frequent flyers. Much of this behavior has been driven by the dictates of the corporate university structure, which seeks to internationalize itself in a competitive bidding war for student numbers, including overseas students, and research funds. This has occurred as governments have reduced funding particularly for public universities.

As primary producers of knowledge, those in professional academic employment (or similar private employment) generally are required to implement research and disseminate research findings. The advent of internationalization has put enormous emphasis on academic conferences, meetings, consulting visits, and field research travel, as well as job interviews. In fact, the need has generated a thriving academic conference industry. Success in an academic career and the international standing of a researcher's institution both hinge in part on travel-enabled activities. Consequently, academics tend to be highly aeromobile.

Air Travel on the Part of Anthropologists

We suspect that while the majority of anthropologists around the world accept the findings of climate scientists and recognize that climate change has already had an adverse impact on many of the participants or interlocutors in their research, and will continue to do so throughout the course of the 21st century, they are not aware—or compartmentalize their awareness—that their flying may be contributing to a 4 degree Celsius (7.2 degree Fahrenheit) warmer world by 2100 if emissions from many sources are not abated in the next few decades. When several years ago, one of the authors of this volume, Hans Baer, asked a world-renowned anthropologist based at a prestigious US university whether he was a "frequent flyer," he replied: "Isn't everyone?" We suspect that many of the people he studies who live on a different continent are not. Many of them have probably never flown.

When the authors of this book became anthropologists during the 1970s, one of the most world-renowned anthropologists at the time was Napoleon Chagnon. He spent many years studying the Yanomamo, a horticultural Indigenous group living in villages scattered across the Amazonian basin which overlaps both Venezuela and Brazil. In his provocative auto-ethnography, he reports that he spent some 35 years studying the Yanomamo who live primarily in the Venezuelan jungle (Chagnon 2013: 3). Chagnon's initial field trip was conducted from November 1964 to March 1966—some 17 months. However, he returned to conduct ethnographic research on the Yanomamo on a further 25 or so occasions from his university base (Chagnon 2013: 3). It is possible to do a very rough calculation of the amount of carbon dioxide produced during these flights. The distance from Columbia, Missouri (where Chagnon was a professor from 2013 until his retirement), to Caracas, Venezuela, is a round trip of about 5,000 air miles. According to BlueSkyModel (n.d.), created by students and staff at Wake Forest University's Babcock Graduate School of Management, the average commercial flight emits 53.3 pounds of carbon dioxide per air mile (0.024 metric tons). So, 25 round-trip flights from

Columbia to Caracas creates over 125,00 pounds (60.44 metric tonnes) of carbon dioxide emissions. However, to give Chagnon his due, he would have been in good company among his fellow anthropologists and he probably was operating in an era in which anthropologists and other scholars had little or no consciousness of the environmental damage resulting from their flying including to their respective field sites.

It is not just sociocultural anthropologists who engage in a lot of flying over the course of their careers to conduct ethnographic research on their "people," but also archeologists and physical anthropologists. A case in the point is the renowned Ardi team which included Donald Johanson and Tim White who discovered numerous early hominid fossils, including "Lucy," in Ethiopia's Afar region. Human origins research has been heavily reliant on philanthropy (Pattison 2020: 197). A member of the Ardi team included Ann Getty, a middle-aged San Francisco socialite who was married to Gordon Getty, an heir to the Getty oil fortune. After enrolling in White's anthropology class at Berkeley, she became so smitten by human fossils that she and her husband became "generous benefactors of human origins research and put their private 727— known as 'the Jetty'—at the disposal of the Ardi team" (Pattison 2020: 197).

Attendance at professional conferences has become an integral component of academic life. Many academics are involved in air travel, a practice that seems to be spreading. Flying to attend conferences and research meetings and to conduct research started quite early among prominent British anthropologists, such as Bronislaw Malinowski, who "used *Imperial Airways* to attend international conferences in Cape Town and Johannesburg, and to visit southern African fieldworkers in the summer of 1934" (Pirie 2012: 100). Undoubtedly, in her later years, Margaret Mead gave anthropology an international profile as a frequent flyer. Academic air travel often increases with seniority, affiliation with elite universities and funding from granting agencies. Parker and Weik (2013: 168) observe:

> Setting aside package tourism to sunny beaches, the elite nomads from the traveling classes then include academics from the elite institutions of the Global North. They have generally travel budgets and something to say at the conferences and symposia that keep chain hotels profitable. This is a mobility that appears to be chosen and is socially valued, not one that is forced and humiliating. It is a mobility that speaks with passport stamps, conversations about different airports and the name-dropping about where you are now and where you are going.

They further argue that while reflexivity is often a valued trait among academics, they often fail to reflect upon the environmental impact of their hypermobility. As anthropologists who have attended and presented at our share of academic conferences, we have witnessed at first-hand the reality that these are events where an academic can present new ideas and insights on his or her own research and even on the state of the world. However, most speakers are only provided with short periods of time in which to speak, perhaps 15–20 minutes—or maybe 30 minutes with a little luck. Probably the main benefit of the conference is the opportunity to network with colleagues, prompting some attendees not to attend many of the presentations, and to spend time in sightseeing and eating at new restaurants. A survey of staff at Aalborg University in Denmark revealed that on average they took two international trips per year, with 22% of their trips being to Scandinavian countries, 56% to other European countries, and 22% to countries outside of Europe (Lassen 2006: 304–305). In a survey of over 300 Australian academics, Andrew Glover (2016) and his research team at the Royal Melbourne Institute of Technology found that the average Australian academic takes 1.7 overseas round-trip flights and three domestic flights per annum for academic purposes.

Anthropologists, at least some of whom are frequent flyers, may therefore be engaging in an activity that the subjects of their ethnographic research have never—or at least seldom— engaged in. Furthermore, given that most anthropologists hail from the Global North and often engage in studies of Indigenous, or peasant and poor urbanites in the Global South, much of their work focuses on people who already have and will continue to be the most adversely impacted by climate change— populations in places such as low-lying islands in the South Pacific, the delta of Bangladesh, the mountainous regions in the Andes and Himalayas, and the semi-arid and arid regions of sub-Saharan Africa and the Middle East (Baer and Singer 2018). Even in countries of the Global North, it is the Indigenous people who are the most adversely impacted by climate change, such as the Inuit and Inupiat in the Arctic, Native Americans in the US Southwest, and Aboriginal Australians residing in remote communities.

Flying on the Part of Other Academics

Many medical researchers are frequent flyers owing to the handsome grants that they tend to receive. At the University of Melbourne, a case in point is Peter Doherty, a 1996 Nobel Prize co-winner for his work in immunology. Over the course of his illustrious career he has been based at the Australian National University, St. Jude Children's Research Hospital in Memphis, Tennessee, and, most recently, the University of Melbourne. Doherty (2021: 3) reports:

Having returned to Australia from the University from the United States to join the Department of Microbiology and Immunology at the University of Melbourne Medical School in mid-2002, I'd continued through the decade with a pattern of direct, though progressively decreasing, involvement in laboratory-based research that initially involved spending nine months each year in Melbourne and three months at St Jude Children's Research Hospital (SJCRH) in Memphis, Tennessee, where I'd relocated from the Australian National University (ANU) in 1988 to head the Department of Immunology. These United States visits from 2002 were generally split into three to four trips a year, with that frequently involving an exhausting round-the-world itinerary (often with my wife, Penny) to give invited lectures in Asia or Europe. That type of travel is part of the unofficial science ambassador/global citizen role that senior Australian scientists fulfil, and in my case at least, all costs were paid from international sources.

In his memoir of the time spent in lockdown during the COVID-19 pandemic, Doherty continues:

Apart from my trips to Memphis and Seattle combined with other lecture and conference obligations, Penny and I were both becoming aware that long-distance international flights were taking an increasing toll on our health and well-being. Even following the precept "Just say no" most of the time, I was still signed up to be at four Northern Hemisphere events in 2020, and was dreading the prospect of most of them. For obvious reasons that didn't happen and, finally connecting with the convenience of Skyping or Zooming across the planet, that frequent-flyer travelling-scientist lifestyle described in my 2018 book, *The Incidental Tourist*, has largely come to an end. If COVID-19 doesn't get me, it may just have saved my life!

(2021: 15)

Ironically, Doherty, who has periodically expressed in various formats his alarm about climate change, probably unwittingly has contributed to the problem with his pattern of academic jet-setting.

Despite the fact that their research systematically records the impact of anthropogenic forces in contributing to climate change, many climate scientists are frequent flyers, perhaps particularly ones serving as leading authors of IPCC climate change periodic assessment reports. Joelle Gergis (2023: 3), a climate scientist based at the Australian National University, in a gripping account of her work, notes that she served as one of the 234 lead authors from 65 countries who worked on the Working Group I's sixth IPCC assessment report. Her first three meetings entailed "long-haul flights to China, Canada and France for incredibly intense five-day meetings" (Gergis 2023: 4). When she attended her third lead IPCC author meeting in Toulouse in southern France in August 2019, Gergis (2023: 142) reports that while "Europe sweltered through one of its hottest summers on record, a burst of late season heat took hold, creating an eerily apt backdrop for our meeting." She is a relatively politically progressive climate scientist who acknowledges capitalism as a generator of economic exploitation and ecological destruction. Indeed, Gergis (2023: 165) observes:

> The world's dominant economic model is capitalism, which rests on the exploitation of the planet's natural resources and the poor for corporate profit, often with scant regard for the collective good or wisdom of First Nations peoples. Since the end of World War II, capitalism has turned humans into consumers and the Earth into a giant quarry to generate wealth for people to live comfortable lifestyles, predominantly in rich nations.

Despite this astute observation, relatively unusual among climate scientists, Gergis makes no mention of airplane flights as one of the drivers of GHG and climate change.

By contrast, Swedish climate activist Greta Thunberg avoided air travel and sailed from Europe to North America in 2019, as part of a month-long climate tour of the Americas. This reflects her commitment to stay grounded and to travel to speak out on climate change by other means. Thunberg, who took a 32-hour train ride from her home to Davos, Switzerland, told delegates at the 2019 World Economic Forum conference: "I think it's very weird that people come here in private jets to discuss climate change and say ... 'Oh we care about this very much' but they obviously don't" (quoted in Kottasova and Macintosh 2019).

Alternative Forms of Flying and Flying Far Less

The growing concern about climate change has prompted discussion about the possible revival of airships powered by a hydrogen-helium mixture or helium, thus circumventing the dangers of disasters such as the explosion of the *Hindenburg*. If perfected, airships could constitute a form of slow travel, given that they travel at speeds of 150–200 kilometers per hour. As passenger transport by ship is not environmentally sustainable, transoceanic travel could make considerable use of wind power. However, within the parameters of existing global capitalism in which "time is money," such slow forms of long-distance travel are not feasible, although they might be within the context of an eco-socialist world system.

In the meantime, while awaiting a global socio-ecological revolution which remains long in the offing—during which time much damage will be done, particularly to many of the peoples many of us as anthropologists have studied and on whose behalf, for better or worse, we have sometimes intervened—what can we do, at both the individual and collective level, to be part of the still-burgeoning climate justice movement, which proclaims that "system change, not climate change," as opposed to the climate action movement, which is largely focused on ecological modernization, exemplified by the adoption of renewable energy sources, energy efficiency, electric vehicles and other techno-fixes—not that some of these are not needed?

Over the past decade or so a campaign has emerged urging academics and their associates, as well as researchers, including climate scientists, to reduce their flying through a number of strategies, including video conferencing and traveling to meetings and conferences by train or coach (Carter 2014). In the introduction to a book on academic flying, Kristian Bjørkdahl and Adrian Santiago Franco Duharte (2022: 2) observe:

> In the face of the climate crisis, however, academics' infatuation with flying may increasingly come across as odd, or even absurd—not to say irresponsible ... How can academics carry on their jet-set lifestyle when, every day the world grows more conscious of the change needed to avert catastrophic global warming? How can academics maintain and even expect their own inclination to fly, when digital alternatives to this "academic tourism" are becoming ever more viable as a means of academic communication—not to mention how these same technologies have been demonstrated their utility during the COVID-19 pandemic?

Fortunately, a small but growing number of academics and research scientists are seeking to grapple in a variety of ways with the contribution made by flying to GHG emissions and climate change. Kevin Anderson (2014), a world-renowned climate scientist based at the Tyndall Centre for Climate Change Research at Reading University in the United Kingdom, no longer flies. He argues that the "[r]eal world has us flying halfway around the world to give banal 20-minute presentations to audiences who know what we're going to say" (Anderson 2014: 11). Anderson, who originally worked in the oil industry, joined the Tyndall Centre because he wanted to work on climate change issues, obviously cognizant of the contribution that fossil fuels have historically made to anthropogenic climate change. He regards "carbon offsetting, the clean development mechanism, and emissions trading as all ruses to maintain the status quo whilst giving the impression of action" (Anderson and Nevins 2016: 211). Anderson advocates a "fly-less" rather than a strict no-fly stance and argues that flying should be restricted for "truly extraordinary and important reasons" (Anderson and Nevins 2016: 214). Anderson and Nevins (2016: 214) view flying as "emblematic of our high-emissions lifestyles and the internationalization agenda of universities—from conferences to students" (Anderson and Nevins 2016: 214). They assert:

> The issue is not flying in and of itself, but the fact that flying is such a carbon-fuel-intensive and emissions-spewing activity. There may come a day when we can substitute high-carbon kerosene for a zero or very low carbon alternative, at which point, from a purely climate change perspective, we can resume flying. So I'm not saying that we have to

eliminate flying forever, but to hugely curtail it until we develop a truly zero carbon aviation sector. But we're very far from that point. Until that day arrives—if it ever does—we have no choice but to radically cut how much we fly.

(Anderson and Nevins 2016: 214–215)

Anderson does not advocate an individualistic approach to flying less but maintains that individual actions may serve as the catalyst for deeper systemic changes that contribute to climate change mitigation (Anderson and Nevins 2016: 216). He took a 20-day return trip by train from the UK to a conference in Shanghai, China, and argues:

Travelling slowly forces us to travel much less, to be much more selective in what events we attend, and to endeavour to get more out of those trips we do take. Fewer trips and potentially long stays: not rocket science—just climate change basics.

(Anderson 2014: 73)

However, an individual may not be able to take 20 days off for one work event. In a capitalist world, time is money, and time is wages, but in an alternative world attuned to social and climate justice with less pressure to be always earning or checking to see if your career is secure, slow travel could be the modus operandi.

This point aside, climate scientist Michael Mann (2021), the world-renowned developer of the "hockey stick hypothesis," takes a swipe not only at flight shame activists but also his colleague Kevin Anderson. He asserts:

While doomism itself might be dismissed as a rather fringe movement, there is some evidence of "seepage" of doomist conspiracy-mongering into the mainstream climate discourse. Consider, for example an exchange between climate experts back in January 2020. It started with Kevin Anderson, a climate scientist who has been critical of the mainstream climate science community for what he perceives to be complacency and a lack of urgency in the face of crisis.

(Mann 2021: 157)

While noting that he flies less than he used to, Mann (2021: 80) asserts: "Flight shaming is a good fit with those who see capitalism itself as the enemy." We suspect that he also would categorize eco-socialists as dividers: people who engage in "class warfare" with global elite polluters, who now claim they seek to take climate action. Is Mann himself, as a privileged academic and former advisor to the Bill Clinton campaign on energy and climate in 2016, perhaps deflecting attention from the considerable contribution of academics to emissions? (see Mkono et al. 2020; Chiambaretto et al. 2021;

Doran et al. 2022). One good point that Mann makes is that corporations have long had an interest in persuading people that it is individual behavior that must change, not the practices or economic system that are behind their mounting profit. BP, a fossil fuel company, he notes, created the first personal carbon footprint calculator. It did not create a corporate carbon footprint calculator, however. The critical question—discussed below—is the role that academics might play in lowing GHG emissions by limiting their flying and participating in the anti-flying movement.

Baer had the opportunity to speak briefly with Mann after he delivered a plenary address entitled "New Climate Reality Check" on February 14, 2020, at the 2020 National Climate Emergency Summit in Melbourne Town Hall, Australia, in which he argued that the new emphasis on climate adaptation rather than climate mitigation "avoids talking about real solutions," and alluded to the need for systemic changes, although he did not spell out what they might be. In his conversation with Baer following the address, Mann admitted that the level of consumption fostered by global capitalism is a problem, but added that this is a difficult issue to address.

The culture of academics is problematic, particularly since it is embedded within an increasingly corporatized structure which places a strong demand on marketing the university as a brand competing with other brands. Giroux (2014: 22) maintains that many universities and colleges "have become unapologetic accomplices to corporate values and power." In a similar vein, Escrigas (2016: 10) maintains that universities have come under increasing pressure to contribute to economic growth to justify the public funding that they receive, and in the process "emphasizing competiveness over collaboration and instrumental over holistic knowledge."

Some academics may assert that refusing to get on a plane to deliver a 20-minute presentation at a far-off conference will not make any difference in the grand scheme of things because the plane is scheduled to take off anyway. However, if a critical mass of academics and other travelers opted to fly less for conferencing, business, or pleasure, airlines would inevitably have to reduce the number of flights that they offer. As Parkinson (2009: 211) so aptly argues, particularly "those who are building careers and reputations on sounding the alarm about the dangers of greenhouse gas emissions should be a great deal more conscientious than many of them seem to be about how much flying they do."

At the individual level, we can join the small but growing movement urging academics to fly less, such as the one mounted by Parke Wilde (2015), an academic at Tufts University in the United States, who has posted a petition urging universities and professional associations to reduce the amount they fly. Some universities have begun to take into consideration the GHG emissions generated by university air travel and have even taken modest steps—generally tokenistic ones—towards mitigating it. While Baer's university proclaims that it is making a "bold commitment to reducing air travel' in the form of 'high quality teleconferencing facilities" (University of Melbourne 2017: 14), he continues to witness numerous colleagues flying hither and thither to conferences, research meetings, consulting meetings, short-term overseas teaching stints and research projects, as well as other academics visiting the campus, sometimes for a few days to deliver a keynote address, or maybe a few weeks to teach an intensive subject, albeit that for a brief period during the COVID-19 pandemic the Australian government restricted air travel.

The commitment of Australian universities to environmental sustainability does not even touch on the matter of overseas students who have become a major source of income now that the Australian government has cut back on funding for public universities. While theoretically overseas students contribute to the cosmopolitanism of any university, whether in a developed or a developing country, sociologist Raewyn Connell (2019: 191) asserts that under the present circumstances the "international market in fee-paying students sucks money out of developing countries to pay universities in richer ones"—another example of unequal economic exchange under the parameters of the capitalist world system.

Given the looming climate crisis, it is imperative that anthropologists, along with other academics, closely reflect upon how they pursue their careers and research. If, for example, anthropologists, as part of their PhD thesis fieldwork, decide to (and can) spend a long-term period overseas (perhaps getting there by slow travel), they should seriously consider not making numerous additional short-term trips overseas, sometimes annually, to the original research site, but confine future research to sites much closer to home. Conversely, if one's PhD thesis was conducted close to home, one might have the option to conduct long-term research in a faraway place perhaps later in one's career.

In terms of conferences, instead of attending international conferences in faraway places, anthropologists should confine their conference attendance to their own countries, or countries that can be easily reached by rail. European anthropologists are much better positioned to travel to conferences in their region because of a dense, international rail network. Perhaps in time North American countries, including the United States, Mexico, and the Central American countries, as well as Australia, could develop a comparable network.

In the case of the United States, rather than focusing on flying to the American Anthropological Association conference, which for many functions as an international conference, there should be a strengthening of regional associations, such as the Northeast Anthropological Society or the Southern Anthropological Society, which hold conferences that can be reached by land. Anthropologists in large urban areas—where a multiplicity of anthropologists are situated in universities, government agencies, nongovernment organizations and other institutions—could organize local conferences, with the option of teleconferencing for distinguished overseas speakers.

However, anthropologists will have to overcome their elitist predilection to avoid local conferences because they are deemed parochial—which does not necessarily have to be the case, particularly if eminent anthropologists from afar are accessible via videoconferencing platforms. Personally, over the years, we have found that such conferences can be very stimulating intellectually and often touch upon the burning issues of the day.

Anthropological postgraduate students often attend conferences, particularly the American Anthropological Association conference, to interview for a position. Such interviews rarely result in being hired, given their rapidity and superficiality. Anthropology departments and anthropological associations need to grapple with strategies by which the interviewing process results in numerous flights, excluding teleconferencing, an approach which already has become more common.

As part of creating a more even playing field in the world, anthropological associations and universities in developed countries can financially support the training of anthropologists from the Global South in their own countries. Anthropologists can both study and provide support to movements to reduce flying and halt the expansion of airports, such as Aviation Justice (United States), Plane Stupid (UK) and the Global Anti-Aerotropolis Movement.

Teleconferencing

While academics often argue that attending conferences provides excellent opportunities to disseminate their research findings, a 20-minute presentation or an hour-long keynote address is arguably not the most efficient means for achieving this goal. Ultimately, journal articles, book chapters, and books are much better and longer-lasting mechanisms for disseminating research findings. As Koch (2012: 118) observes, new communication technologies have "made knowledge diffusion widely independent of an individual scientist's travel activity." Even prior to the emergence of numerous e-conferences around the world as a result of the COVID-19 pandemic, at least until the end of 2021, e-conferences occurred as early as 2010 in New Zealand (Krumdieck 2014). In 2016, the Environmental Humanities Initiative at the University of California, Santa Barbara, United States, convened an online "Climate Change Views from the Humanities" conference (Society for Cultural Anthropology 2018).

In the field of anthropology, a breakthrough in the promotion of environmentally sustainable conferencing occurred in 2018. That year, the Society for Cultural Anthropology's biennial conference on displacements, in collaboration with the Society for Visual Anthropology, was billed as an "international experiment in carbon-conscious conferencing and radically distributed access":

> Air travel is one of the fastest growing sources of greenhouse emissions worldwide, and one of the chief ways that an academic livelihood

contributes to carbon pollution. Our format is also meant to enable broader geographic access and participation, most especially in a political climate of intensified restrictions on international travel. We have set the conference registration fee at a flat $10 with expanded access to anthropological knowledge and dialogue in mind. The conference brings together pre-recorded presentations from anthropologists, filmmakers, artists and activities in 46 countries, with participants tuning in from many other places as well.

(Society for Cultural Anthropology 2018)

The conference entailed 55 local nodes, including 15 in the Global South, which functioned as "decentralized, affinity-based forms of collaboration and exchange" (Society for Cultural Anthropology 2018).

It is difficult to say what will be the long-term effect of teleconferencing given that countries around the world have lifted COVID-19 restrictions on air travel. For the moment, what appears to have emerged are hybrid conferences with some of the speakers presenting online and some of the audience tuning in remotely. Research needs to be done on how to make video conferencing as satisfactory for networking and relationship-building as face-to-face conferencing. Strengers (2015: 602) asserts that

[T]elepresence meetings avoid much of the bodily 'work' that is needed to conduct meetings enabled by air travel, including the arduous and unhealthy toll that air travel can take on the body and the impact of the body's physical absence in family life.

Given the dangers that climate change poses for people around the world, particularly in the Global South, it is imperative that academics begin to grapple seriously with the implications of their flying behavior.

Conclusion

Pinpointing various social categories who engage in frequent flying and—knowingly or unknowingly—a form of environmentally irresponsible behavior, is perhaps a touchy topic among relatively progressive people, such as academics, environmentalists, and climate activists, who at some level are aware that flying results in GHG emissions and that it is a growing and significant driver of anthropogenic climate change. Raising the topic among colleagues, friends, and acquaintances can be an even touchier matter. Airplane flights have become an integral part of doing business, socializing, touring, and holidaying in the modern world and have become a hegemonic aspect of everyday life, one that all too often is ignored, or is acknowledged but put into the "too hard basket."

While individuals may opt not to fly or to reduce their flying—an option that some people have pursued—work and career demands have made aeromobility central to the logic of the capitalist world system and an increasingly corporatized university sector. As a species, we need to move beyond airplanes as much as possible, but such an effort will have to be part and parcel of creating an alternative world system, one that preserves both human life and biodiversity. Perhaps in time, solar or hydrogen-powered flights—beyond the short, experimental ones that presently exist—will become a reality, but such a scenario looms on a distant horizon at best. The global socio-economic, ecological, and climate crises that are the by-products of global capitalism require that we much more quickly re-examine much of what we do in terms of work and leisure, what we eat and consume in general, what sort of dwellings we reside in and how we move around on our planet. A simpler way for the affluent world would entail a disposal or minimizing of the use of airplanes and motor vehicles.

Anthropologists and other social scientists need to contribute in various ways by participating in a growing, but still disparate, climate justice movement, which is particularly strong in the Global South. This differs from the narrower climate movement prevalent in the Global North, which tends to focus on technological solutions, particularly renewable energy sources as the major means for decarbonization, while downplaying social justice issues. Just as anthropology went through an effort to reinvent itself in the late 1960s and early 1970s, it once again needs to reinvent itself by moving beyond current impasses and shift towards becoming a more ethical, socially just, localized, postcolonial, and ecologically sustainable endeavor, while maintaining a global vision that recognizes that all human beings, particularly those living in the Global South—whom many of us continue to study—face the threat of catastrophic climate change if emissions are not drastically reduced over the next few decades.

References

Anderson, Kevin. 2014. Slow and low—the way to go: A systems view of emission. In *Beyond Flying*. Chris Watson, ed. pp. 66–81. Cambridge: UIT Cambridge

Anderson, Kevin and Nevins, Joseph. 2016. Planting seeds so something bigger might emerge: The Paris Agreement and the fight against climate change. *Socialism and Democracy* 30(2): 209–218

Baer, Hans A. 2020. *Airplanes, the Environment, and the Human Condition*. London: Routledge

Baer, Hans A. and Singer, Merrill. 2018. *The Anthropology of Climate Change: A Critical Integrated Perspective* (2nd edn). London: Routledge

Bauman, Zygmunt. 1998. *Globalization: The Human Consequences*. Cambridge: Polity Press

Berners-Lee, Mike. 2010. *How Bad Are Bananas? The Carbon Footprint of Everything*. London: Profile Books

BlueSkyModel. n.d. 1 air mile. https://blueskymodel.org/air-mile?&_ga=2.248671175. 1503921652.1700592510-1045964116.1700592510#:~:text=To%20the%20best%20of %20our,short%20tons%20of%20carbon%20dioxide

Bjørkdahl, Kristian and Duharte, Adrian Santiago Franco. 2022. Introduction: Ending the romance of academic flying. In *Academic Flying and the Means of Communication.* Kristian Bjørkdahl and Adrian Santiago Franco Duharte, eds. pp. 1–18. Singapore: Palgrave Macmillan

Bows, Alice. 2009. *Aviation and Climate Change: Lessons for European Policy.* New York: Routledge

Bridger, Rose. 2013. *Plane Truth: Aviation's Real Impact on People and the Environment.* London: Pluto Press

British Broadcasting Corporation. 2016. *City in the Sky* [TV documentary]. https:// www.imdb.com/title/tt5820022/

Carter, Chris, ed. 2014. *Beyond Air Travel: Rethinking Air Travel in a Globally Connected World.* Cambridge: UIT Cambridge

Chagnon, Napoleon. 2013. *Noble Savages: My Life among Two Dangerous Tribes— the Yanomamo and the Anthropologists.* New York: Simon & Schuster

Chiambarreto, Paul, Mayenc, Elodie, Chappert, Hervé, Engsig, Juliane, Fernandez, Anne-Sophie, and Le Roy, Frédérick. 2021. Where does flygskam come from? The role of citizens' lack of knowledge of the environmental impact of air transport in explaining the development of flight shame. *Journal of Air Transport Management* 93: 102–149

Clark, Duncan. 2009. *The Rough Guide to Green Living.* London: Rough Guides

Connell, Raewyn. 2019. *The Good University: What Universities Actually Do and Why It's Time for Radical Change.* Melbourne: Monash University Press

Doherty, Peter. 2021. *An Insider's Plague Year.* Melbourne: Melbourne University Press

Doran, Rouven, Pallesen, Stale, Böhm, Gisela, and Ogunbode, Charles A. 2022. When and why do people experience flight shame? *Annals of Tourism Research* 92: 103254

Escrigas. C. 2016. A higher calling for higher education. *Great Transitions Initiative,* June. https://www.greattransition.org

Gergis, Joelle. 2023. *Humanity's Moment: A Climate Scientist's Case for Hope.* Melbourne: Black, Inc.

Giroux, Henri. 2014. *Liberalism's War on Higher Education.* Chicago: Haymarket Books

Glover, Andrew. 2016. *Opportunity and obligation: Air travel practices of Australian academics.* Australian Sociological Association conference, Australian Catholic University, Melbourne campus, 1–4 December

Gössling, Stefan, Ceron, Jean-Paul, Dubois, Ghislain, and Hall, Michael C. 2009. Hypermobile travelers. In *Climate Change and Aviation: Issues, Challenges and Solutions.* Stefan Goessling and Paul Upham, eds. pp. 131–150. London: Earthscan

Goodal, Chris. 2010. *How to Live a Low-Carbon Life: The Individual's Guide to Stopping Climate Change.* London: Earthscan,

Gough, Kathlen. 1968. Anthropology and imperialism. *Monthly Review* April: 12–27

Grimalda, G. 2023. Refusing to fly has lost me my job as a climate researcher: It's a price worth paying. *The Guardian.* https://www.theguardian.com/commentisfree/ 2023/oct/12/fly-climate-breakdown-germany-climate-change-papua-new-guinea

Intergovernmental Panel on Climate Change (IPCC). 1999. Aviation and the global atmosphere. http://www.ipcc.ch/index.htm

Intergovernmental Panel on Climate Change (IPCC). 2007. Transport and its infrastructure. https://archive.ipcc.ch/publications_and_data/ar4/wg3/en/ch5.html

International Air Transport Association. 2022. Air passenger numbers to recover by 2024. March 1. https://www.iata.org/en/pressroom/2022-releases/2022-03-01-01

Koch, Max. 2012. *Capitalism and Climate Change: Theoretical Discussion, Historical Development and Policy Responses.* New York: Palgrave Macmillan

Kottasova, I. and Mackintosh, E. 2019. Teen activist tells Davos elite they're to blame for climate crisis. CNN, January 25. https://edition.cnn.com/2019/01/25/europe/greta -thunberg-davos-world-economic-forum-intl/index.html

Krumdieck, Susan. 2014. The no-flying conference: Signs of change. In *Beyond Flying: Rethinking Air Travel in a Globally Connected World.* Chis Watson, ed. pp. 114–121. Cambridge: Cambridge UIT

Lassen, Claus. 2006. Aeromobility and work. *Environment and Planning A* 38(2): 301–312

Mann, Michael. 2021. *The New Climate War: The Fight to Take Back the Planet.* Melbourne: Scribe

McBurney, Stuart. 1990. *Ecology into Economics Won't Go, or Life is Not a Concept.* Cambridge: UIT Cambridge

Millman, O. 2022. A 17-minute flight? The super-rich who have "absolute disregard for the planet." *The Guardian.* https://www.theguardian.com/environment/2022/jul/ 21/kylie-jenner-short-private-jet-flights-super-rich-climate-crisis

Mkono, Mucha, Hughes, Karen, and Echentille, Stella. 2020. Hero or villain? Responses to Greta Thunberg's activism and the implications for travel and tour- ism. *Journal of Sustainable Tourism* 28: 2081–2098

Parker, Martin and Weik, Elkie. 2013. Free spirits: The academic on the aeroplane. *Management Learning* 45: 167–181

Parkinson, C. 2009. *The Coming Climate Crisis: Consider the Past, Beware the Big Fix.* Lanham, MD: Rowman & Littlefield

Pattison, Kermit. 2020. *Fossil Men: The Quest for the Oldest Skelton and the Origins of Humankind.* New York: William Morrow

Pirie, Gordon. 2012. *Cultures and Caricatures of British Imperial Aviation: Passengers, Pilots, Publicity.* Manchester: Manchester University Press

Pitron, Guilaume. 2023. *The Dark Cloud: How the Digital World Is Costing the Earth.* Bianca Jacobsohn, trans. Melbourne: Scribe

Society for Cultural Anthropology. 2018. *Displacements: The Biennial Meeting of the Society for Cultural Anthropology.* https://www.displacements.jhu.edu

Strengers, Yolande. 2015. Meeting in the global workplace: Air travel, telepresence and the body. *Mobilities* 10: 592–608

Tickell, Oliver. 2008. *Kyoto2: How to Manage the Global Greenhouse.* London: Zed Books

University of Melbourne. 2017. *Sustainability Plan 2017–2020.* https://www.unimelb.edu.au .

Watson, Chris. 2014. Introduction. In *Beyond Flying.* Chris Watson, ed. pp. 16–21. Cambridge: Cambridge UIT

Wilde, Parke. 2015. Calling for universities and professional associations to greatly reduce flying. https://www.aireform.com/wp-content/uploads/2015102cpy .

9
TWO GENRES OF THE CLIMATE MOVEMENT

Climate Action vs. Climate Justice

Case Study

Recognizing that they have been born into a world facing climate turmoil, youth have been key players and leaders of public discourse on climate change. One of them is Leah Namugerwa, born in 2004, a climate activist who gained attention by leading a tree-planting campaign—notably she celebrated her 15th birthday by planting 200 trees instead of having a birthday party—and starting a petition to enforce the plastic bag ban in her home country of Uganda in East Africa. Her Birthday Tree project provides seedlings to youth who celebrate their birthdays by planting trees. Inspired by the efforts of the Swedish environmental activist Greta Thunberg, and motivated by watching local news reports about mudslides and flooding in rural sections of her country, she joined the school strikes in 2019 under the slogan "Fridays for Future Uganda." Namugerwa has lobbied the Ugandan government to fully implement the Paris Agreement, or the Paris Climate Accords, to cut carbon dioxide emissions. She asserts, "Most people do not care what they do to the environment … I noticed adults were not willing to offer leadership and I chose to volunteer myself. Environmental injustice is injustice to me" (quoted in Hope 2022). Namugerwa was a youth delegate at the United Nations (UN) Climate Change Conference (COP25) in 2019 and spoke at the World Urban Forum, a conference on pressing urban issues run by the UN Human Settlements Programme in 2020. In 2023, Namugerwa was selected to serve on the UN Secretary-General's Youth Advisory Board, designed to provide the UN with practical and outcome-focused advice, diverse youth perspectives, and concrete recommendations on accelerating the Secretary-General's climate action agenda. She proclaims, "I want to raise a generation that cares about the environment … At least if the leaders can't make a

DOI: 10.4324/9781003469940-10

difference, we can make a difference. We, as kids, we're not too young to make a positive difference" (quoted Crowe 2019).

Climate Movements

Given that it can be said that corporations and most governments along with the UN have not been acting in a responsible and effective manner in terms of serious climate change mitigation, despite considerable rhetoric to the contrary, much of the collective effort will have to be spurred by anti-systemic movements, including a burgeoning international climate movement that is quite variable in terms of addressing social justice or equity issues. The climate movement, both internationally and nationally, is a broad and disparate phenomenon that draws in part on earlier movements, particularly the environmental movement but also the anti-corporate globalization movement and even the labor movement. Countries with the largest climate movements include the United States, Germany, the United Kingdom, Canada, Australia, New Zealand, South Africa, Brazil, and India (Dietz and Garrelts 2014). But smaller movements can be found across the Global South. Manuel Castells (2012: 325) observes: "The internet has played an increasingly important role in the global movement to prevent global warming" (see also Chen et al. 2023).

The global environment movement that emerged in the late 1960s and early 1970s has evolved in many directions, ranging from the emergence of mainstream environmental organizations to radical groups and environmentalists of different sorts, including deep ecologists, eco-socialists, and eco-anarchists. Over the past two decade or so, many environmental groups and Green parties around the world have in one way or another become involved in the climate movement as have many other political actors, including grassroots climate action groups, socialist political parties and groups, and labor unions. Charles Derber (2010: 205) maintains that in order to address the climate crisis, social movements must acknowledge the existential reality of climate change, struggle for regime change within nation-states and internationally, allow the labor movement to take the lead, integrate personal and systemic changes, and cooperate with heads of state while pushing them to adopt a more radical stance on the systemic problems at hand. He argues that the "labor movement has begun to advocate strongly for green economic solutions to the current crisis, as well as to remake the economy and ensure full employment for the long term" (Derber 2010: 209). Unfortunately, at the present time and for the foreseeable future labor appears to be too weak to take the lead in the climate movement (Reed 2023).

However, many environmentalists and climate activists do not want put labor in a position of leadership because many labor union leaders are committed to providing their members with a high material standard of living with little thought of the environmental consequences associated with it. In the United States, many coalminers, for example, voted for Donald Trump, a climate change denier, for president because he promised (falsely as it turned out) to bring back coal production and coal-related jobs. Conversely, environmental groups all too often have been willing to cooperate with corporate interests, even the more progressive ones. As James Anderson (2006: 252) observes,

> National and international NGOs [nongovernmental organizations], such as Friends of the Earth, Greenpeace, Oxfam, and other international aid and development organizations, play a crucial role in mobilizing scientific expertise, and some are important players in anti-globalization. But despite this progressive role, they generally occupy an ambiguous and often highly constrained position in the power game and hence are vulnerable to incorporation and ideological subversion by established powers.

Tendencies in the International Climate Movement

As depicted in Table 9.1, the climate movement both at the international level and within specific countries is quite divided, with some segments seeking to work within capitalist parameters and others challenging capitalism to a greater or lesser degree.

Eve Croeser (2020: 118) juxtaposes differences between how climate action activists and how climate justice activists seek to "connect the dots":

> The 2012 climate action movement's 350.org-initiated campaign aimed to "connect the dots" between climate change and the increasing number of

TABLE 9.1 Tendencies in the International Climate Movement

Climate Action Tendency	Climate Justice Tendency	In Between Tendency
Tends to be more pronounced in the Global North	Radical and anti-capitalist	Recognizes social justice issues but is not explicitly anti-capitalist
Seeks to regulate capitalism	Contains vocal eco-socialist and eco-anarchist elements	
Places strong emphasis on ecological modernization	May be most pronounced in the Global South	
Sympathetic to green capitalism	Contains pockets in the Global North	
Focuses on lobbying politicians and electing ones favorable to climate action	Views ecological modernization as important but is insufficient to mitigate climate change	
Muted on social equity and social justice issues	Calls for the transcendence of global capitalism with an alternative social system committed to social parity and justice and environmental sustainability	

"extreme weather events" around the world, thereby narrowly focus on the physical and technical aspects of GHG [greenhouse gas] emissions and the scientific and technical aspects of their physical effects. Climate justice activists argue that a genuine understanding of the global warming crisis entails "connecting the dots" between the scientific evidence of anthropogenic climate change and other anthropogenic threats on the Earth's biosphere; the social, economic and political systems created by humans (and therefore subject to being changed by humans) causing the current crisis; and the ethical implications of different courses of action.

The climate movement both at the international level and within specific countries is quite fragmented, with some of its segments, such as the Climate Action Network, Greenpeace, and Al Gore's Climate Reality Project, trying to work within the parameters of global capitalism, and others, such as Friends of the Earth, Rising Tide, Climate Justice Action, and various socialist and anarchist groups, challenging it and even calling for efforts to transcend it, thus prompting the mantra, "System change, not climate change."

While the climate movement has tended to be a secular phenomenon, some faith-based people, including proponents of eco-theology, have become part of it. Nita (2016: 28) observes that "some national Christian grassroots networks preoccupied with the environment and climate change have developed, particularly in the Anglophone world, the UK, the USA, Canada and Australia." Faith-based concern about climate change was boosted by Pope Francis's 2015 Encyclical Letter entitled "On Care for Our Common Home," *Laudato Si*, in which Francis spoke forcefully about the gravity of climate change: "Climate change is a global problem with grave implications: environmental, social, economic, political and for the distribution of goods. It represents one of the principal challenges facing humanity in our day."

Unfortunately, based on an analysis of Gallup annual surveys on the environment, Konisky (2018) found that US Christians, whether Catholic, Protestant, or members of other denominations, tend to exhibit less concern about climate change than the broader US population.

The Climate Action Movement

Many climate action groups in North America, Europe, and Australia tend to focus on ecological modernization as their primary climate change mitigation strategy, thus either ignoring or downplaying social justice issues. The international climate action movement appears to have started around 1989 with the formation of the Climate Action Network (CAN), which by mid-2008 had grown into an alliance of some 430 NGOs (Camilleri and Falk 2010: 309). CAN lobbied heavily in support of the 1997 Kyoto Protocol and supported emissions trading schemes or their passage, such in Australia. Some of its leaders have had connections with carbon-trading companies (Bond 2009: 215).

350.org

Perhaps the largest contingent of the climate action movement is 350.org, an organization spearheaded by Bill McKibben (1989), a renowned US environmentalist whose book *The End to Nature* (one of the 20 books he has written) propelled him to fame as a significant environmental writer, in large part because it was the first book about climate change targeted at a general audience. He has advocated carbon trading and until 2012 was open to nuclear energy as a purported climate change mitigation strategy. McKibben is a staunch advocate of renewable energy and believes that people need to shift away from meat consumption, develop public transport networks to replace automobiles, densify cities, and engage in regenerative farming to restore carbon to the soil. He supports labor unions, voting rights, social safety nets, possibly state banks, and public-owned utility companies (McKibben 2019).

In 2010, McKibben and others formed 350.org in an effort to create a global climate movement aimed at returning the planet to an atmospheric carbon dioxide concentration of 350 parts per million in contrast to the 390 parts per million recorded at the time. 350.org utilizes demonstrations, marches, and other forms of protest, such as locking onto infrastructure as forms of climate action. It also relies heavily on local volunteers using digital media and online platforms (Gunningham 2019). In 2011 and 2012, McKibben led a massive environmental campaign against the proposed Keystone XL pipeline project. He likes to serve as a catalyst for jumpstarting climate action campaigns, leaving the groundwork to "young, diverse, and remarkable organizers around the world" (McKibben 2019: 222).

350.org served as the catalyst for an international divestment movement from fossil fuels which is based on the ethical argument that it is morally wrong to benefit from the climate crisis. Unfortunately, the divestment movement has not fully challenged the economic growth paradigm and has diverted attention away from more radical actions by focusing on the management of stock portfolios. Whereas the first phase of the climate movement lacked a leader or figurehead, Malm (2021: 33) asserts that McKibben fulfilled this role during its second phase. He argues that McKibben manifested an ethic of moral pacifism which prompts him to fret "about cracks in discipline that might allow 'adventurers' to spoil the movement" (Malm 2021: 35). Gunningham (2019) suggests that 350.org's impact may have peaked and that it has been replaced by other climate action groups, particularly School Strike 4 Climate and Extinction Rebellion.

School Strike 4 Climate

At the age of 15, Greta Thunberg single-handedly staged the first school climate strike in front of the Swedish Parliament in Stockholm in August 2018, an event that sparked an international phenomenon and provided a much-needed catalyst for the flagging international climate movement spread across northern Europe, North America, and Australia. She carried a sign that read *Skolstejk foer klimatet* (School Strike for Climate). As a result, a petite teenage girl who had been diagnosed as being autistic was propelled to international fame and became not only a frequent speaker at climate rallies around the world, but also at mainstream international conferences, including the World Economic Forum in Davos, Switzerland, in both January 2019 and January 2020, as well as the European Parliament in April 2019 and the US Congress on September 23, 2019 (Thunberg 2019). In her speech at the 24th COP meeting in Katowice, Poland, on December 15, 2018, she boldly declared:

> You only speak of green, eternal growth because you are too scared of being unpopular … We are about to sacrifice our civilization for the opportunity of a very small number of people to make enormous amounts of money. We are about to sacrifice the biosphere so that rich people in countries like mine can live in luxury. But it is the suffering of the many which pay for the luxuries of the few.
>
> *(Thunberg 2019: 14–15)*

The School Strike 4 Climate led to a global strike on March 19, 2019, in which over one million school-age strikers rallied in some 2,200 actions in 125 countries. Another global school strike occurred on May 24, 2019, timed to coincide with the 2019 European Parliament election. The September 2019 climate change strikes, billed as the Global Week for Future, occurred around the world for three days before the UN COP meeting in New York City. More school strikes ensued on September 27, 2019. European youth engaged in "Fridays 4 Future" (F4F) rallies in which thousands of students skipped school every Friday to protest for climate action. Public reaction to the rallies were mixed, varying from strong approval and support to outright condemnation of young students for being absent from school and for their alleged failure to be obedient to authority figures (Misch et al. 2021). Conservative German newspapers, such as the *Frankfurter Allgemeine Zeitung* and the *Tageszeitung*, opted to "align with the exclusive hegemony of an established environmental governance regime," such as the Commission for Growth, Structural Change and Employment (the coal commission) assigned in 2018 to advise Germany to phase out coal production (Bergmann and Ossewarde 2020: 267). Conservative politicians around the world viciously asserted that children should remain in school and learn rather than engaging in climate change rallies. Even Angela Merkel, who is often depicted as a more enlightened conservative politician, "hinted that children like Greta should not themselves organize such a large strike movement—there may be some dark force pulling the strings behind them, such as Putin" (Žižek 2020: 83). In their survey of 19 political parties in Austria, Germany, and Sweden, Berker and Pollex (2023: 768) report three distinct responses to the Fridays 4

Future movement: "Left parties uniformly support F4F, PRRP [populist radical right parties] fiercely oppose it and centre-right parties are divided— either supporting F4F or showing an at best cautious attitude by ignoring the movement." In contrast to Extinction Rebellion rallies, Fridays 4 Future rallies tend to be moderate. In the United States, Jane Fonda (2020), the Hollywood actor who had been active in the anti-Vietnam War movement in the late 1960s, became part of the "Fire Drill Fridays" rallies and teach-teams during 2019 in Washington, DC, teaming up with Greenpeace, Bill McKibben, and author Naomi Klein at these events.

While by and large the Fridays 4 Future movement has adopted a moderate stance on climate action, often beseeching climate policymakers to take heed of the findings of climate science, Thunberg at times has adopted explicitly radical rhetoric as exemplified below:

> Corporate elite are destroying the planet for profits and to give rich people a better lifestyle. Politicians and media let this exploitation happen without acting and therefore are responsible. The adults are responsible through lifestyles and everyday life consumption. They must make choices in the name of the youth and a future generation that cannot vote.
>
> The youth, future generations, ecosystems, and other victims are hailed as the true people, who will rise up against the system, demanding a revolution.
>
> *(Quoted in Nordensvard and Ketola 2022: 870)*

She adds: "Leaving capitalist consumerism and market economics as the dominant stewards of the only known civilization in the universe will most likely seem, in retrospect, to have been a terrible idea" (Thunberg 2023).

The deeper question is why have elites in various forums, such as the COP meetings, the World Economic Forum, the UK Parliament, and the US Congress have allowed Thunberg to speak and scold her for her failure to take deeper climate actions given that other prominent climate activists have not been provided airtime of the same sort. Was it an elite strategy to appease young people around the world who will inherit the worst impacts of the climate crisis? Perhaps part of the reason lies in Thunberg's tendency to avoid addressing the climate crisis as an explicitly left/right issue, despite her condemnation of capitalism, corporations, banks, governments, and other social institutions. Only time will tell whether in the future elites will allow her to address them in the bold manner that she has in the past.

Extinction Rebellion

Extinction Rebellion (XR) was launched by Gail Bradbrook, a coal miner's daughter, Simon Bramwell, a proponent of regenerative culture, and Roger Hallam, an organic farmer and proponent of civil disobedience. XR and climate activists affiliated with Rising Up! rallied in London in October 2018 when they engaged in various acts of civil disobedience shortly afterwards. In April 2019, an estimated 6,000 people blocked five major bridges across the River Thames, supergluing themselves to commuter trains and buildings such as the Stock Exchange and planting trees in Parliament Square (Gunningham 2019. In its "Declaration of Rebellion," XR proclaims:

> The science is clear: we are in the sixth mass extinction event and we will face catastrophe if we do not act swiftly and robustly. Biodiversity is being annihilated around the world … The breakdown of the climate has begun. There will be more wildfires, unpredictable super-storms, increasing famine and untold drought as food supplies and freshwater disappear.
>
> *(Extinction Rebellion 2019: 1)*

XR made three demands on the UK government and soon afterwards on other governments around the world, namely to "[t]ell the truth and act as if that truth is real on the climate and ecological emergency" and given the prospect of extinction due to the ecological crisis to "take radical collective action, which means engaging in a rebellion against the government" (Hallam 2019: 19–20). XR uses a circle hourglass as an extinction symbol to warn that time is running out for many species on the planet. Hallam argues that 3.5% of the population in rebellion is needed to orchestra a climate revolution. He drew the ire of many, including XR members, however, for having down-played the gravity of the Holocaust. XR claims that it entered its international phase in April 2019. Sam Knights (2019: 10) reports:

> In Pakistan, we marched through the capital. In the US, we glued ourselves to a bank. In the Netherlands we occupied The Hague. In Austria, we blocked roads. In Chile, we lay down in the middle of the street. In Ghana, we blew whistles to sound the climate alarm.

XR established some 3,130 branch groups across the UK and quickly spread to 45 countries with about 650 local groups around the world. It has proposed the creation of Citizens' Assemblies on climate and ecological justice in the various countries where it is based. In the case of the UK, XR envisions the Citizens' Assembly as a forum for deciding which climate policies need to be promoted, in essence functioning as a "jury-like structure, chosen by a lot to get a cross-section of society" with the Parliament remaining in place but in essence playing an "advisory role to the Citizens' Assembly" (Knights 2019: 112). XR asserts that some 30 years of climate activism thus far has achieved little in terms of getting governments and corporations to act in a meaningful manner on climate change. While Malm (2021: 126) concedes that XR has achieved some considerable achievements, he argues that it has "remained persistently aloof from the factor of class and race, remaining based in a white middling strata" and has refrained from anti-capitalist rhetoric, in contrast to climate justice activists.

The Climate Justice Movement

The international climate justice movement has existed since the late 1990s. The first reference to the notion of climate justice appeared in a report entitled *Climate Gangsters vs. Climate Justice* published in 1999 by the Corporate Watch group in San Francisco, United States (Bruno et al. 1999). Although the report primarily targeted the oil industry, it outlined an approach to climate justice that includes the following premises:

- Holding corporations accountable for the root causes of global warming;
- Opposition to the distinctive impacts of oil development and support for communities most adversely affected by weather-related disasters;
- Drawing on the struggles of environmental justice communities and organized labor to encourage a just transition away from fossil fuels;
- Confronting corporate globalization and the powerful influence of international institutions in the struggle for climate justice.

(Tokar 2018: 16)

At COP4 in Buenos Aires, Argentina, in 1998, the Climate Action Network, the principal climate action organization, "drew media attention to its 'Fossil of the Day Award' and organized seminars on justice and equity with regard to emissions of greenhouse gases" (Pattberg and Stripple 2010: 137). The climate justice movement emerged in full force in 2000 at the Climate Justice Summit that convened outside the COP6 meeting held in The Hague, Netherlands, when it declared that the "causes of climate change are the production and consumption patterns in industrialised countries." At the Climate Justice Summit held in October 2002 in New Delhi the India Climate Justice Forum stated:

> We, the representatives of the poor and marginalised of world, representing fishworkers, farmers, Indigenous Peoples, Dalits, the poor and the youth resolve to actively build a movement from … a human rights, social justice and labour perspective… We reject the market-based principles that guide the current negotiations to solve the climate crisis: Our World is Not for Sale!

In 2007, at the COP13 meeting in Bali, Indonesia, representatives of climate change-impacted communities unified under the slogan "Climate Justice Now!" They called for more radical actions on climate change, including the following:

- Leaving fossils in the ground and investing in appropriate energy efficiency and clean and community-led renewable energy projects;
- Drastically reducing wasteful consumption, particularly in the Global North, but also on the part of elites in the Global South;

- Making large financial transfers from the Global North to the Global South based on the repayment of climate debts;
- Implementing rights-based conservation that "enforces Indigenous land rights and promotes people's sovereignty over energy, forests, land, and water,"
- The development of sustainable family farming and food sovereignty.

(Tokar 2018: 16–17)

The ensuing Climate Justice Network formed in 2007, consisting of some 750 organizations in both the Global North and the Global South. These included the Rainforest Action Network (established in 1985), the Indigenous Environmental Network, the Citizens Climate Lobby, Rising Tide, and the Climate Reality Project (Spears 2019: 201–202). Mobilization for Climate Justice is a North American-based network of organizations and individuals that espouse nonviolent direct action and public education in order to counteract climate change (Tokar 2018). It included the Indigenous Environmental Network, the Global Justice Project, and Rising Tide. Various other Indigenous groups have become involved in the climate justice movement, including the Inuit Circumpolar Council.

In April 2010, Evo Morales, the president of Bolivia, convened the World People's Conference on Climate Change and Mother Earth in Cochabamba. Over 35,000 people from 142 countries attended the conference at which Morales asserted that "either capitalism dies or Mother Earth dies." As the passages delivered on April 22, 2010 below indicate, the rhetoric of the conference was radical and anti-systemic:

> Today, our Mother Earth is wounded and the future of humanity is in danger. Under capitalism, Mother Earth is converted into a source of raw materials, and human beings into consumers and a means of production, into people that are seen as valuable only for what they own, and not for what they are.
>
> It is imperative that we forge a new system that restores harmony with nature and among human beings. And, in order for there to be balance with nature, there must first be equity among human beings. We propose to the peoples of the world the recovery, revalorization, and strengthening of the knowledge, wisdom, and ancestral practices of Indigenous peoples, which are affirmed in the thought and practices of "Living Well," recognizing Mother Earth as a living being with which we have an indivisible, interdependent, complementary, and spiritual relationship. To face climate change, we must recognize Mother Earth as the source of life and forge a new system based on the principles of:
>
> - Harmony and balance among all and with all things: complementarity, solidarity, and equality;
> - Collective well-being and the satisfaction of the basic necessities of all: people in harmony with nature:

- Recognition of human beings for what they are, not what they own;
- Elimination of all forms of colonialism, imperialism and inter-ventionism: peace among the peoples and with Mother Earth.
(World People's Conference on Climate Change and Mother Earth 2010)

The People's Agreement drafted at the conference called on countries in the Global North to take the lead in returning the atmosphere's carbon dioxide levels to 300 parts per million, thereby limiting the increase in the average global temperatures to a maximum of 1 degree Celsius (1.8 °F). It also called for the creation of an International Climate and Environmental Justice Tribunal with the legal capacity to judge states, industries, and people with regard to their contribution, either though commission or omission, to climate change. Unfortunately, since the World People's Conference, carbon dioxide emissions have risen to 422 parts per million, the average global temperature has risen to 1.2 °C (2.16 °F) since 1900, and the proposed tribunal has not been established.

Adopting the Quechua expression *sumac kawsay* (living well), the Bolivian Platform against Climate Change highlights how climate change adversely impacts the right of Indigenous people to conserve their livelihoods and culture (Lazaro et al. 2023: 130). In Peru, Asociacion Econsistemas Andinos "works with 21 communities in the Andean highlands outside Cusco to conserve the endangered Polylepis forests directly affected by climate change" (Lazaro et al. 2023: 130).

In light of the disappointing outcomes of COP15 in Copenhagen in 2009—and the feebleness of the climate change deal agreed there—many climate activists, particularly climate justice activists, struggled with how effective their efforts at COP meetings had been in terms of pushing for meaningful climate action. Based on observations in both online and offline meetings leading up to and following the 2015 COP in Paris, de Moor (2018: 1084) argues that in contrast to the Copenhagen mobilization, the "Paris mobilization stood out for its degree of cooperation across the movement's political and strategic spectrum." The two broad contingents at Paris were Coalition Climat 21 and Climate Justice Now! The former not only primarily represented the traditional moderate groups such as CAN and Avass, but also some of the more radical groups such as Attac and La Via Campesina. It closely collaborated with Climate Justice Now! and Climate Justice Action, which included other radical groups such Reclaim the Power (UK) and Ende Gelände (Germany).

Dana Risher and Quinn Renaghan (2023) refer to the climate justice campaigners as forming part of the growing "radical flank" of the climate movement, reflecting a type of splintering common in social movements when disagreement emerges over tactics, targets, and timetables for change. These radical activists have taken to the streets in protest, blocking traffic, marching slowly (again to block traffic), smearing paint (e.g., on the casing of the statue of Degas in the National Gallery of Art in Washington, DC), using chairs to block bank doors, throwing food, disrupting sports and social events, and other forms of nonviolent civil disobedience to bring public and policymaker attention to the severity of the climate crisis and the need for more aggressive, faster-paced climate action. Based on a survey of 60 such activists, the majority of whom were predominantly female, Risher and Renaghan (2023) found that their primary motivations were climate change (83%), racial justice (58%), and income and wealth inequality (46%). The top four organizational affiliations of their participants were XR, Scientist Rebellion, Declare Emergency and Third Act. XR, as noted above, has been engaged in coordinating civil disobedience to disrupt business-as-usual behavior and raise awareness about the climate crisis since it began in the UK in 2018. Inspired by XR, Scientist Rebellion was organized in 2020 and its members gained attention by wearing white laboratory coats during protests. In December 2022, two members, Peter Kalmus and Rose Abramoff, disrupted the American Geophysical Union's annual meeting in Chicago by unfurling a banner on stage that read "Out of the labs and into the streets" during a lunchtime plenary. Declare Emergency is known for its action at the National Gallery of Art, while Third Act has organized seniors to sit on hand-painted rocking chairs in front of bank branches.

Despite internal differences, climate activists of varying political orientations tend to converge on specific campaigns, such blockades against the Northern Gateway Pipeline in North America, the construction of the Keystone Pipeline (which was slated to run from the oil tar sands in northern Alberta to the Gulf of Mexico until President Barack Obama blocked its completion), and what was reportedly the world's largest climate change rally attended by an estimated half-million people from some 166 countries on September 21, 2014 in New York, two days before the UN Climate Summit there. Giacomini and Turner (2015: 30) observe that two broad tendencies were represented at the New York rally, namely the supporters of green capitalism who view climate markets as being essential to mitigating climate change and the defenders of life-affirming solar communing who "seek an end to capitalist relations (meaning the demise of corporate value chains)" and an elaboration of "commoners' value chains (better described as a web or matrix)."

Climate justice activists have mounted numerous campaigns on a wide array of issues. For example, some campaigns have targeted the Clean Development Mechanism, a voluntary carbon market, and various carbon offsetting schemes, arguing that these practices constitute a form of carbon colonialism, or a "means by which rich consumers in the West merely displace their high-carbon consuming practices by buying offset for emissions cheaply from the South" (Newell and Patterson 2010: 32).

The Climate Movement in the Global South

In contrast to the climate movement in the Global North, particularly North America, the UK, Germany, and Australia, other than South Africa (Bond 2012), in-depth examinations of the climate movement in the Global South are few. Below, we provide vignettes of the climate movement in five countries, namely Brazil, India, the People's Republic of China, Kiribati, and sub-Saharan Africa, drawing on some of the case studies presented in *The Routledge Handbook of the Climate Movement* edited by Matthias Dietz and Heiko Garrelts (2014).

Brazil

The following vignette is based upon research conducted by Segebart and Koenig (2014):

- Large sections of the Brazilian climate movement are professionalized and diverse, including campaigns targeting deforestation in the Amazon Basin and the vulnerability of large cities to climate change.
- The Environmental Justice Network in Brazil consists of some 1,000 NGOs, scientific organizations, and grassroots groups which are vocal about climate change issues.
- All actors in the Brazilian climate movement in one way or other are committed to joining climate justice and social justice issues, particularly a segment that expresses a strong critique of capitalism and market mechanisms designed to mitigate climate change.

China

The following vignette is based on research conducted by Schroeder (2014):

- The Chinese climate movement is in large part embedded in civil society organizations (CSOs) which work on a variety of environmental issues.
- CSOs initiated a climate change campaign termed "C+ Actions – Beyond Government Commitment, Beyond Climate, Beyond China."
- CSOs have sought to form links with the international and climate movements.

Kiribati

The following vignette is based upon research conducted by Klepp (2014):

- Kiribati, located in the Micronesia in the central Pacific Ocean, is a central player in the Alliance of Small Island States which has been a vocal voice in UN Framework Convention on Climate Change negotiations.

- Various environmental NGOs, such as Friends of the Earth, have initiated climate justice campaigns in the South Pacific, including in Kiribati.
- Kiribati has served as a focal point in the global media as an iconic example of "sinking island states."

Sub-Saharan Africa

The following vignette is based upon research conducted by Bond (2014):

- The international justice movement emerged in Africa during the early 2000s.
- The African climate justice movement has strongly targeted emissions trading schemes, particularly the Climate Development Mechanism and Reducing Emissions from Deforest Degradation; land grabs for various purposes, including dam construction and minerals extraction which power coal-generated and oil-generated electricity.
- The African climate justice movement highlights the notion of climate debt that the polluting Global North and South Africa need to repay to Africans.

India

The following vignette is based upon the research conducted by Harms and Powalla (2014):

- Greenpeace has served as a significant player in the Indian climate movement since it emerged in 1995 and has promoted the notion of energy justice which highlights that India's energy supply is unjust in that a small, wealthy elite has reliable access to it in contrast to the masses who face energy insecurity.
- "Barefoot scientists" in rural areas provide testimonies on climate change to professional climate scientists.
- The Indian climate movement consists of a wide array of civil society groups and campaigns.

Grassroots Resistance to Coal Mining in the Mahan Forests in Central India

Coal constitutes India's most abundant fuel source and has been a central plank in India's economic development. Climate justice activism in India is in large part framed in terms of protecting forest areas from encroachment of coal mining projects. In 2015, activists from the Indian branch of Greenpeace and forest dwellers in Mahan district of the Central Indian state of Madhya Pradesh managed to invoke a Supreme Court order that cancelled development of a coal mine in the area (Talukdar and Pillai 2023: 26). However, the campaign opposing coal mining development continues. While the Indian government has made a commitment to increasing its reliance on renewable energy sources, in 2020 it auctioned 40 new coal mines (Talukdar and Pillai 2023: 24). Talukdar and Pillai (2023: 33) report:

> [I]n comparison to well-educated urban activists, environmentalism of the poor is composed of largely rural populations that can often lack a scientific understanding of the issue of climate change, even though they are attuned to changing weather patterns. The South Asian People's Action on Climate Crisis (SPACC), formed in 2019, is a very recent and unique collaboration between livelihood-focused people's movements. Indigenous groups, trade unions, and farmers across South Asia. It marks an emergent space in mass activism in South Asia, making climate change central to people's movements and attempting to link the existent malcontents of ecosystem-dependent subsistence communities with the broader problem of climate change.

Transforming the Climate Movement into a Climate Justice Movement

Daniel Tanuro (2009: 263–266) calls for a multipronged movement to fight climate change, one that includes struggles for peace, women's rights, jobs, public provision of land, water, natural resources, and Indigenous rights, as well the eradication of poverty, economic insecurity, privatization, and the globalization and liberalization of agricultural markets. To this list, one could add public provision of education, health care, housing, and demilitarization, particularly denuclearization.

Andreas Malm (2014: 29–38), a Swedish professor, delineates the following five theses on the role that climate change may play in fostering revolutionary change, of which the climate justice movement could be part and parcel.

- Thesis 1: Climate change increases the likelihood of revolution.
- Thesis 2: Climate change will cause victorious revolutions to degenerate.
- Thesis 3: Revolution improves the prospects for adaptation to climate change.
- Thesis 4: No revolution can survive business-as-usual, because no one can.
- Thesis 5: The only revolution that can save humanity is the climate revolution.

With regard to Thesis 1, he acknowledges that climate change cannot serve as a "sufficient cause for revolution but it can be one ingredient in a powder keg, and it can, at least potentially light the fuse" (Malm 2014: 29). For example, between 2010 and 2012, a series of extreme weather events very likely related to climate change adversely impacted the global food system in places such as the Russian Federation, Ukraine, Canada, Australia, Argentina, the Middle East, the United States, and China. With regard to thesis 2, there is the danger that "victorious revolutions in the era of climate change will give birth to bulky bureaucracies or even blood-soaked dictatorships" (e.g., the Syrian Arab Republic, Egypt in the aftermath of the "Arab Spring") because climate change is likely to create shortages of a wide variety of basic goods, starting with food (Malm 2014: 33). With regard to thesis 3, a progressive revolutionary regime could potentially ensure that everyone would be granted enough resources to survive the ravages of climate change. With regard to thesis 4, "any social formation, whatever the character of its relations, will succumb to the forces of global warming, because it cannot do without a biophysical resource base" (Malm 2014: 37). Finally, With regard to thesis 5, Malm (2014: 41)—who began his activist career deflating thousands of tires on polluting SUVs in wealthy neighborhoods— highlights that any serious effort to mitigate climate change before humanity embarks beyond the 2 °C threshold will require nothing less than a "war on the accumulation of fossil capital," but unfortunately at the moment no one has yet taken the lead in instigating a climate revolution that will move humanity beyond business-as-usual.

Climate justice and other anti-systemic activists face an incredibly daunting task as Malm (2018: 175–176) observes:

> The fact that the movement is still nowhere approaching the critical mass required for taking down the fossil economy does not diminish its status as benchmark for theoretical utility. It remains weak and scattered— which may have something to do with the fact that most other social movements are too, including, centrally, the organised working class—but there are so many reasons to join and assist it in every way possible. For if anything is ever going to turn in a better direction, a lot of action will be needed.

Malm (2021) has criticized the climate movement for its overemphasis on pacifism, particularly exemplified in the activism of Bill McKibben and XR. Despite numerous climate rallies—some of which have been huge—around the world, the three largest assessment managers globally continued to pump money into oil, gas, and coal at an alarming rate. Malm draws on Verity Burgmann's observation:

> [T]he history of social movement activity suggests that reforms are more likely to be achieved when activists behave in extremist, even confrontational ways. Social movements rarely achieve everything they want, but they secure important partial victories [when activists resort to violent measures].
>
> *(2018: 10)*

Malm (2021: 71)—responding to the deadly heat waves and wildfires sweeping across Europe in 2019—goes on to argue that pipelines, which are used to transport oil and natural gas, can easily be sabotaged, an action he calls "intelligent sabotage," adding: "There is a long and venerable tradition of sabotaging fossil fuel infrastructure, for other reasons than its impact on the climate." He maintains that the climate movement naively has believed that there is strength in numbers, by and large dismissing not only the access that the capitalist class and its political allies have access to state coercive violence but also "overwhelmingly superior capabilities in virtually all fields, including media propaganda, institutional coordination, logistical resources, political legitimacy and above all, money" (Malm 2021: 112). Most climate activists probably would not agree with Malm's suggestion that sections of the climate movement engage in property sabotage [without physically harming people], but given the gravity of the climate crisis, it is a strategy that might warrant consideration. He points out that the French energy and petroleum company TotalEnergies, for example, is constructing the world's longest heated crude oil pipeline (approximately 900 miles or 1443 kilometers) in Uganda and Tanzania, displacing 100,000 smaller farmers in order to carry more oil on the world market and pour fuel on the global fire. If people were to attack construction equipment and blow up the pipeline, Malm would support the action and does not see how such property damage could be seen as morally illegitimate in light of the human and environment costs the pipeline brings, including causing the forced removal of farmers, food insecurity, household debt, school loss, and likely devastating ecosystem effects (Remnick 2021).

Finding Cracks in the System and Dilemmas Facing the Climate Movement as an Anti-Systemic Movement

The prospect of getting a critical mass of humanity to engage in anti-systemic movements and efforts is a most daunting goal. Anti-systemic social movements are a crucial component of moving humanity to an alternative social system but the process is a tedious and convoluted one with no guarantees. Chase-Dunn (2005: 184) argues: "One of the big challenges is how the different kinds of progressive social movements can work together to struggle against capitalist globalization." In his view, major transnational anti-systemic movements consist of the labor, women's, environmental and Indigenous movements (Chase-Dunn 2005: 187). Chase-Dunn (2005: 187) adds that a "truly democratic global peacemaking government should be the eventual goal of the family of anti-systemic movements." Given the potential for climate wars along with the danger of mass extinction from nuclear annihilation, rising tensions between the United States and China in the form of a new Cold War, the Russian invasion of Ukraine, and the Israeli–Palestinian conflict, it is essential that the climate movement enter into an alliance with the peace movement. By and large, the two movements operate in isolated silos.

With respect to climate change, as Victor Wallis (2010: 52) so forcefully states:

> Until large numbers of people are well organised and thoroughly aware of their long-term interests, the idea of reducing carbon emissions by 80% will appear totally unreal. Only with the social transformation well underway will everyone be able to see through the false dilemma— "either" protect the economy, "or" preserve the environment— propounded by those who resist even the most minimal international accords on global warming.

Thus, John Holloway (2010: 11) maintains that the "only way to think of changing the world radically is as a multiplicity of interstitial movements" situated within mainstream society. Ordinary people constitute the carriers of social movements and they often come in and out of them and move from one movement to another.

In trying to survive in the cracks of the capitalist world system, people have developed alternative social relations and even communities, as was the case in El Alto, an Indigenous settlement on the outskirts of La Paz that played a central role in the Bolivian response to neoliberalism under the leadership of Evo Morales. A crack provides an opening for deeper systemic changes and the transcendence of global capitalism and will require the "recognition, creation, expansion and multiplication here and now of cracks in the structure of domination" (Holloway 2010: 35).

To a greater or lesser degree, social movements have historically focused on relatively limited objectives, be they better wages and working conditions in the case of the labor movement, voting rights and better economic and educational opportunities in the case of the women's movement and the civil rights or ethnic rights movements, environmental protection in the case of the environmental movement, and so on. It is important to note that social movements, including anti-systemic movements, often undergo a process of institutionalization, in which they evolve into social movement organizations (SMOs) in order to "act with greater efficiency in the areas of recruitment, fundraising, political lobbying, and, of course, mobilization" (Johnston 2011: 72). As they become professionalized, bureaucratized, and more efficient, SMOs, including those in the climate movement, often become less democratic, less radical, more domesticated, and even co-opted to the point that they make concessions to the capitalist world system, something that is manifested in many NGOs that to a large degree have their roots in various types of social movements. Indeed, many of the largest environmental NGOs, which by and large emerged from the environmental movement, receive core funding from some of the world's largest and most polluting multinational corporations and obtain contributions from very wealthy individuals. For example, the Nature Conservacy has partnered with Boeing, BP, Shell,

Monsanto, and Walmart; the Sierra Club has accepted millions of dollars from the gas industry, especially Chesapeake Energy, one of the world's biggest gas drillers; the Environmental Defense Fund receives money from McDonalds; the World Wildlife Fund is the beneficiary of Coca-Cola and Ikea money; and Conservation International has ties with Starbucks and Walmart.

Nevertheless, a strong anti-systemic orientation appears to have resurfaced, albeit in disparate ways, even as old social movements become institutionalized and corporatized. Social movements struggle internally with how much hierarchy and centralization to adopt in an effort to be efficient, coordinated, and capable of rapid action or how decentralized and democratic to be in order to maintain a sense of commitment among their followers.

Although anti-systemic movements have not come close to displacing capitalism with some sort of global democracy, aside from whether it is based on socialist or anarchist principles or some other notions of democracy, they have not been inconsequential in getting the attention of the powers-that-be. As Buechler (2014: 278) observes,

> The best indicator is that national governments and corporate interests can no longer ignore them and have been forced to respond. The responses have included violence against protesters and repression of dissent was well as attempts to co-opt and pre-empt movement initiatives without substantive change.

Conversely, one of the problems that social movements have is that many of their members, including organizers, lack a concrete theory of social activism and "turn out to be sprinters rather than long-distance runners" (Lynd and Grubacic 2008: 42) who move from one protest to another protest, sometimes around the world (and in the process burn up many carbon miles.) While certain social movements have manifested an anti-systemic quality, in today's world, social movements, including the labor movement, are not necessarily anti-systemic per se or may be only partly anti-systemic or have anti-systemic wings. However, they are the site where the greatest potential to transcending global capitalism lies. While anti-systemic movements often espouse democratic decision-making, all too often a small clique of professional activists "can quickly dominate an organisation and exclude alternative means of communication" (Ruiz 2014: 36).

In recent times, social movements, including the climate movement, have come to rely heavily on the internet and social media for organizing and mobilizing their activities. Whereas as in the past social movements relied on word of mouth, pamphlets, leaflets, newspapers, and conferences for organizing protest rallies, they have increasingly come to rely on multimodal, digital networks for this purpose. Castells (2012: 233) maintains that contemporary social movements are "largely made by individuals living at ease with digital technologies in the hybrid world of real virtuality." Of course, one of the dangers that social movements face in relying too heavily on digital networks is that the powers-that-be can shut them down. China has created sophisticated techniques to censor material on the internet that potentially could provoke social dissent. Thus, it is imperative that social movements continue to rely on more traditional forms of communication. Furthermore, digital networks can destroy or interfere with convivial relationships among individuals who isolate themselves in their homes with their laptops, smartphones, and so forth, rather than interacting with each other face-to-face, whether in each other's homes, cafés, classrooms, or a wide assortment of public places. Ultimately, anti-systemic movements need structures, albeit democratic ones, to coordinate their efforts, to follow up on protest rallies, which all too often fail to go anywhere concretely in terms of creating another world.

Perhaps the biggest challenge that anti-systemic movements face is how they can unite in their struggle against a common foe, namely global capitalism. Ethnic, national, religious, and cultural differences all too often serve to divide subalterns around the world and prevent them from forming a united revolutionary movement. Chase-Dunn (2005: 184) maintains that humanity "needs both more democratic global governance and more local autonomy, and that the globalization-from-above movements should work together with the local-autonomy movements, or at least with those who are progressive and willing." In contrast, the corporate class and its political allies are able to use a number of transnational organizations, such as the World Bank, the International Monetary Fund, the World Trade Organization, the Group of 7 and the Group of 20, the World Economic Forum, and even the UN, to further their various agendas.

Harvey (2012: 127) maintains that the formation of a viable anti-capitalist movement will have to re-evaluate many past as well as present anti-capitalist or anti-systemic movements in terms of "what can and must be done, and who is going to do it where" and needs to address three critical questions. He asserts that the "first is that of crushing material impoverishment for much of the world's population," which requires an "anti-wealth politics and to the construction of alternative social relations to those that dominate within capitalism" (Harvey 2012: 127). We often hear parties ranging from the World Bank to sporting superstars making an appeal for "eradicating global poverty" or "making poverty history." However, following Harvey, we can make a case for "making wealth history" and the eradication of poverty will follow. For him, the "second question derives from the clear and imminent dangers of out-of-control environmental degradations and ecological transformations," which will require "major shifts in consumerism, productivism, and institutional arrangements" (Harvey 2012: 127). And the third question for Harvey (2012: 127–128) "derives from a historical and theoretical understanding of the inevitable trajectory of capitalist growth," which "requires the abolition of the dominant class relation that underpins and mandates the perpetual expansion of surplus value production and realization."

Conclusion

Unfortunately, as De Lucia (2014: 66) comments, the term climate justice "has been stretched very thin." For example, Mary Robinson (2018), a former president of Ireland and the head of the Mary Robinson Foundation—Climate Change who has worked very closely with the UN COP process—wrote a book entitled *Climate Justice*. However, it is far from clear how she reconciles a desire for environmental sustainability with her membership in a group of business leaders (formed at Davos) that includes Richard Branson whose Virgin Galactica project promises space tourism for the very wealthy. Robinson acknowledges the need to reduce social inequality but fails to confront the growing concentration of wealth in most countries in the world—a phenomenon that became even more pronounced during the COVID-19 pandemic— and the persistence of major social inequalities. Ultimately, her conception of climate justice reduces it to a moral problem that can be solved by persuading the rich to do better without attending to much deeper systemic changes, namely the transcendence of global capitalism and the transition to an alternative world system based on social justice and parity, deep democracy, environmental sustainability, and a safe climate.

It is likely that the next two or three decades, if not the next decade, will bring much hardship for much of humanity, particularly for the poor in the Global South and even the Global North, as we have seen and continue to see not only in the ravages of climate change but also the COVID-19 pandemic. This hardship includes the rise of authoritarian regimes and reactionary social movements and political parties that claim that climate change is a hoax. As climate change increasingly impacts human and nonhuman beings, the powers-that-be will seek to construct a Fortress World to protect their privileges, borders, and market systems. There are no easy fixes for these harsh realities, but it is imperative that climate activists become climate justice activists who are part of a global force consisting of a wide array of anti-systemic movements that challenge and endeavour to transcend capitalism. The alternative, sadly, is a very grim future for a growing number of Earth's inhabitants.

References

Anderson, James. 2006. Afterword: Only sustain the environment, "Anti-Globalization," and the runaway bicycle. In *Nature's Revenge: Reclaiming Sustainability in an Age of Corporate Globalization*. Josee Johnston, Michael Gismondi, and James Goodman, eds. pp. 273–425. Toronto, ON: Broadview

Bergmann, Zoe and Ossewaarde, Ringo. 2020. Youth climate activists meet environmental governance: Ageist depictions of the FFF movement and Greta Thunberg in German newspaper coverage. *Journal of Multicultural Discourses* 15: 267–290

Berker, Lars E. and Pollex, Jan. 2023. Explaining differences in party reactions to the Fridays for Future movement: A qualitative comparative analysis (QCA) of parties in three European countries. *Environmental Politics* 32: 755–792

Bond, Patrick. 2009. Carbon crusade. In *People First Economics*. David Ransom and Vanessa Bird, eds. pp. 211–216. Oxford: World Changing

Bond, Patrick. 2012. *Politics of Climate Justice: Paralysis Above, Movement Below*. Scottsville: University of KwaZulu-Natal Press

Bond, Patrick. 2014. Climate justice in, by, and for Africa. In *Routledge Handbook of the Climate Change Movement*. Matthias Dietz and Heiko Garrelts, eds. pp. 205–221. London: Routledge

Bruno, Kenny, Karliner, Joshua, and Brotsky, China. 1999. *Greenhouse Gangster vs. Climate Justice*. San Francisco, CA: Transactional Resource & Action Center

Buechler, Steven M. 2014. *Critical Sociology* (2nd edn). Boulder, CO: Paradigm Publishers

Burgmann, Verity. 2018. The importance of being extreme. *Social Alternatives* 37(2): 10–13

Camilleri, Joseph A. and Falk, Jim. 2010. *Worlds in Transition: Evolving Governance across a Stressed Planet*. Cheltenham: Edward Elgar

Castells, Manuel. 2012. *Networks of Outrage and Hope: Social Movements in the Internet Age*. London: Polity

Chen, Kaiping, Molder, Amanda L., Duan, Zenig, Bouliane, Shelley, Eckart, Christopher, Malleri, Prince, and Yange, Diyi. 2023. How climate movement actors and the news media frame climate change and strike: Evidence from analyzing Twitter and new media discourse from 2018 to 2021. *International Journal of Press Politics* 28: 413–884

Chase-Dunn, Christopher. 2005. Social evolution and the future of world society. *Journal of World-Systems Research* 11(2): 171–192

Chomsky, Noam and Pollin, Robert. 2020. *Climate Crisis and the Global Green New Deal*. London: Verso

Cooke, Sophie. 2010. "Leave it in the ground": The growing global struggle against coal. In *Sparking a Worldwide Energy Revolution: Social Struggles in the Transition to a Post-Petrol World*. Kolya Abramsky, ed. pp. 424–438. Oakland, CA: AK Press

Croeser, Eve. 2020. *Ecosocialism and Climate Justice: An Ecological Neo-Marxian Analysis*. London: Routledge

Crowe, P. 2019. As Greta Thunberg inspires a world revolution, one young Ugandan is bringing the climate fight home. *The Independent*. https://www.independent.co.uk/ climate-change/news/climate-change-leah-namugerwa-greta-thunburg-activism-pro test-uganda-a9261326.html

De Lucia, Vito. 2014. The climate justice movement and the hegemonic discourse of technology. In *Routledge Handbook of the Climate Change Movement*. Matthias Dietz and Heiko Garrelts, eds. pp. 83–89. London: Routledge

De Moor, Joost. 2018. The "efficacy dilemma" of transnational climate activism: The case of COP21. *Environmental Politics* 27: 1079–1100

Derber, Charles. 2010. *Greed to Green: Solving Climate Change and Remaking the Economy*. Boulder, CO: Paradigm Publishers

Dietz, Matthias and Garrelts, Heiko. eds. 2014. *Routledge Handbook of the Climate Change Movement*. London: Routledge

Extinction Rebellion. 2019. *This Is Not a Drill: An Extinction Rebellion Handbook*. London: Penguin

Fisher, D. and Renaghan, Q. 2023. Understanding the growing radical flank of the climate movement as the world burns. Brookings Institute. https://www.brookings.edu/articles/ understanding-the-growing-radical-flank-of-the-climate-movement-as-the-world-burns/

Fonda, Jane. 2020. *What Can I Do? The Truth About Climate Change and How to Fix It*. London: HarperCollins

Francis, Pope. 2015. *Laudato si*. https://www.vatican.va/content/francesco/en/encyclicals/documents/laudato si/papa-francesco-2015024-encilica-laudato-si.html

Giacomini, Terran and Turner, Terisa. 2015. The 2014 People's Climate March and Flood Wall Street Disobedience: Making the transition to a post-fossil capitalist, communing civilization. *Capitalism Nature Socialism* 26(2): 27–45

Gunningham, Neil. 2019. Averting climate catastrophe: Environmental activism, Extinction Rebellion and the Coalitions of Influence. *King's Law Journal* 30(2): 194–202

Hallam, Roger. 2019. *Common Sense for the 21st Century: Only Nonviolent Revolution Can Now Stop Climate Breakdown and Social Collapse*. White River Junction, VT: Chelsea Green Publishing

Harms, Arne and Powalla, Oliver. 2014. India: The long march to a climate movement. In *Routledge Handbook of the Climate Change Movement*. Matthias Dietz and Heiko Garrelts, eds. pp. 179–193. London: Routledge

Harvey, David. 2012. *Rebel Cities: From the Right of the City to the Urban Revolution*. London: Verso

Holloway, John. 2010. *Crack Capitalism*. London: Pluto Press

Hope, B. 2022. Get to know the world's top five youth climate activists. *Sustainability*. https://sustainabilitymag.com/sustainability/faces-of-change-the-top-five-youth-climate-activists-named-greta-emissions-change-pledge

India Climate Justice Forum. 2002. *Delhi Climate Justice Declaration*. https://www.indiaresource.org.

Johnston, Hank. 2011. *States & Social Movements*. London: Polity

Klepp, Silja. 2014. Small island states and the new climate change movement: The case of Kiribati. In *Routledge Handbook of the Climate Change Movement*. Matthias Dietz and Heiko Garrelts, eds. pp. 308–318. London: Routledge

Knights, Sam. 2019. Introduction: The story so far. In *This Is Not a Drill: Extinction Rebellion Handbook*. Clare Farrell, Alison Green, Sam Knights, and William Skeaping, eds. pp. 9–13. London: Penguin

Konisky, David M. 2018. The green of Christianity? A study of environmental attitudes over time. *Environmental Politics* 27: 267–291

Kottosova, I. and Macintosh, E. 2019. Teen activist tells Davos elite they're to blame for climate crisis. CNN, January 25. https://edition.cnn.com/2019/01/25/europe/greta-thunberg-davos-world-economic-forum-intl/index.html

Lazaro, Lira Luz Benites, Lauda-Rodriguez, Zenaida Luisa, Boerner, Susanne, Lampis, Andrea, and Giatti, Leandro Luiz. 2023. Climate justice in Latin America. In *Climate Debate in the Majority World*. Neil J.W. Crawford, Kavya Michael, and Michael Kikulewicz, eds. pp. 124–147. Abingdon: Routledge

Lynd, Staughton and Grubacic, Andrej. 2008. *Wobblies & Zapatistas: Conversations on Anarchism, Marxism and Radical History*. Oakland, CA: PM Press

Malm, Andreas. 2014. Tahrir submerged? Five theses on revolution in the era of climate change. *Capitalism Nature Socialism* 25(3): 28–44

Malm, Andreas. 2018. *The Progress of Storms: Nature and Society in a Warming World*. London: Verso

Malm, Andreas. 2021. *How to Blow Up a Pipeline*. London: Verso

McKibben, Bill. 1989. *The End of Nature*. New York: Random House

McKibben, Bill. 2019. *Falter: Has the Human Game Begun to Play Itself Out?* Melbourne; Black, Inc.

Misch, Antonia, Kristen-Antonow, Susanne, and Paul, Markus. 2021. A question of morals? The rise of moral identity in support of the youth climate movement Fridays4Future. *PLoS One* 16(3): 1–16

Newell, Peter and Patterson, Matthew. 2010. *Climate Capitalism: Global Warming and the Transformation of the Global Economy*. Cambridge: Cambridge University Press

Nita, Marta. 2016. *Praying and Campaigning with Environmental Christians. Green Religion and the Climate Movement*. London: Palgrave Macmillan

Nordensvard, Johan and Ketola, Markus. 2022. Populism as an act of storytelling: Analyzing the climate change narratives of Donald Trump and Greta Thunberg as populist truth-tellers. *Environmental Politics* 31: 861–882

Pattberg, Philip and Stripple, Johannes. 2010. Agency in global climate governance: Setting the stage. In *Global Climate Governance Beyond 2012: Architecture, Agency and Adaptation*. Frank Biermann, Philip Pattberg, and Fariborz Zelli, eds. pp. 137–145. Cambridge: Cambridge University Press

Reed, R. (2023). Unions' extension into politics was necessary—and contributed to their decline, says Harvard Law expert. Harvard Law School. https://hls.harvard.edu/today/unions-extension-into-politics-was-necessary-and-contributed-to-their-decline-says-harvard-law-expert/

Remnick, D. 2021. Should the climate movement embrace sabotage? *The New Yorker* [podcast]. https://www.newyorker.com/podcast/the-new-yorker-radio-hour/should-the-climate-movement-embrace-sabotage

Risher, Dana and Renaghan, Quinn. 2023. The rising radical flank of the climate movement. August 1. http://www.givingcompass.com

Robinson, Mary. 2018. *Climate Justice: Hope, Resilience, and the Fight for a Sustainable Future*. London: Bloomsbury

Ruiz, Pollyanna. 2014. *Articulating Dissent: Protest and the Public Sphere*. London: Pluto Press

Schroeder, Patrick. 2014. China's emerging climate change movement: Finding a place to stand. In *Routledge Handbook of the Climate Change Movement*. Matthias Dietz and Heiko Garrelts, eds. pp. 194–204. London: Routledge

Segebart, Doerte and Koenig, Claudia. 2014. Out of the forest: The climate movement in Brazil. In *Routledge Handbook of the Climate Change Movement*. Matthias Dietz and Heiko Garrelts, eds. pp. 194–204. London: Routledge

Spears, Ellen Griffith. 2019. *Rethinking the American Environmental Movement*. New York: Routledge

Talukdar, Ruchira and Pillai, Priya. 2023. Southern climate justice activism in the context of the energy transition: Forest rights over coal in Mahan, Central India. In *Climate Debate in the Majority World*. Neil J.W. Crawford, Kavya Michael, and Michael Kikulewicz, eds. Pp. 19–41. Abingdon: Routledge

Tanuro, Daniel. 2009. Climate crisis: 21st century socialists must be ecosocialists. In *The Global Fight for Climate Justice: Anticapitalist Responses to Global Warming and Environmental Devastation*. Ian Angus, ed. Pp. 237–282. London: Resistance Books

Thunberg, Greta. 2019. *No One is Too Small to Make a Difference*. London: Penguin

Thunberg, Greta. 2023. *The Climate Book: The Facts and Solutions*. London: Penguin Press

Tokar, Brian. 2018. On the evolution and continuing development of the climate justice movement. In *Routledge Handbook of Climate Justice*. Tahseen Jafry, Michael Mikulewicz, and Karin Helwig, eds. pp. 13–25. London: Routledge

Wallis, Victor. 2010. Socialism and technology: A sectoral overview. In *Eco-Socialism as Politics: Rebuilding the Basin of Our Modern Civilization*. Q. Huan, ed. pp. 45–61. New York: Springer

World's Peoples Conference on Climate Change and Mother Earth. 2010. Cocha-bamba, Bolivia, April 4–April 22. https://www.rightsandresources.org/event/world-peoples-conference-on-climate-change-and-the-rights-of-mother-earth

Žižek, Slavoj. 2020. *A Left That Dares to Speaks Its Name.* Oxford: Polity

10

TOWARDS A CRITICAL ANTHROPOLOGY OF THE FUTURE

Climate Change and Future Scenarios

Case Study

Some climate scientists have been forced by their research findings to consider worst-case climate change futures. They ask themselves some difficult questions: could anthropogenic climate change cause worldwide societal collapse or even eventual human extinction? Unfortunately, there are ample reasons to suspect that climate change could result in global catastrophe for humans and other species and in the not too distant future. Some say that analyzing the mechanisms for these extreme outcomes could help to sharpen awareness of the impending situation and encourage the provision of funding needed to galvanize action, improve resilience, and inform policy, including identifying emergency responses. It is time, some climate scientists assert, for the scientific community to grapple with the challenge of better understanding the bleakest of futures that could result from catastrophic climate change.

One climate scientist who has stressed the feasibility of such a scenario is Gaya Herrington, a Dutch sustainability researcher and advisor to the Club of Rome, a Swiss think tank. In 1972, based on a study conducted at the Massachusetts Institute of Technology, the Club of Rome suggested that industrial civilization was on track to collapse sometime in the 21st century, due to overexploitation of limited planetary resources. When the Club of Rome study was first published, it was hotly debated and condemned by various pundits and politicians. Out of pure scientific curiosity, Herrington decided to undertake a data check of the Club of Rome's computer model using up-to-date empirical observations of environmental and climate changes and impacts. Herrington concluded that a global collapse of contemporary societies and a decline in human populations and welfare standards could set in as soon as the 2040s. She comments, "The key finding of my study is that

DOI: 10.4324/9781003469940-11

we still have a choice to align with a scenario that does not end in collapse" (quoted in Helmore 2021). She adds,

> [T]o get to a sustainable society, technological innovation is going to be absolutely imperative. So that's part of the solution. But the idea that it's going to deliver us, because that's how it feels, that people think that it's just going to deliver us all and we don't have to do anything, is ridiculous because technology is just a tool. And in a system that is geared towards growth, the tools will typically just be geared towards that goal.
>
> *(Accidental Gods n.d.)*

Playing with Time

The overwhelming majority of climate scientists have come to the conclusion that the warming of the planet and the associated climatic events are largely anthropogenic, or the result of human activities in particular. The pace and effects of warming have been increasing, and this change in the world inhabited by humans has had significant, if not severe, consequences for human wellbeing as well the wellbeing of numerous fauna and flora, large and small.

Historically, anthropologists have concerned themselves with either human societies of the distant past—the domains of archeology and paleoanthropology—or the recent past or present—the domain of sociocultural anthropology. When we consider how long humans have lived as farming and herding communities—some 10,000 years—or how long numerous numbers of humans have lived in stratified state societies—some 6,000 years—our presence in such social arrangements represents just a tiny fraction of the already brief timeline of our species compared with the age of the planet Earth. The Intergovernmental Panel on Climate Change and prognosticators often speak of the state of humanity in 2050 or 2100, but generally not much beyond that. Sheila Jasanoff (2010: 241) suggests that "[c]limate change invites humanity to play with time," including projecting the mind's eye into the future as we might imagine it will unfold.

Over the past few decades, anthropologists and other social scientists have often alluded to a cavalcade of "posts," such as post-modernism, post-colonialism, post-Fordism, post-structuralism, and post-feminism. Anthropologists might entertain the possibilities of two other "posts," namely post-capitalism and post-anthropogenic climate change. While in the 1970s and 1980s various anthropologists grappled with imagined future scenarios for humanity, the demise of the Soviet bloc countries and the disillusionment with grand theory or metanarratives under the guise of post-modernism appear to have predisposed a younger generation of anthropologists to steer away from seemingly grandiose projects of attaining a better world based upon social justice and environmental sustainability.

Yet a revival of the anthropology of the future strikes us as long overdue. Fortunately, John Bodley (2012) in the six editions of his book *Anthropology and Contemporary Human Problems* concludes with a chapter on "The future' in which he envisions a *sustainable planetary system*. Climate change compels us to engage with what Immanuel Wallerstein (1998: 1) terms *utopistics*, which he defines as "serious assessment of historical alternatives, the exercise of our judgment as to the substantive rationality of alternative possible historical systems." In seeking to assess possible future scenarios with respect to climate change one must consider the possibility of a dystopian future in the hope that this will contribute to the realization that serious climate change mitigation efforts will require an alternative to the capitalist world system, one that is based on both social justice and environmental sustainability, and that will allow humanity to reach a steady state for itself and for biodiversity on the planet.

In this chapter, we outline three possible scenarios for the future of humanity:

1. A dystopian future characterized more or less as "business-as-usual," with ongoing economic growth, and increasing social stratification. This is a Fortress World in which the affluent attempt to protect their privileged lifestyles, amid environmental degradation and runaway climate change, although ultimately they may find themselves in situations analogous to the experiences of elites during both the French and Bolshevik Revolutions, albeit on a much more global scale. This is a future stuck in the myths, and modes of being, of contemporary capitalism, incapable of adequately acting on, or accepting, any other view of life.
2. A future of "reflexive modernization," which emphasizes *ecological modernization* (renewable energy sources, energy efficiency, electric vehicles, etc.) and *sustainable development*, with ongoing social inequality coupled with some amelioration of global poverty, and some mitigation of and adaptation to climate change. This represents a world society with a slightly more flexible political economy, namely green capitalism, that at present is still unable to imagine a significant alternative to the current world order.
3. An socio-ecological or eco-socialist revolution which would entail public or social ownership of the means of production, highly democratic processes, increasing social and economic equality, a focus on the wellbeing and flourishing of all species, a steady-state world economy, environmental sustainability and a safe climate. This assumes a society capable of radical imagination and readiness to explore alternative modes of being.

The Road to Dystopia

Despite pledges by governments and increasing numbers of corporations to achieve net zero emissions by 2050, the grim reality is that emissions continue to rise along with increases in global temperatures, heat waves, droughts, wildfires, hurricanes, torrential rains and floods, melting glaciers and ice caps, etc. Climate change scenarios prompt us to imagine dystopian visions of the future, if for no other reason than to forewarn us to take serious measures to counteract possible doomsday events.

Mark Lynas

In his book *Six Degrees*, journalist Mark Lynas (2020), based on his perusal of numerous climate scientific reports, vividly portrays climate change scenarios in the event of increases in global temperature of between 1 and 6 °C (33.8–42.8 °F), which will have a negative impact on human populations. Given that many climate scientists are envisioning that global warming will increase by 4 °C (39.2 °F) or more by 2100 (see Christoff 2014), if humanity does not seriously begin to curtail its greenhouse gas (GHG) emissions, Lynas envisions a hotter world resulting in the following:

1. the loss of one-third of Bangladesh's land area, including in the displacement of millions from the Ganges-Brahmaputra-Meghna delta;
2. flooding of low-lying islands and deltaic cities such as Shanghai, Mumbai, Alexandria, Boston, New York, London, and Venice;
3. massive shrinking of Greenland's ice cap into the center of the landmass;
4. slowing and shutdown of the North Atlantic Conveyor Belt, which would have a significant cooling impact on northwestern Europe.
5. spreading of new deserts in southern Europe;
6. possible July and August temperatures of 48 °C (118.4 °F) in Switzerland, accompanied by wildfires and diminished water supplies;
7. a completely ice-free summer in the Arctic Ocean;
8. the release of methane contained in frozen Arctic soils.

Moreover, in a 5 °C warmer world, human populations would be greatly restricted in terms of habitable areas due to drought and flooding; northern Europe and Siberia would possibly constitute crowded refugee areas; and Patagonia, Tierra del Fuego, Tasmania, the South Island of New Zealand, and the ice-free Antarctica Peninsula could serve as other refugee areas, In a 6 degree Celsius world, the eruption of oceanic methane might result in massive human extinction. We would like to add that if the world were to warm by between 4 degrees and 6 °C this would possibly lead to climate wars as people try to control useable land areas or to seize such areas from those occupying them.

The Possibility of Eco-Authoritarian Regimes

Perhaps frustrated by the cumbersome nature of global and national climate governance processes, including in liberal democracies, various scholars have argued that democratic processes are moving too slowly to contain climate changes, suggesting the need for eco-authoritarian, or even eco-fascist regimes. James Anderson (2006: 245; emphasis in the original) suggests that the "radical changes necessary to sustain capitalism could indeed turn out to be an extremely authoritarian *counter-revolution.*"

James Lovelock, the inventor of the Gaia hypothesis, has identified portions of the Earth that may be inhabitable in what he regards to be the dystopian future given the inevitability of ongoing climatic and ecological crises. These include the northern regions of the United States and the Russian Federation, Canada, Scandinavia, Siberia, Patagonia, southern Chile, and various island nations or states, such as Japan, New Zealand, and the British Isles, and smaller islands, such as Hawaii, Taiwan, Tasmania, and the Philippines (Lovelock 2009: 11). Lovelock (2009: 57) refers to such places as "lifeboats for humanity" and concurs that the various continents will have well-watered oases to sustain agriculture. Later, Lovelock (2014) advocated the creation of high-rise, high-density, climate-controlled cities like Singapore where people can escape the hot world outside in their artificial environment, a development that will require the suspension of democracy. In the event that humanity does not sufficiently mitigate climate change, he predicts:

> It is most unlikely that the number of humans surviving will be less than a million: this is more than enough needed for the survival of our species ... [O]ur present choice of city life would be a powerful benign force moving us to a future existence in city organisms.
>
> *(Lovelock 2014: 122–123)*

Andreas Malm and the Zetkin Collective (2021) refer to the possibility of one genre of climate dystopia that they term *fossil fascism*. They report that "all European far-right parties of political significance in the early 21st century expressed climate denial" (Malm et al. 2021: 4). While some of them have retreated slightly from climate denialism, it lurks in their backgrounds. Indeed, climate denialism marked the Trump Administration (2017–2021) and remains embedded in the fabric of the Republican Party. Furthermore, it was part and parcel of significant sectors of successive coalition governments in power in Australia. Malm et al. (2021: 20) assert that while "white people have ascended the evolutionary ladder in the comfort and affluence" afforded by fossil fuels, black people "have stayed behind in the fossil fuel bottom to break their own backs." While capitalist climate governance regards global warming as a reality with capital positioned as providing a way out of the climate crisis, Malm et al. (2021: 37–38) posit that the far right "objectively worked as the defensive shield of fossil capital as a totality and primitive fossil capital in particular, even if—or, rather, precisely because—it was not set up or financed by them." Climate denialism is now firmly entrenched in countries often portrayed as progressive paragons of climate change mitigation strategies, such as Austria, the Netherlands, Denmark, Sweden, and

Germany. In the wake of the closure of the Swedish border in 2015 by a government of social democrats and Greens, the far-right Sverigedemokraterna (Sweden Democrats) called for the remigration of refugees who had previously been admitted into the country.

Norway, a country of some five million inhabitants, is often viewed as a progressive nation on various counts, including environmental policies. However, this is a chimera in that, as of 2016, Norway was the 14th largest producer of oil and the seventh largest producer of natural gas in the world, fossil fuels which "were under the control of the Ministry of Petroleum and Energy" (Malm et al. 2021: 119), a body which was, between 2013 and 2020, headed up by four leaders from the far-right Fremskrittspartiet (Progress Party). Norway's ability to juxtapose its purported environmentalism with fossil fuel extraction was developed in the 1990s by an:

> [I]deological state apparatus—here truly centred on the state—consisting of the Ministry of Finance, state-owned Statoil, the social democratic and conservative parties and a cohort of journalists, working in concert to inculcate in Norway trust in its fossil fuels.
>
> *(Malm et al. 2021: 121)*

This grim reality suggests that Norway constitutes an example of creeping fascism. Elsewhere, various right-wing leaders such as Marine Le Pen in France along with the likes of Paul Kingsnorth (one of the founders of the Dark Mountain Project) in the United Kingdom, have come to embrace green nationalism which regards national borders as ecological protection structures. As Connolly (2019: 61) observes, "The danger of fascism now is bound immediately to cultural resentment in several old capitalist states against growing refugee pressures, some of which will accelerate due to the close relation between growing regional droughts and civil wars."

Eve Darian-Smith (2022) continues the discussion of authoritarian patterns associated with the climate crisis, using the recent wildfires in Australia, Amazonia, and California in the United States as omens of ecological collapse. She defines *free-market authoritarianism* as the "collusion between political governance and corporate sectors in banking, energy, agribusiness, technology, and pharmaceuticals" (Darian-Smith 2022: 25). Darian-Smith focuses on three specific instances of extractivism in which corporations call the shots in legislative settings. The first is Pacific Gas and Electric Company, a utility that provides natural gas and electricity in northern California which has been complicit in climate denialism. The second is the mining industry, another bastion of climate denialism, which "played a direct and indirect role in creating the environmental conditions for Australia's catastrophic bushfires of 2019 and 2020" (Darian-Smith 2022: 53). The third is Brazil's agribusiness which found support in Jair Bolsonaro, a Brazilian business leader, during his presidency. While Scott Morrison, the Australian coalition prime minister, and

Bolosonaro, the Brazilian president, were deposed in 2022, both Australian and Brazilian agribusiness continue to exert a strong influence on political processes. This reality is borne out by the fact that the Australian Labor Party government under the leadership of Prime Minister Anthony Albanese continues to support the expansion of fossil fuel projects, despite its rhetoric of being stronger on climate action than the previous coalition governments. Regardless of where it occurs, Darian-Smith (2022: 67) maintains that free market authoritarianism that conflates neoliberalism and anti-democratic processes exhibits three common features: ultranationalism, multilateralism, and anti-environmentalism.

The Potential for Climate Wars

Geographer Christian Parenti (2011: 7) maintains that climate change could interact with existing crises of poverty and violence in a *catastrophic convergence* in which political, economic, and environmental problems "compound and amplify each other, one expressing itself through the other. He identifies a *Tropic of Chaos* situated between the Tropic of Capricorn and the Tropic of Cancer which is being particularly adversely impacted by climate change:

> The societies in this belt are also heavily dependent on agriculture and fishing, thus very vulnerable to shifts in weather patterns. This region was also on the front lines of the Cold War and of neoliberal economic restructuring. As a result, in this belt we find clustered most of the failed and semifailed states of the developing world.
>
> *(Parenti 2011: 9)*

Welzer (2017) identifies three broad tendencies driving interstate wars: (1) international commodities markets and supply infrastructure, such as pipelines; (2) conflicts over basic resources such as water and potentially new resources being uncovered due to the melting Arctic and Antarctic ice cover; and (3) climate change as a factor contributing to social system breakdown, including civil war or genocide. He delineates various social conflicts, wars, and security measures resulting from environmental and climatic factors. These include:

- Soil degradation, flooding, water shortages, or hurricanes resulting from climate change, which place serious constraints on subsistence;
- Violence in societies most adversely impacted by climate change, which will increase the "flow of refugees and migrants, both within and between countries";
- "The terrorism that has grown in step with global modernization and processes is legitimated and strengthened by inequalities and injustices resulting from climate change."

(Welzer 2017: 160–161)

Various defense agencies, including those in the United States, the UK, and Australia, along with other organizations around the world, have made note of the "security risks" associated with climate change. In its recognition that global warming or climate change may pose a "security threat" to the United States, the Pentagon commissioned CNA (n.d.), a nonprofit national security organization, to write a report on this issue. CNA convened a panel of retired military officers and national security experts as part of its effort to assess the security implications of global warming. In its report, CNA (n.d.) asserts that global warming "acts as a threat multiplier for instability in some of the most volatile regions of the world" and "will seriously exacerbate already marginal living standards in many Asian, African, and Middle Eastern nations, causing widespread political instability and the likelihood of failed states." CNA stresses that global warming poses the possibility of an even greater number of people attempting to emigrate, either legally or illegally, from Mexico to the United States, and more turbulent seas could adversely affect US naval operations in the North Atlantic.

Peter Schwartz and Doug Randall (2003: 1) also authored another Pentagon-commissioned report entitled *An Abrupt Climate Change Scenario and Its Implications for United States National Security*, outlining worst-case scenarios that, "although not most likely," are "plausible" and thus would "challenge United States security in ways that should be considered immediately." While climate scientists may very well argue with their abrupt change scenarios, Schwartz and Randall (2003: 2) envisage the possibility, over the next few decades, of an annual average temperature increase of up to 2.75 °C (36.95 °F) in Asia and North America and 3.3 °C (37.94 °F) in northern Europe; an annual average temperature increase of up to 2.2 °C (35.96 °F) in key areas of Australia, South America, and southern Africa; long-term drought in "critical agricultural resource regions for major populations in Europe and eastern North America"; and intense winter storms and winds in western Europe and the North Pacific region. Due to these climatic changes, they envisage the possibility of food shortages due to diminished "net global agricultural production," "decreased availability and quality of fresh water in key regions," and "disrupted access to energy supplies due to extensive sea ice and storminess." Schwartz and Randall (2003) have outlined possible conflict scenarios for Europe, Asia, and the United States for 2010–2020 and 2020–2030. They envisage the possibility of conflicts in the Persian Gulf and Caspian Sea regions in 2020 due to increased oil prices; a civil war in the People's Republic of China and border wars with adjacent countries in Southeast Asia; and also an "internal struggle in Saudi Arabia" in 2025, which "brings Chinese and U.S. naval forces to [the] Gulf in direct confrontation" (Schwartz and Randall 2003: 17). According to Schwartz and Randall (2003: 2),

As global and local carrying capacities are reduced, tension could mount around the world, leading to two fundamental strategies: defensive and offensive. Nations [such as the United States and Australia] with the resources to do so may build virtual fortresses around their countries, preserving resources for themselves. Less fortunate nations, especially those with ancient enmities with their neighbors, may initiate in struggles for access to food, clean water, or energy.

Miller et al. (2021) report that world's top ten historic GHG emitters since 1850 spent much more on border spending per year during the period 2013–2018 than they did on climate financing. For example, the United States as the leading historic emitter had a ratio of 10.9:1 and Australia as the ninth leading historic emitter had a ratio of 13.5:1, thus exhibiting a strong Fortress World mentality in both (Miller et al. 2021: 11).

Dan Smith and Janani Vivekananda (2007: 3) wrote a report in which they assert that many of world's poorest countries and communities face a "double-headed" dilemma: global warming and the potential for violent conflict. Like others, they maintain that global warming "could compound [the] propensity for violent conflict, which in turn will leave communities poorer, less resilient and less able to cope with [the] consequences of climate change" (Smith and Vivekananda 2007: 46). Smith and Vivekananda submit that the 46 countries where the "effects of global warming interacting with economic, social and political problems could create a high risk of violent conflict" are home to some 2.7 billion people and that the 56 countries "where governments will have great difficulty in taking the strain of climate change on top of all their other current challenges" are home to some 1.2 billion people. Their report includes case studies of Algeria, Peru, Bangladesh, Mali, Chad, Liberia, and Nepal. The Christian Science Monitor has identified six potential flash points that could erupt into conflict as a result of global warming: Nepal, Indonesia, Lagos (Nigeria), the United States, the Arctic, and East Africa (Shapley 2007).

Thus, due to the pressures and stresses induced by climate change, states may be willing to resort to war in order to protect resources, including keeping climate refugees out of their territories, and to acquire new resources given the contraction of resources, such as diminishing food supplies due to climate change, within their own borders. Alvarez (2017: 80) reports:

> In many places, something as basic as providing enough food for a population will present challenges as drought and flooding decrease agricultural productivity. Even land itself, especially arable land, will be seen as a resource worth fighting and killing for. In some parts of Ethiopia, local subsistence farmers are already finding steel fences erected around vast swathes of land in their communities. These segregated tracts of farmland have been bought by Saudi Arabia. How will those local farmers react when the struggle to meet their own need for food intensifies? Violence has already broken out as Ethiopian police and military personnel have cracked down on local protests organized by local villagers and farmers.

Reflexive Modernization

The prominent German sociologist Ulrich Beck in his various books about "risk society" tended to eschew any specific reference to capitalism, repeatedly referring to it as "industrial modernity" or "Western modernity" (Beck 2007). Nevertheless, he recognized that:

> [I]n the light of climate change, the apparently independent and autonomous system of industrial modernization has begun a process of self-destruction and self-transformation. This radical turn marks the current phase in which modernization is reflexive, which means: we have to open up to global dialogues and conflicts about redefining modernity ... It has to include multiple extra-European voices, experiences and expectations concerning the future of modernity.
>
> (Beck 2010: 264)

Beck called for a form of "cosmopolitanism," which transcends national interests and has the potential to create a green modernity.

Australian political scientists Peter Christoff and Robyn Eckersley (2013) argue for a conflation of reflexive modernization and reflexive globalization. The globalization aspect is inherent, given that modernization, which has entailed population growth and technological innovation, has been the primary driver of global environmental change, a process that has been accentuated under the neoliberal phase of economic globalization. Christoff and Eckersley (2013: 189–204) posit that these dual processes need to undergo change to incorporate the following:

- Reflexive governance which entails adapting the established set of good governance principles, such as the rule of law, the absence of corruption, transparency, participation, accountability, efficiency, and effectiveness, to globalization;
- Extension of the notion of accountability to responsibility that includes a "critical understanding of the historical conditions and the social structures that systematically produce environmental injustices across space and time";
- Adoption of the "precautionary principle" which reduces exposure to and adding to global risks, such as climate change and biodiversity loss;
- Reflexive consumption which incorporates the principle of ecologically responsible consumption as a form of ecological citizenship in the global system;
- Reflexive production which "requires the manufacturer to take environmental responsibility for its product throughout its entire life cycle," including resource extraction, use, disposal, reclamation, or recycling;
- Greening of both national and international governance.

Christoff and Eckersley believe that governments should impose "ecological ceilings or safe sustainability boundaries" on economic activity that will eventually result in absolute, rather than relative, decoupling from resource depletion. However, in the long run, such a decoupling process may constitute wishful thinking that prevents contemplating post-capitalist alternatives. Indeed, Christoff and Eckersley (2013: 206) concede that "[t]he globalisation and transnationalization of capitalism has [*sic*] made reflexive governance increasingly difficult," particularly in the "advancing and new economies," presumably such as China, where notions of ecological governance and citizenship remain weak.

In keeping with this emphasis on *reflexive modernization*, at least among those who take the findings of climate science seriously, ecological modernization has become a virtually hegemonic imaginal stance that asserts that environmental sustainability and effective climate change mitigation and adaptation can be implemented by adopting more efficient, environmental sustainable, and low-carbon-emitting energy sources and manufacturing processes. Ecological modernization is part and parcel of what goes under the rubric of *green capitalism* or *climate capitalism*. Newell and Patterson (2010: 9) assert:

> So the challenge of climate change means, in effect, either abandoning capitalism, or seeking a way to find a way for it grow while gradually replacing coal, oil, and gas. Assuming that the former is unlikely in the short term, the questions to be asked are, what can growth be based on? What are the energy sources to power a decarbonised economy? ... What kind of climate capitalism do we want? Can it be made to solve desirable social, as well as environmental, ends?

Essentially, like Christoff and Eckersley (2013), they seek to imagine a decoupling of increasing GHG emissions from economic accumulation and growth.

Various parties have been calling for a Green New Deal (Simms 2009). In one version of the Green New Deal, Tim Jackson (2017: 7), economics commissioner of the Sustainable Development Commission in the UK, argues that the global financial crisis of 2008 offered humanity a "unique opportunity to address financial and ecological sustainability together" by questioning the "underlying vision of prosperity built on continual growth." In keeping with the Jevons paradox, or rebound effect, which arises when technological progress or government policy increases the efficiency with which a resource is used, but the falling cost thereof then leads to rising consumption, he acknowledges that improvements in energy and carbon intensity tend to be offset by increased economic growth and consequently greater usage of fuels. Jackson calls upon governments to invest in public infrastructure, reduce social inequality, redistribute existing jobs and reduce work hours, reverse the culture of consumption, implement resource/emissions caps, and shift to alternative energy sources. As a green Keynesian economist, he appears to assume that global capitalism can function as a steady-state or non-growth economic system, despite the fact that history has repeatedly demonstrated that, by its very nature, it must grow or die out.

Geoengineering

Various parties, including technocrats, climate scientists, and even some climate activists, have proposed a wide range of geoengineering remedies, or what might be called *climate engineering*, as climate change mitigation strategies. These include placing a gigantic sunshade in space so as to drastically cool the Earth's average temperature; injection of sulfate aerosols into the atmosphere to increase cloud cover, thus increasing solar reflectivity; removal of carbon dioxide from the atmosphere through the use of photosynthesis by growing ocean plants such as plankton; and enrichment of some ocean areas with iron particles to better absorb carbon dioxide. Edward Teller, a renowned physicist, has proposed placing "billions of tinfoil strips in orbit around the Earth to reflect up to 2% of the incoming sunlight and cool down the planet" (Gore 2009: 314).

Perhaps the most highly touted geoengineering proposal is solar radiation management (SRM), also called albedo modification or solar geoengineering (SGE). It has become very popular due to the speed in which one version of it, namely the spraying of sulfuric acid aerosols into the stratosphere, could be deployed on a large scale, quickly, and supposedly cheaply, in order to reduce or slow global warming. SGE is being actively researched and promoted, particularly in the United States. David Keith (2013: x), a Harvard University-based scientist, refers to SGE as a "cheap tool that could green the world." Geoengineering advocates tend to argue that mitigation will not be sufficient to stave off the worst impacts of climate change. The implementation of this controversial measure at both the global level and as an economic opportunity for the corporate sector remains problematic. Bill Gates, who has in recent years become alarmed about the gravity of climate change, has spent millions of dollars on geoengineering research and has invested in Intellectual Ventures, a firm developing the StratoShield, a 19-mile-long hose designed to spray sulfur dioxide particles into the atmosphere and another device designed to curtail hurricanes (Klein 2019: 106–107). Jeremy Baskin (2019: 246–250) posits five factors that thus far have constrained SGE's normalization:

- Uncertainties and risks associated with its deployment;
- Its association with a dystopian view of the future;
- The widespread view on the part of people and politicians that climate change is a slow process that either is not a serious threat or one to which humanity can slowly adapt;
- The absence of a powerful advocate, with the possible exception of Russia; and

- The reality that is opens up the possibility of unforeseeable and extremely disruptive geopolitical consequences.

SGE has not been mentioned in the Paris Agreement, or the Paris Climate Accords, and by and large it has encountered antipathy from the United Nations Framework Convention on Climate Change community.

SGE has the backing of fossil fuel interests and involves spraying sulfate aerosols into the atmosphere to reflect solar rays away from Earth, providing a kind of chemical sunshade. Critics point out that even if it were possible to add enough aerosols to counteract the amount of warming from daily carbon pollution, it would not halt the various adverse effects of past GHG emissions. Furthermore, many geoengineering simulations produce undesirable phenomena such as tropical overcooling, which entails land near the equator cooling while areas near the poles remain much hotter. Meanwhile, research by Carlson et al. (2022) found that SRM will cause a significant geographic shift in malaria cases that in some scenarios could put an additional billion people at risk of malaria. Ord (2020: 113) cautions:

> A central problem with geoengineering is that the cure may be worse than the disease. For the very scale of what it is attempting to achieve could create a risk of massive unintended consequences over the entire Earth's surface, possibly posing a greater existential risk than climate change.

Limitations of Reflexive Modernization

Some aspects of ecological modernization, such as renewable energy sources (solar, wind, and geothermal), improved efficiency and building construction and design and a massive shift from private vehicles to energy-efficient public transport systems, have the potential to serve as important climate change mitigation strategies. However, as anthropologist Alf Hornborg (2001: 25–26) argues:

> What ecological modernization has achieved is a neutralization of the formerly widespread intuition that industrial capitalism is at odds with global ecology … The discursive shift since the 1970s has been geared to disengaging concerns about environment and development from the criticism of industrial capitalism as such. But the central question about capitalism should be the same now as it was in the days of Marx: Is the growth of capital to everybody, or only a few at the expense of the others?

Thus, ultimately, technological innovations that on the surface appear to be more environmentally sustainable or energy-efficient must be part and parcel of a shift to a steady-state or zero-growth global economy if they are to circumvent the Jevons paradox.

Geoengineering represents a more extreme genre of ecological moderniza-tion, namely ecomodernism, which has found support from various capital-ists, such as Bill Gates, and may border on a more climate dystopian stance than a reflexive modernization one in that it is presented as the ultimate techno-fix when others not have not sufficiently mitigated climate change.

Reflective modernization is largely synonymous with green capitalism which fails to address the treadmill of production and consumption that contributes to the depletion of natural resources, environmental degradation, and climate change. It tends either to be oblivious to social justice issues or at best to downplay them or pay them lip service, and ignores the strong possi-bility that capitalists or the corporate sector will resist changes which imply a limiting of profit-making.

Eco-Socialism as an Alternative World System

While the powers that be around the world are seeking to address climate change within the parameters of global capitalism, as Simms (2009: 184) observes, "global warming probably means the death of capitalism as the dominant organising framework for the global economy." Thus, it is impera-tive to think outside the box and develop an alternative global capitalism as the ultimate climate mitigation strategy, even though it will not be achieved any time soon—if ever. As humanity enters an era of catastrophic climate change, it will have to consider alternatives to global capitalism in order to circumvent dystopia scenarios.

Proposals for an alternative system have come under various terms, including *global democracy, Earth democracy, eco-socialism,* and *eco-anar-chism.* Despite all the baggage associated with the term *socialism* and the desire of various progressive thinkers to substitute other terms for it, it is important for people oriented toward a socially just and environmentally sustainable world to grapple with ideals of socialism and the social experi-ments that have been labelled socialist or even communist at national and local levels. As Samir Amin (2009: 22) so aptly asserts:

> The expression counterculture is fraught with difficulty—because socialist culture is not there in front of our eyes. It is part of a future to be invented, a project of civilization, open to the creativity of investigation.

In other words, socialism remains very much an active vision, continually being revisited and redefined, and a way of imagining a world without unconstrained capitalism with which various individuals and groups continue to grapple, often by seeking to frame it in new guises. As Stilwell (1992: 211) argues, "liberated from Stalinist legacy, it now makes sense to start asking what a progressive socialism involves."

In the 19th century, various revolutionaries and reformers sought to develop alternatives to an increasingly globalizing capitalist world system. Efforts at the national level to create such an alternative began with the 1848 Paris Commune and the Bolshevik Revolution in Russia in 1917 and later encompassed subsequent revolutions in various other countries, including China in 1949, Vietnam in 1954, Cuba in 1959, and Nicaragua in 1979. Unfortunately, as Wright (2010: 106) observes, "these attempts at ruptural transformation ... have never been able to sustain an extended process of democratic experimental institution-building" for a variety of complex reasons. Scholars have spilled much ink trying to determine whether these societies constituted examples of "state socialism," "actually-existing socialism," "degenerated workers' states," "post-revolutionary societies" which had to undergo democratic revolutions in order to transition into socialism," or "new class societies." They also asked why some of these societies, particularly the Soviet Union or more precisely Russia but also China, later became fully incorporated into the capitalist world system. Suffice it to say that the failure to achieve authentically democratic socialist societies was related to both internal forces specific to each of these societies and external forces that created a besieged environment. According to Wright (2010: 106),

> [p]erhaps the failure of sustained democratic experimentalism in the aftermath of revolution was because revolutionary regimes always faced external pressure, both economic and military, from powerful capitalist countries, and felt a great urgency to consolidate power and build institutions of sufficient strength to withstand that pressure ... Or perhaps the problem was mainly the low level of economic development of the economies within which revolutionary movements succeeded in seizing political power.

In 1997, the authors of this volume, along with Ida Susser, coined the term *democratic eco-socialism* as a vision for creating a healthy world in our medical anthropology textbook, which went through two subsequent editions (Baer et al. 1997, 2003, 2013). We simply conflated the two previously existing terms *democratic socialism* and *eco-socialism.*

Democratic Socialism

Numerous Marxists have asserted that socialism is inherently more democratic than capitalist societies could ever be, and thus democracy is an inherent component of authentic socialism.

Miliband (1994: 51) delineates three core propositions that define socialism: (1) democracy; (2) egalitarianism; and (3) socialization or public ownership of a predominant portion of the economy. Socialist democracy would not be identical to total state ownership and centralized planning but could include collective, cooperative, and even individual property. Tariq Ali (2009: 88) argues that 21st-century socialism should include political pluralism, freedom of speech, access to the media, the right to form trade unions, and cultural liberty. In recent years, various leftists have been revisiting the concept not only of socialism but also of communism. Alan Badiou argues that "more than ever, we can, we must, and we will reactivate the communist hypothesis" (Badiou and Gauchet 2016: 48). Jodi Dean (2016: 1), another proponent of the communist hypothesis, argues that an alternative goal to capitalism is needed in "light of the planetary climate disaster and ever intensifying class war as states distribute wealth to the rich in the name of austerity." Most scholars who have embraced the communist hypothesis have not discussed the global ecological crisis and how communism, let alone socialism, might serve to solve it. For example, Badiou and Žižek have dismissively asserted that "ecology has become the new opium for the masses" (quoted in Foster and Clark 2016: 9).

Eco-Socialism

In the past, Marxian political economy has tended to give, at best, passing consideration to environmental factors. However, various Marxist theorists, including Herbert Marcuse, Erich Fromm, E.P. Thompson, Raymond Williams, André Gorz, Barry Commoner, and Rudolf Bahro served as precursors to present-day eco-socialism (Wall 2010: 82–89). A growing number of neo-Marxian scholars, as well as other radical scholars, have been attempting to integrate ecological considerations into their analyses of various types of social formations and societies. Such endeavors have been referred to as the political economy of ecology, green socialism, eco-socialism, radical ecology, socialist ecology, social ecology, eco-anarchism, and Earth democracy. Foster (2014: 57) maintains:

> Although contributions to ecological thought within the Marxist tradition have existed since the beginning—going back to Marx himself—ecosocialism, as a distinct tradition of inquiry, arose primarily in the late 1980s and early '90s under the hegemony of green theory (and in the context of the crisis of Marxism following the downfall of Soviet-societies). The general approach adopted was one of grafting Marxian conceptions onto already existing green theory—or, in some cases, grafting green theory onto Marxism.

However, the earliest use of the term eco-socialism may harken back to a pamphlet entitled *Ecosocialism in a Nutshell* published in 1980 by the Socialist Environment and Resources Association (Albritton 2019). Albritton (2019: 5) argues that "since the publication of this pamphlet, 'ecosocialism' has come to be seen by large numbers of people as the theoretical and action concept most appropriate for mobilizing against capitalism in the twenty-first century."

Ecological Marxism or eco-socialism has made some headway among Chinese Marxist scholars. Wang et al. (2014: 49) believe that ecological

Marxism "can help remind the Chinese Communist Party (CCP) of its ecological responsibility, and because environmental issues involve the vital interests of the people, it can remind the Party of its social responsibility as well." However, they admit that the "mainstream in China still tends to rely on the modernistic, technologically determinist, developmentalist way of thinking to address the problems facing China," despite the official CCP declaration of commitment to the notion of *ecological civilization.*

Democratic Eco-Socialism

Democratic eco-socialism embraces the following principles:

- An economy oriented to meeting basic social needs, namely food, clothing, shelter, education, health, and dignified work;
- A high degree of social equality;
- Public ownership of the means of production;
- A blend of representative and participatory democracy;
- Environmental sustainability.

Democratic eco-socialism rejects a statist, growth-oriented, productivist ethic and recognizes that humans live on an ecologically fragile planet with limited resources that must be sustained and renewed as much as possible for future generation.

The vision of democratic eco-socialism closely resembles what world system theorists Terry Boswell and Christopher Chase-Dunn (2000) term *global democracy*, a concept that entails the following components:

- An increasing movement towards public ownership of productive forces at the local, regional, national, and international level;
- The development of an economy oriented towards meeting social needs, such as basic food, clothing, shelter, and healthcare and environmental sustainability rather than profit-making;
- The eradication of health and social disparities and the redistribution of human resources between developed and developing societies and within societies in general;
- The curtailment of population growth that in large part would follow from the previously mentioned conditions;
- The conservation of finite resources and the development of renewable energy resources;
- The redesign of settlement and transport systems to reduce energy demands and greenhouse gas emissions;
- The reduction of waste through recycling and transcending the reigning culture of consumption.

Democratic eco-socialism constitutes what Erik Olin Wright (2010) terms a *real utopia*, a vision that is achievable but only through much theorizing and social experimentation. As the existing capitalist world system continues to self-destruct due to its socially unjust and environmentally unsustainable practices, many of which result in GHG emissions that contribute to anthropogenic climate change, democratic eco-socialism seeks to provide a vision to mobilize human beings around the world, albeit in different ways, to prevent ongoing human socioeconomic and environmental degradation, including catastrophic climate change.

For the purposes of this book, in the spirit of the larger eco-socialist project, we drop the adjective "democratic," operating under the premise that socialism as global endeavor by definition must be democratic. Many have argued that socialism has been tried in places like the Soviet Union and China, and even Cuba for that matter, and has proven wanting. Of these societies, Cuba is the closest example of a existing post-revolutionary country that embodies socialist ideals and practices. Yet socialism along with eco-socialism remain visions rather than existing social systems per se. As Foster (2009: 276) so aptly argues:

> It is important to recognize that there is now an *ecology* as well as a political economy of revolutionary change. The emergence in our time of sustainable human development in various revolutionary interstices within the global periphery, could mark the beginning of a universal revolt against both world alienation and human self-estrangement. Such a revolt, if consistent, could have only one objective: the creation of a society of associated producers rationally regulating their metabolic relation to nature, and doing so not only in accordance with their own needs but also those of future generations and life as a whole. Today, the transition to socialism and the transition to an ecological society are one.

While at the present time or for the foreseeable future, the notion that eco-socialism may be eventually implemented in any society, either in the Global North or the Global South, or in a bloc of countries may seem utterly ridiculous, history tells us that social changes can occur very quickly once economic, political, and social structural forces have reach a tipping point, one that would result in a socio-ecological revolution.

Radical Perspectives with which Eco-Socialism Can Dialogue

Bearing this thought in mind, we now explore various radical perspectives, namely eco-anarchism, eco-feminism, the de-growth perspective, and the Indigenous perspective on the socio-ecological crisis from which eco-socialists can draw and hopefully vice versa, particularly in the Australian context.

Eco-Anarchism

In terms of environmental issues, Anderson (2006: 256) asserts that "[a]narchism has a long history of environmentalism, from early anarchist thinkers such as Peter Kropotkin to the influential social ecologist Murray Bookchin (1991), who links the exploitation of nature to the exploitation of human beings, in much the same vein as eco-socialists. Greek eco-anarchist Takis Fotopoulos (1997) maintains that "modern hierarchical society," which for him includes both the capitalist market economy and "socialist" statism, is highly oriented towards economic growth, which has glaring environmental contradictions. Ted Trainer is an Australian eco-anarchist, who Baer visited briefly in 2008 at the University of New South Wales where he was an honorary fellow in late 2009 and with whom Baer has corresponded periodically since. In the November 3, 17, and 24, 2020, issues of *Green Left* Trainer and Baer engaged in a dialogue comparing eco-socialism and eco-anarchism. While Baer does not agree with Trainer on all issues, including the role of the state and social movements, in bringing about a socially just and environmentally sustainable society, his thinking has influenced Baer in various ways. He is a scholar whose work eco-socialists should consider. Although Trainer has written several books, we turn to some of the principal points made in his book entitled *The Transition of a Sustainable and Just World*, in which he explores the following themes about capitalist society in general:

- It is highly unsustainable;
- It has a highly unjust global economy;
- It fosters over-consumption;
- Scarcity is the defining feature of the coming era;
- It cannot fix itself;
- It requires massive and radical changes;
- The alternative entails a "Simpler Way."

(Trainer 2010: 2–12)

Trainer delineates the following five core principles of the Simpler Way:

- Material living standards, on the whole, must diminish;
- There is a need for small-scale, highly self-sufficient local economies and communities;
- Local communities based on cooperative and participatory principles control their own affairs, largely independent of international and global social structures;
- A radically different economic system needs to be developed, one that is socially controlled, oriented towards meeting needs as distinct from maximizing profits and not driven by market forces and a growth paradigm;

- These changes require a radical shift in values and worldview, especially away from competition, greed, and acquisitiveness.

(Trainer 2010: 6–13)

Trainer's first principle applies primarily to people who are relatively affluent in both the Global North and the Global South and particularly the super-rich. It clearly does not apply to people who are abjectly poor in developing countries, particularly those in sub-Saharan Africa, or the indigent poor, particularly among Indigenous peoples, ethnic minorities, and homeless people in developed countries, including Australia. Many of these people need to undergo "appropriate development," which for Trainer

> [S]hould be about improving all aspects of society, including the quality of food and water and health services, the opportunities for leisure and cultural activity, the level of debate and discussion, the processes for government and administration, the moral standards, the geographic and aesthetic conditions in which people live, citizenship and social responsibility, openness and accountability, social cohesion, equity, concern for the less fortunate, the quality of life, security, the conditions of the poorest, and especially ecological sustainability.

(2010: 129–130)

Trainer (2010: 139) has a high regard for Indigenous and peasant lifestyles, which are "typically highly collectivist, and whose economies are governed by customs and traditions, not profit and gain."

Terry Leahy, another Australian eco-anarchist, envisions a "gift economy" in which

> [p]roducers in all sectors should decide what to produce and how to distribute it based on the needs of other groups. The status of the givers would depend on genuine needs being met by the gift. Production in each sector would be aware of their dependence on the services of other sectors.

(2011: 114)

He argues that "hybrid strategies" or practices employed by producers or community members could serve as transitions from the existing system to a gift economy. Existing examples of hybrids include support for environmental taxes, willingness to pay a higher price for more environmentally sustainable products, and volunteering to work in community gardens, which already exist in abundance in Australia (Leahy 2011: 122). Leahy (2011: 122) argues that a "revolution by stealth" might occur as the pervasiveness and influence of hybrid organizations come to dominate the economy, nationally, regionally, and internationally. Conversely, any gift economy would probably "have to achieve either a near universal spread or the military capacity to defend attacks from capitalist forces" (Leahy 2011: 125).

Eco-Feminism

Eco-feminism emerged out of the feminist, peace, and environmental movements in the late 1970s and 1980s. It seeks to eradicate not only injustice against women and the environment, but all forms of social injustice. Eco-feminism assumes various genres, including liberal eco-feminism, cultural ecofeminism, social ecofeminism, and socialist ecofeminism (Merchant 2005: 200–211). Many eco-feminists advocate some form of an environmental ethic that deals with the twin oppressions of the domination of women and nature through an ethic of care and nurture that arises out of women's culturally constructed experiences. Liberal eco-feminists seek to liberate women within the contours of mainstream society while nurturing the environment. Spiritual eco-feminists focus on the female principle as cardinal to Mother Earth, who nurtures and sustains all human beings. For instance, Australian spiritual eco-feminist Freya Matthews (1991) draws inspiration from her claim that Indigenous people view the planet Earth as the great Mother and womb of life, which they view as sacred. Radical eco-feminists challenge the North/South divide associated with capitalist patriarchy, in which "nature is subordinated to man; woman to man; consumption; and the local to the global, and so on" (Mies and Shiva 2014: 5). They maintain that patriarchal societies are built on five interrelated pillars: sexism, racism, classicism, capitalism, and environmental degradation.

Socialist eco-feminism makes the category of reproduction as opposed to production central to the notion of achieving a socially just and environmentally sustainable world system. Ariel Salleh (1997: 12–13) argues that the ecological crisis stems from a Eurocentric capitalist patriarchal culture built on the domination of nature and domination of women "as natural." She argues that global justice and sustainability demand that the "North will have to review its high-tech consumption in favor of more species-egalitarian models by the South provisions itself (Salleh 1997: 142). Salleh (1997: x) argues that eco-feminism reaches for an "Earth democracy" that "reframes environment and peace, gender, socialist, and post-colonial concerns beyond the single-issue approach fostered by bourgeois right and its institutions." Nature must benefit all human beings rather than just privileged ones as is the case under capitalist parameters. For Salleh (2009: 291), eco-feminism constitutes an "environmentalism of the poor," arguing that even in the Global North women, as a result of their regenerative labors, experience poverty and pain while men do not.

De-Growth

The de-growth perspective has developed more or less in tandem with eco-socialism since roughly the late 1970s or early 1980s. At times eco-socialism and de-growth have conversed with each other, but at others they have talked past each other. For example, Serge Latouche, a French economist and anthropologist who popularized de-growth, drew on the work of various radical thinkers, including French Marxist André Gorz (1980), who challenged the growth paradigm by arguing that perpetual growth was impossible. Vincent Leigey, a spokesperson for the French and international de-growth movement, and Anitra Nelson, an Australian social scientist who at times has referred to herself as an eco-socialist, state:

> Degrowth is an invitation to go on the inevitably long journey of the decolonisation of our growth imaginaries, moving from cultural awareness to a systemic and material transformation changing our everyday practices. Degrowth insists on the deconstruction and re-evaluation of beliefs within, and relations between, capitalism and productivism, consumerism and materialism, development and the quasi-religion of economism, science and technology.
>
> *(2021: 12)*

Anthropologist Jason Hickel (2020: 204) has pinned his hope for a pathway to a post-capitalist world in the de-growth paradigm which proposes "reducing the material and energy throughput of the economy to bring it back into balance with the living world, while distributing income and resources more fairly, liberating people from needless work, and investing in the public goods that people need to thrive." In order to achieve a global post-capitalist economy, he notes a need to democraticize the World Bank, the International Monetary Fund, and the World Trade Organization; a debt jubilee freeing poor countries to invest in needed services; ending corporate land grabs; distributing land back to small farmers; and reforming regimes that provide high-income countries an unfair advantage in the global agricultural industry. Other steps that Hickel suggests for moving humanity to an "ecological civilisation" include ending planned obsolescence; cutting advertising; shifting from ownership to usership; eradicating food waste; scaling down ecologically destructive industries; eliminating "unnecessary" jobs and shifting workers to jobs in renewable energy production, public services, maintenance, and so on; shortening the workweek; capping wage disparities; and expanding the commons. However, his book, which appears to have drawn considerable attention, suggests that he is not conversant with the burgeoning literature on eco-socialism.

We fear that the notion of de-growth is a Eurocentric one, more applicable to the Global North than to the Global South. Obviously, there are large sectors of developed societies and smaller sectors of developing societies that need to undergo de-growth, but the abjectly poor of the developing countries in particular, along with homeless people or Indigenous peoples living on reservations in North America and Australia, need to undergo some sort of growth or development in terms of access to nutritious food, decent housing and sanitation, healthcare, and education. While most de-growth theorists would probably agree, our reading of the de-growth literature is that it does not place enough emphasis on social justice issues.

Indigenous Voices

Indigenous people created a sociocultural system in which they "understood that nature would generously provide for all of their material needs in return for minimal human effort" (Bodley 2017: 32). In order to survive in a wide array of eco-niches, Indigenous people around the world have developed traditional ecological knowledge based on their observations of local ecological conditions that have allowed them to adapt to environmental changes. Some Indigenous Australians have been involved in the Caring for Country movement, which seeks to preserve, protect, and rejuvenate traditional lands reclaimed through native title laws, many of which had been designated Indigenous Protected Areas (Altman and Kerins 2012). Indigenous peoples in both the Global North and the Global South have been on the front lines of struggles for land rights and fighting environmental devastation due to land grabs and corporate-based extractivism. Historically, they have struggled against colonization and continue to struggle against racism, neocolonialism, and imperialism around the world. Without romanticizing them, particularly given that sectors of the communities have been seduced by corporate promises of prosperity, their views on communalism and collective responsibility are highly congruent with eco-socialism.

Half-Earth Socialism

Troy Vettese and Drew Pendergrass (2022) in *Half-Earth Socialism* maintain that environmentalists and socialists need a boost to regain political momentum. They draw upon E.O. Wilson's notion of *half-Earth* which asserts that humanity needs to rewild half of the planet to halt the severe biodiversity loss that is presently under way. Vettese and Pendergrass (2022) also argue that the pursuit of global social equality must be part and parcel of half-Earth socialism. While they cannot say how a "Half-Earth socialist coalition might come to power', they argue that the dire future that current socio-ecological conditions presage makes "it is all the more pressing to imagine utopian alternatives to motivate and mobilize the dispirited masses" (Vettese and Pendergrass 2022: 17–188). For them, half-Earth socialism would entail a massive global planning system which would include the following dimensions:

- Supplying "everyone with the material foundation for a good life—sustenance, shelter, education, art, health—while protecting the biosphere from destabilization" (Vettese and Pendergrass 2022: 96);
- Setting "half the Earth aside for rewilding to limit the ecocide of the Sixth Extinction" (Vettese and Pendergrass 2022: 101), a measure that would require shifting food production drastically away from livestock toward veganism;
- The manufacture of solar panels, wind turbines, super-efficient insulation and railways;
- Massive investment in public transit and renewable energy, including a clean hydrogen industry;
- An "almost complete abolition of private vehicles" (Vettese and Pendergrass 2022: 110);
- Stabilizing global population at a maximum of 10 billion people;
- Retrofitting buildings to conserve energy and adapting private mansions and private headquarters to communal use;
- Rewilding private lawns and golf courses;
- Wide-ranging improvements to industrial processes to reduce pollution, fuel use, and wastewater;
- Grappling with the "failures of past socialist societies" (Vettese and Pendergrass 2022: 130);
- A serious commitment to democracy and meaningful work.

Despite the numerous shortcomings of the Soviet Union, Vettese and Pendergrass (2022: 126) argue that it was a "crucial player in the development of climate science." Ultimately, they view their concept of half-Earth socialism as a "starting point for a deeper discussion of how socialism should function in an age of ecological crisis" (Vettese and Pendergrass 2022: 133).

Conclusion

Ongoing global warming and associated climatic and other anthropogenic environmental changes raise the question of how long humanity can thrive into and beyond 2100. While a large section of the international elite has come to recognize the seriousness of climate change, the solutions they propose under the guise of ecological modernization, green capitalism, and existing climate regimes are insufficient to contain catastrophic climate change. As a result, perhaps more than any other environmental crisis, anthropogenic climate change forces us to examine whether global capitalism needs to be transcended and humanity needs to develop a new approach, as some would see it, along eco-socialist lines. Indeed, Dawson (2017: 299) suggests that climate change increases the likelihood of a revolution in the future and argues that "humanity needs a global people's movement to battle climate chaos while generating work for the disenfranchised masses of the world."

Noam Chomsky, a world-renowned linguist and political analysist, and Robert Pollin, a political economist at the University of Massachusetts, United States, have written a short book entitled *Climate Crisis and the Global New Deal* (2020). Previously, Chomsky had only fleetingly touched upon climate change in short commentaries, but in this new book he engages more deeply with the topic. Chomsky's co-author Pollin is particularly critical of William Nordhaus, who received the Nobel Prize in Economics in 2018 for his research on the economics of climate change, describing his approach as being "utterly sanguine about accepting the risks we would face allowing the global mean temperature to rise by 4°C by 2150" (Chomsky and Pollin 2020: 23). Indeed, many climate scientists are now arguing that if the world does not rapidly reduce GHG emissions, a 4 °C warmer world will be reached in 2100, not 2150.

In commenting on the Australian bushfires in late 2019 and early 2020, Chomsky observes in the book that, after the Coalition Prime Minister Scott Morrison "returned grudgingly from a vacation to assure his constituents that he felt their pain," the "opposition labor leader toured the coal plants, calling for expansion of Australia's role as world champion coal exporter and assuring the country that this was quite consistent with Australia's serious comment to combating global warming" (Chomsky and Pollin 2020: 12).

Chomsky argues that a revival of the labor movement is essential for a variety of reasons, including addressing the environmental crisis, noting that Tony Mazzocchi, head of the US Oil, Chemical and Atomic Workers International Union, had been a "harsh critic of capitalism as well as committed environmentalist" (Chomsky and Pollin 2020: 50). However, while sympathetic to efforts to transcend capitalism, Chomsky asserts that the immediacy of the climate crisis requires addressing it within the "framework of state-capitalist systems" (Chomsky and Pollin 2020: 58), hence his endorsement of a Green New Deal as an interim strategy. Pollin argues that a viable Green New Deal would require large-scale public investment, public ownership, and the stringent regulation of emissions.

On the question of eco-socialism, Chomsky does not view it as a viable political project at the present time, seeing it rather as providing a forum for "sharpening ideas" about what a future society might look like. Differently, his co-author Pollin asserts that "eco-socialism and the Green New Deal are fundamentally the same project" (Chomsky and Pollin 2020: 146). Contrasting with the relative dynamism of climate movements in advanced capitalist societies, he argues that the movements in most low- and middle-income countries operate at more modest levels, although conceding that this situation could quickly change.

In terms of the foreseeable future, Immanuel Wallerstein (2007: 382) observes:

> I do not believe that our historical system is going to last much longer, for I consider it to be in a terminal structural crisis, a chaotic transition to some other system (or systems), a transition that will last twenty-five

to fifty years. I therefore believe that it could be possible to overcome the self-destructive patterns of global environmental change into which the world has fallen and establish alternative patterns. I emphasize however my firm assessment that the outcome of the transition in inherently uncertain and unpredictable.

As Wallerstein's remarks suggest, as much as we social scientists would like to forecast future scenarios, as we have done in this chapter, our imaginings are filled with contingency and unpredictability.

Nevertheless, anthropologists and other social scientists can play a small but critical role in contributing their analytical skills and insights to a much larger struggle for the future—one in which we as a species learn to live in harmony with each other and with nature, including the rich biodiversity within it. Perhaps more so than any other issue, climate change allows critical social scientists, including anthropologists, to illuminate the contradictions of the capitalist world system and to envision the creation of an alternative world system based upon eco-socialist principles—and to test whether they have a real understanding of social action. Anthropology as a discipline has the potential to present the "big picture" while drawing on their observations of "small pictures" in different societies and cultures around the world. It has been very good at looking at human societies in the present and the past, both the immediate past and even the distant past. With a few exceptions, anthropology has been rather weak in attempting to look at the future but hopefully new imaginings, prompted by the seriousness of anthropological climate change and the ecological crisis will inspire anthropologists to engage with an anthropology of the future and be part of the larger project of developing a critical, and future directed, social science.

References

Accidental Gods. (n.d.) Five insights for avoiding global collapse with Gaya Herrington, episode 189 (podcast). https://accidentalgods.life/five-insights-for-avoiding-global-collap se-with-gaya-herrington/

Albritton, Robert. 2019. *Eco-Socialism for Now and the Future*. Cham: Palgrave Macmillan

Ali, Tariq. 2009. *The Idea of Communism*. London: Seagull Books

Altman, Jon and Kerins, Séan. eds. 2012. *People on Country: Vital Landscapes, Indigenous Futures*. Annadale: Federation Press

Alvarez, Alex. 2017. *Unstable Ground: Climate Change, Conflict, and Genocide*. Lanham, MD: Rowman & Littlefield

Amin, Samir. 2009. Capitalism and the ecological footprint. *Monthly Review*, November: 19–22

Anderson, James. 2006. Afterword: Only … the environment, "anti-globalization", and the runaway bicycle. In *Nature's Revenge: Reclaiming Sustainability in an Age of Corporate Globalization*. Josee Johnston, Michael Gismondi, and James Goodman, eds. pp. 245–273. Toronto, ON: Broadview

Badiou, Alan and Gauchet, Marcel. 2016. *What Is to Be Done? A Dialogue on Communism, Capitalism, and the Future of Democracy.* Susan Spitzer, trans. London: Polity

Baer, Hans A., Singer, Merrill, and Susser, Ida. 1997. *Medical Anthropology and the World System: A Critical Perspective.* Westport, CT: Bergin & Garvey

Baer, Hans A., Singer, Merrill, and Susser, Ida. 2003. *Medical Anthropology and the World System: A Critical Perspective* (2nd edn). Westport, CT: Praeger

Baer, Hans A., Singer, Merrill, and Susser, Ida. 2013. *Medical Anthropology and the World System: A Critical Perspective* (3rd edn). New York: Praeger

Baskin, Jeremy. 2019. *Geoengineering, the Anthropocene and the End of Nature.* Cham: Palgrave Macmillan

Beck, Ulrich. 2007. *Environment and Social Theory* (2nd edn). London: Routledge

Beck, Ulrich. 2010. Climate for change, or how to create a green modernity. *Theory, Culture and Society* 27: 254–266

Bodley, John. 2012. *Anthropology and Contemporary Human Problems* (6th edn). Lanham, MD: AltaMira Press

Bodley, John. 2017. *Cultural Anthropology: Tribes, States, and the Global System* (6th edn). Lanham, MD: Rowman & Littlefield

Bookchin, Murray. 1991. *The Ecology of Freedom* (2nd edn). Montreal, ON: Black Rose Books

Boswell, Terry and Chase-Dunn, Christopher. 2000. *The Spiral of Capitalism and Socialism.* Boulder, CO: Lynne Rienner Publishers

Chomsky, Noam and Pollin, Robert. 2020. *Climate Crisis and the Green New Deal.* London: Verso

Christoff, Peter, ed. 2014. *Four Degrees of Global Warming: Australia in a Hot World.* London: Earthscan

Christoff, Peter and Eckersley, Robyn. 2013. *Globalization & Environment.* Lanham, MD: Rowman & Littlefield

CNA. n.d. National security and the threat of climate change. https://www.cna.org

Connolly, William E. 2019. *Climate Machines, Fascist Drives, and Truth.* Durham, NC: Duke University Press

Darian-Smith, Eve. 2022. *Global Burning: Rising Antidemocracy and the Climate Crisis.* Palo Alto, CA: Stanford University Press

Dawson, Ashley. 2017. *Extreme Cities: The Peril and Promise of Urban Life in the Age of Climate Change.* London: Verso

Dean, Jodi. 2016. *Crowd and Party.* London: Verso

Foster, John Bellamy. 2014. Paul Burkett's *Marx and Nature* fifteen years after. *Monthly Review,* December: 56–62

Foster, John Bellamy and Clark, Brett. 2016. Marxism and the dialectics of ecology. *Monthly Review* 68(5): 1–17

Fotopoulos, Takis. 1997. *Towards an Inclusive Democracy: The Crisis of the Growth Economy and the Need for a New Liberatory Project.* New York: Cassell

Gore, Al. 2009. *Our Choice: A Plan to Solve the Climate Crisis.* London: Bloomsbury

Gorz, André. 1980. *Ecology as Politics.* Boston, MA: South End Press

Helmore, E. 2021. Yep, it's bleak says expert who tested 1972 end-of-the-world-prediction. *The Guardian.* https://www.theguardian.com/environment/2021/jul/25/gaya-herrington-mit-study-the-limits-to-growth

Hickel, Jason. 2020. *Less Is More: How Degrowth Will Save the World.* London: Heinemann

Hornborg, Alf. 2001. *The Power of the Machine: Inequalities of Economy, Technology, and Environment.* Walnut Creek, CA: AltaMira Press

Jackson, Tim. 2017. *Prosperity without Growth: Foundations for the Economy of Tomorrow* (2nd edn). London: Routledge

Jasanoff, Sheila. 2010. A climate for society. *Theory, Culture & Society* 27 (2–3): 233–253

Keith, David. 2013. *A Case for Geoengineering*. Cambridge, MA: MIT Press

Klein, Naomi. 2019. *On Fire: The Burning Case for a Green New Deal*. London: Allen Lane

Leahy, Terry. 2011. The gift economy. In *Life Without Money: Building Fair and Sustainable Economies*. Anitra Nelson and Frans Timmerman, eds. pp. 111–135. London: Pluto Press

Liegey, Vincent and Nelson, Anitra. 2021. *Exploring Degrowth: A Critical Guide*. London: Pluto Press

Lovelock. James. 2009. *The Vanishing Face of Gaia: A Final Warning*. Melbourne: Allen Lane

Lovelock. James. 2014. *A Rough Ride to the Future*. London: Allen Lane

Lynas, Mark. 2020. *Our Final Warning: Six Degrees of Climate Emergency*. London: Fourth Estate

Malm, Andreas and the Zetkin Collective. 2021. *White Skin, Black Fuel: On the Danger of Fossil Fascism*. London: Verso

Matthews, Freya. 1991. *Ecological Self*. London: Routledge

Merchant, Carolyn. 2005. *Radical Ecology: The Search for a Livable World*. New York: Routledge

Mies, Maria and Shiva, Vandana. 2014. *Ecofeminism*. London: Zed Books

Miliband, Ralph. 1994. *Socialism for a Skeptical Age*. London: Courage Press

Miller, Todd, with Buxton, Nick, and Akkerman, Mark. 2021. *Global Climate Wall: How the World's Wealthiest Nations Prioritise Borders Over Climate Action*. October. Amsterdam: Transnational Institute

Newell, Peter and Patterson, Matthew. 2010. *Climate Capitalism: Global Warming and the Transformation of the Global Economy*. Cambridge: Cambridge University Press

Ord, Tony. 2020. *The Precipice: Existential Risk and the Future of Humanity*. London: Bloomsbury

Parenti, Christian. 2011. *Tropic of Chaos: Climate Change and the New Geography of Violence*. New York: Nation Books

Salleh, Ariel. 1997. *Ecofeminism as Politics: Nature, Marx and the Postmodern*. London: Zed Books

Salleh, Ariel. 2009. From eco-sufficiency and global justice. In *Eco-Sufficiency & Global Justice: Women Write Political Ecology*. Ariel Salleh, ed. pp. 297–312. London: Pluto Books and Melbourne: Spinifex Press

Schwartz, Peter and Randall, Doug. 2003. An abrupt climate change scenario and its implications for United States national security. October. https://www.gbn.com

Shapley, Dan. 2007. Global warming and war: 6 flashpoints. *The Daily Green*, December 6. https://www.thedailygreen.com/environment-news/latests/national-security-4712607

Smith, Dan and Vivekananda, Janani. 2007. A climate of conflict: The links between change, peace and war. *International Alert*, November

Stilwell, Frank. 1992. *Understanding Cities and Regions*. Leichhardt: Pluto Australia

Trainer, Ted. 2010. *The Transition to a Sustainable and Justice World*. Canterbury: Envirobook

Vettese, Troy and Pendergrass, Drew. 2022. *Half-Earth Socialism: A Plan to Save the Future from Extinction, Climate Change, and Pandemics*. London: Verso

Wall, Derek. 2010. *The Rise of the Green Left: Inside the Worldwide Ecosocialist Movement*. London: Pluto Press

Wallerstein, Immanuel. 1998. *Utopistics: Or Historical Choices for the Twenty-First Century*. New York: New Press

Wallerstein, Immanuel. 2007. Climate disasters: Three obstacles to doing anything. Commentary No. 2005, Fernand Braudel Center, March 15. https://www.binghampton.edu/fbc/archive/205en.htm .

Wang, Zhihe, He, Hulli, and Fan, Meijun. 2014. The ecological civilization debate in China: The role of ecological Marxism and constructive postmodernism—beyond the predicament of legislation. *Monthly Review*, November: 37–59

Welzer, Harald. 2017. *Climate Wars: Why People Will Be Kills in the 21st Century*. London: Polity

Wright, Erik Olin. 2010. *Envisioning Real Utopias*. London: Verso

11

ECO-SOCIALISM AS THE ULTIMATE CLIMATE CHANGE MITIGATION STRATEGY

Fighting Climate Change: Reformist Reforms and Nonreformist Reforms

Historically socialists have engaged in intense debates as to whether the transition from capitalism to socialism will occur through revolutionary change or more gradual change in the form of reforms in various parts of the world. On this point, Marx himself was clear stating in one of his early writings that "without revolution, socialism cannot be made possible" (Marx 1975: 420). Revolutions involve more sudden and radical social transformations and are often associated with armed insurrection, as was the case with the American, French, Chinese, and Cuban revolutions, but ironically was not the case for the Bolshevik Revolution in Russia in October 1917. As (Cochran 2021) notes, "What is noteworthy … is how infrequent and limited … bloodshed was across such a tumultuous year." Reforms of the prevailing socioeconomic system, as has been advocated by those who wish to make capitalism more humane or more environmentally aware, despite the best of intentions, are often problematic in that they may serve to stabilize capitalism, as has repeatedly been the case around the world. Exemplary is the New Deal promoted by President Franklin D. Roosevelt during the period 1933–1940 in response to the misery and potential rebelliousness brought on by the Great Depression. As Ross-Nazzal (2018) remarks, "President … Roosevelt went to work to bring immediate relief, long-term recovery and to save capitalism." In light of this pattern, André Gorz (1973) differentiates between *reformist reforms* and *nonreformist reforms*. He uses the term reformist reform to designate the conscious implementation of minor material improvements that avoid any alteration of the basic structure in the existing social system. Between the opposite poles of reformist reform and complete and radical structural transformation, Gorz identifies a category of

DOI: 10.4324/9781003469940-12

applied work that he labels nonreformist reform. Here he refers to efforts aimed at making permanent emancipatory changes in the social alignment of power. In other words, implementing changes that are more than bandages on a festering system, changes that move social arrangements and popular consciousness forward towards socialism. In reality, these two types of reforms are sometimes hard to distinguish, but one way might be whether they are initiated by the powers-that-be or are pushed by the working class or various other subaltern groups or anti-systemic social movements. Gorz (1994: 40) argues that the welfare state particularly characteristic of developed societies constitutes "more or less humanized capitalism" rather than "democratic socialism" as the Scandinavian countries are often described as having (this contrast is unclear). Reforms like social security, veteran's benefits, or improved health care coverage—all of which reflect socialist thought—are hard for conservative opponents to roll back because they move the majority of the population to a new level of understanding and expectation about the role of government in supporting social wellbeing.

Transitioning to Eco-Socialism

The transition towards an eco-socialist world system is not guaranteed and will require a tedious, even convoluted, path, one in which anti-systemic movements must play a central role. Marx viewed blueprints as a distraction from the political tasks that needed to be undertaken in the present moment and it is important to note that these efforts are indeed paramount. But history tells us that there always will be immediate struggles that must be addressed. Perhaps in this spirit we seek to delineate various proposals for system-challenging or radical transitional reforms. We often find that when people ask what it would take to make a transition to an eco-socialist world system, they are seeking some basic guidelines on how to move forward beyond merely bumbling along haphazardly one step at a time.

More and more, scholars and activists have been exploring how to move beyond the existing capitalist world system and towards an alternative world system based on social justice, democratic processes, and environmental sustainability. In this spirit, sociologist Erik Olin Wright (2010: 8) makes a case for envisioning *real utopias* as a necessary component of a "general framework for systematically exploring alternatives" to capitalism. By this, he means the elaboration of "clear-headed, rigorous, and viable alternatives to existing social institutions that both embody our deepest aspirations for human flourishing and take seriously the problem of practical design" (Wright 2010: 37).

British eco-socialist Derek Wall (2010: 68) delineates ten transitional policies that he hopes will promote discussion:

- Supporting Indigenous peoples' control of rainforests and other vital ecosystems;

- Empowering workers to assume control of bankrupt businesses;
- Utilizing government bailouts to support mutualistic resources;
- Converting arms and SUV production for public programs;
- Promoting open-source patenting;
- Land reform;
- Large-scale funding for libraries and other social services;
- A tax and welfare system to support common ownership;
- "Competition reform to transform ownership";
- Public ownership of pharmaceuticals and healthcare.

Whereas the achievement of eco-socialism constitutes a long-term vision, the struggle over the short term will entail more modest nonreformist reforms that pave the way for deeper systemic changes. As Victor Wallis (2009: 96–97) so astutely observes:

> The long-term dimension will encompass the total reorganization of society, including a dramatic change in the definition of both individual and common goals. The short-term dimension will reflect the need to immediately arrest the headlong drive to environmental collapse. It will inescapably require steps that can engage capitalist participation: the challenge will lie in not letting corporate interests define the bigger picture. This will entail developing a high level of mass consciousness, so that when ecological advances are attained, popular constituencies will be ready to claim the credit for them, and not allow corporate adaptations or concessions—such as renewable energy and "green jobs"—to be repackaged as a triumphant achievement of capital.

Thus, the trick becomes the extent to which radical activists work within the system, be it, for instance, with private renewable energy companies or government agencies, such as the US Environmental Protection Agency, that espouse progressive reforms without being co-opted by them. For purposes of discussion, and debate, this chapter focuses on the following systemic-challenging or radical transitional reforms:

1. The creation of new progressive, anti-capitalist parties designed to capture the state;
2. The implementation of greenhouse gas (GHG) emissions taxes at the sites of production that include measures to protect people on low incomes;
3. Increasing public ownership, socialization, or nationalization in various ways of the means of production;
4. Increasing social equality within nation-states and between nation-states;
5. Achieving a sustainable global population;
6. The implementation of workers' democracy;

7. The creation of meaningful work and shortening the workweek;
8. The creation of a new zero-growth world economy; renewable energy sources, energy efficiency, and green jobs;
9. The expansion of public transportation and massive diminution of a reliance on private motor vehicles and air travel;
10. The implementation of sustainable food production and forestry;
11. Resistance to the culture of consumption and adoption of sustainable and meaningful consumption;
12. The implementation of sustainable trade;
13. The implementation of sustainable settlement patterns and local communities;
14. Large-scale demilitarization.

These transitional steps constitute loose guidelines for shifting human societies or countries towards eco-socialism and a safe climate, but it is important to note that both of these phenomena will entail a global effort, including the creation of a progressive global climate governance regime. Our litany of proposed transitional reforms is a modest effort to contribute to an ongoing dialogue and debate as to how to move forward from the present impasse in which the world finds itself today. The application of these suggested transitional reforms will have to be adapted for countries around the world, both in the Global North and the Global South. Furthermore, our suggested transitional reforms do not exhaust other possible ones that will be needed to implement an alternative world system.

New Progressive Parties Designed to Achieve Democratic Capture of the State

Multinational or transnational corporations to a greater or lesser extent have come to dominate governments and politicians in countries around world. National governments have found much of their power taken over by transnational organizations in which corporate elites and politicians often confer, such as the United Nations, the World Bank, the International Monetary Fund, the World Trade Organization, the Group of 7, the Group of 20, the World Economic Forum, and the European Union (EU). These bodies constitute components of what Robinson (2014: 77–95) terms the "transnational state," a structure that facilitates the accumulation efforts of the "transnational capitalist class."

Althusser (2014: 109) maintains that revolution is "unthinkable without the capture of state power." Obviously the shift to an eco-socialist world will require a revolution of some sort that will have to be played out in various ways depending on the national context. Without doubt, the capitalist class and its political allies around the world will be fully and even viciously resistant to such a revolution. The larger question is whether an eco-socialist-oriented revolution can be achieved largely through peaceful measures or whether it will entail violence or perhaps a mixture of both, depending on the country. As Arrighi et al. (1989: 32) observe, regardless of whether state power is achieved through a "legal path of political persuasion" or through the "illegal path of insurrectionary force," in reality the terms reform and revolution "have become so overlaid with polemic and confusion that today they obscure more than they aid analysis." Althusser (2014: 107) asserts that achieving socialism through parliamentary means has been a nonexistent path thus far, although various attempts, such as in Chile in the early 1970s when Salvador Allende was elected president, have been pursued. This effort, however, failed as the result of a US-supported military coup that brought the repressive government of Gen. Augusto Pinochet to power. Indeed, as support for the Popular Unity movement increased between 1970 and the 1973 in Chile, "so did the *reaction* to these advances" (Hoffman 1984: 161; emphasis in the original). Nevertheless, while Marx and Engels did indeed envisage an armed overthrow of capitalism in some situations, they also gave attention to achieving reforms within the bowels of capitalist societies and viewed such efforts as a "school for evoking the political consciousness of the proletariat and building up its political organization," a strategy for "weakening the class domination of the bourgeoisie in all of its aspects," and a means for "potential peaceful transformation of capitalist society into a socialist one" (Wlodzimierz 1967: 87). Marx believed that socialist-oriented revolutions might occur in countries such as the United Kingdom, the Netherlands, and even the United States through peaceful and presumably parliamentary means.

Ultimately, achieving most of the transitional reforms in our listing may require that new left parties or socialist-oriented parties come to power and in a sense democratically "capture the state" and ensure that there is a political will that should guarantee their implementation. For example, nationalization of the means of production would be difficult to achieve without a progressive political party in power. Martin Ryle (1988) maintains that an environmentally friendly progressive state will be necessary to challenge corporate power. Until the election of the Coalition of the Radical Left—Progressive Alliance (known as Syriza) in Greece in early 2015, the possibility of new left parties coming to power in developed capitalist countries had appeared to be remote. But given the gravity of both the global economic and ecological crisis, including climate change, one should not rule out the possibility of political tipping points in something of the same sense that climate scientists speak of tipping points that had set off a number of irreversible climatic events. Due to the machinations of the European Central Bank, illegal practices by Greek banks, and internal divisions within its ranks (a common weakness of broad-based coalitions), the Syriza government came to a rapid demise. Nevertheless, its brief existence demonstrated the possibility that a socialist party could capture state power in a developed country.

Emissions Taxes

Along with various types of eco-taxes in general, emissions taxes and various other eco-taxes potentially can serve as progressive climate change mitigation strategies given the seriousness of the climate crisis (Loewy 2015: 11). Various climate scientists, such James Hansen, have argued that a realistic safe level of carbon dioxide would be in the order of 300–350 parts per million. Larry Lohmann (2006: 72; emphasis in the original) delineates five reasons why emissions trading constitutes a flawed approach to mitigation:

> *First*, in order to work, greenhouse gas trading has to create a special system of property rights in the earth's carbon-cycling capacity. This system sets up deep political conflicts and makes effective climate action exceedingly difficult. *Second*, pollution trading is a poor mechanism for stimulating the social and technical changes needed to address global warming. *Third*, the attempt to build new carbon-cycling capacity is interfering with genuine climate action. *Fourth*, global trading systems for greenhouse gases can't work without much better global enforcement regimes than are likely in the near future. And *fifth*, building a trading system reduces the political space available for education, movement-building and planning around the needed fair transition away from fossil fuels.

Various emissions trading schemes, including those in the United States and the EU, initiated in 2005, essentially grant corporations and developed countries property rights to emit GHG emissions. The carbon permit-prices under the EU over time and the EU scheme does not cover many sources of emissions. Conversely, a carefully crafted emissions tax has the potential to serve as a radical transitional reform. However, emissions taxes or carbon taxes are fraught with danger given the tendency on the part of corporations to find loopholes and pass the burden of the tax onto working-class people, thus compounding the primary injustice of climate change with secondary injustices. An emissions tax has several advantages over an emissions trading scheme, including that "taxes are generally less complicated, more transparent and less expensive to administer than emissions trading" and it "does not allow speculators to create financial instruments that would provide money-making opportunities and hence increase the costs of the carbon price" (Diesendorf 2014: 216–217). James Hansen, a retired NASA climate scientist, has called for a steep carbon tax as a strategy for quickly reducing GHG emissions. He has proposed a "fee and dividend" strategy for the United States through which fossil fuel companies would be charged an easily implemented carbon fee imposed at the well head, mine shift, or source of energy extraction (see Foster 2013).

We believe we must pursue taxation of the rich by closing the offshore wealth loophole and other accounting gimmicks—and use the money generated to fund effective climate policy. A glaring loophole in the heart of US tax law, for example, enables corporations to avoid paying tax on profits made in foreign countries until this income is brought home. Known as "deferral," this lobbied for privilege provides a powerful incentive to keep profits offshore as long as possible. Many corporations, in fact, choose not to bring their profits home and never pay US taxes on them. Tax avoidance through offshore tax haven loopholes is a significant reason why corporations, which paid one-third of federal revenues 60 years ago, now only pay one-tenth of federal tax revenues.

Public Ownership of the Means of Production

In an era of increasing privatization of social and health services, and even military activities and prisons, raising the specter of public ownership, nationalization, or socialization of the means of ownership is taboo in conventional economic and political circles. Incredibly, even disasters caused by climate change, such a Hurricane Katrina in the United States, have been used to implement unpopular privatization. As Naomi Klein (2007) asserts, when a society experiences a major shock there is widespread desire for a rapid, decisive response to rectify the problem. This clamor for bold action creates the opportunity for unscrupulous actors to take advantage of the situation to implement disaster capitalist policies while people are too distracted, both emotionally and physically, to develop effective resistance. In New Orleans, in the aftermath of Hurricane Katrina, for example, a well-financed movement, supported by a network of right-wing think tanks, swiftly launched a push to fire thousands of teachers, privatize the school system, and build a network of charter schools.

Privatization is often justified in terms of economic efficiency. But are such claims supportable? In the 21st century thus far, the world has witnessed three financial and fiscal crises caused by private companies. The global recession of 2007–2009, for example, was triggered by major private investment banks that overleveraged themselves using profitable but financially unsound mortgage-backed "junk" securities. Falling housing prices created a chain reaction of defaults and bank failures. What stopped the bleeding: government bailouts. In the process, wealth inequality increased. While the wealthy did suffer temporary losses in the short run as the financial crisis disrupted the stock market, middle class homeowners lost a great deal of unrecovered wealth, especially the emerging black and Latinx families who bought their first homes in the mid-2000s.

While state or government enterprises or services can be terribly inefficient for complex reasons, this does not necessarily have to be the case. There are numerous examples of publicly owned enterprises that operate relatively efficiently. As Tanuro (2010: 124) observes, ultimately eco-socialism requires "expropriation of monopolies in the energy sectors and confiscation of their assets" as well as a radical extension of the public sector in general, particularly in the "domains of transport and housing" and social services, which would be free. In calling for socialization of the ownership of monopolies, Amin (2013: 136) maintains that "[m]onopolies are institutional bodies that must be managed according to the principles of democracy, in direct conflict with those who sanctify private property." Public ownership could consist of a number of social arrangements, including state ownership, collective ownership, worker-owned enterprises, and cooperatives. State ownership, in turn, can and already does occur at various levels of government, including at the federal or national, state or provincial, and the municipal or local level. It is important to note that public ownership or nationalization of the means of production does not in and of itself constitute socialism, despite the fact that people have often assumed that it does. For example, after World War II the British state nationalized heavy industry that had been in decline for over 50 years, but retained previous owners in managerial positions. According to Rogers (2014: 106),

> [I]t appears that the government socialized the costs of post-war reconstruction and British relative economic decline, before returning industries to private hands, which in turn allowed those who were already at the top end of the income pyramid to reap the financial benefits of government's investment in declining industries while sharing the cost.

Wall (2010: 57) maintains that "eco-socialism is founded on the principle of common property rights, which allow individuals [and groups] free access to a resource as long as they don't damage it." What is also needed is preservation and extension of the common property system that still exists throughout much of the world. According to Burkett (2006: 314),

> The range of resources that are still being managed as common property is impressive: Southeast Asian wetlands, forests in parts of India and Nepal, fisheries in Japan and other Pacific Island nations, dams in Sri Lanka, canals in India, rangelands in various African countries, grazing lands in Britain and continental Europe, and groundwater in numerous countries.

Increasing Social Equality

While organizations such as UN Development Programme and the World Bank claim to be committed to the eradication of global poverty, they rarely, if ever, call for a redistribution of wealth or increasing social equality. Instead, such organizations, along with mainstream economists and most of the governments in the world, call for more economic growth, never acknowledging that wealth and poverty, development and underdevelopment, are interwoven. While some redistribution of wealth has been achieved under capitalism at various historical junctures and particularly in the Global North societies with strong labor unions and left-of-center governments, social inequality is an inevitable dimension of the capitalist world system. Ultimately, a shift towards greater social equality or parity will require transcending global capitalism and moving towards an eco-socialist world system.

Socialists have over the years engaged in intense debates about what sort of wage differentials should exist under socialism. Stilwell (2000: 130) argues that a 3:1 ratio of the highest to lowest incomes would be a tolerable standard for a socialist society. In reality, there are other compensations for work than material rewards, such as the intrinsic rewards of intellectual and even physical stimulation and the sense that one has contributed to the greater good. So long as rich people and corporations exist, progressive taxation that does not allow for tax loopholes constitutes an important mechanism for redistributing wealth.

Achieving a Sustainable Global Population

Population growth is largely a symptom of two deeper problems, namely poverty and patriarchy, both of which would be eradicated in an eco-socialist world. As Fred Magdoff (2013: 22) argues, while population can be an environmental problem, it is "usually not the main one, given that economic growth generally outweighs population growth and environmental degradation arises mainly from the rich rather than the poor." Many middle-class environmentalists who posit population growth as the principal environmental problem appear to want to maintain more or less their present material standard of living, albeit on a planet inhabited by far fewer people. However, in reality bringing down population growth will entail an improvement in quality of life for the poor, or essentially the elimination of poverty, which from an eco-socialist perspective should go hand-in-hand with creating a high degree of social equality. Authoritarian measures to reduce population growth have historically either met with "open rebellion, as in the case of Indira Gandhi's India, or evasion and subversion, as in the case of Deng's China under the one-child family policy" (Shue 2014: 74). What has worked to slow population growth: improved social and economic equality.

Workers' Democracy and Socialist Planning

Workers' economic or participatory democracy would constitute an integral component in a shift towards democratic eco-socialism. As Joel Kovel (2007: 143) asserts, "[w]e call ecosocialism that society in which production is carried out by freely associated labor [workplace democracy] with consciously eco-centric means and ends." Workers' democracy entails democratic and open debate at all levels of the worksite. Wolff (2012: 122–124) maintains that workers' self-directed enterprises (WSDEs) could play an important role in transitioning away from capitalism. Each WSDE would allow its workers to collaborate with the enterprise's board of directors in decision-making and would use part of its surplus to pay taxes to various governmental sectors that have supported its operation and would also consider the impact of its operations on the environment. Enterprise managers and directors could be elected so that they represent various levels of workers within an organization or directly by the entire workforce.

Since the 1960s, a modern wave of collective and egalitarian ownership among workers in factories and fields has spread around the world. Studies of these cooperatives show that they are more likely to avoid closure, have slightly lower wages but higher worker retention (i.e., lowering collected wages rather than firing workers during hard times), fewer wage differentials among workers/managers, are more productive, and have higher job satisfaction, trust, worker health, and commitment, and tend to use natural resource inputs more efficiently and are less growth-oriented than privately owned companies (Pérotin 2017).

Socialist planning is "nothing other than radical democratization of the economy" (Loewy 2015: 27). As part of socialist planning, "[e]very single fact of industrial production—energy production most urgently, but also transportation, housing, trade, agriculture, manufacture of commodities, and waste production and treatment—all require gigantic systemic change and complete structural reorganization" (Williams 2010: 217). Democratic planning needs to be central to the production process, such as in deciding what goods are needed and whether they are environmentally sustainable.

Meaningful Work and Shortening the Work Week

Socialism is committed to the notion of un-alienated and fulfilling or meaningful work. Satisfying work contributes to positive self-esteem and a sense that one is contributing to society and one's fellow human beings. Unemployment for most people, at least those in early adulthood and middle age, can be psychologically devastating. Even work or employment for people over the traditional retirement age of 60–65, depending on the country in question, can be a fulfilling and meaningful activity. A shorter workweek would permit everyone to be employed and thus eliminate the "industrial reserve army," which is an inherent feature of capitalist economies but should not occur in a socialist system. In his conception of eco-socialism, Sarkar (1999: 208) maintains that there would exist no unemployment:

> Firstly, labour-intensive technologies would be preferred, not only to provide jobs, but also because such technologies reduce resource consumption and, consequently, have less negative environmental impact. Secondly, even in a steady-state economy, there would be a lot of necessary work to be done. Food, clothing, housing, and services like education, health, postal service, and so on, would have to be produced. This would demand much labour, which would be equitably distributed among

all who can work. Thirdly, an eco-socialist government would pursue a policy of stabilising and then reducing the population ... Fourthly, ecologically benign technologies such as repairing, recycling, reusing, manual weeding instead of using pesticides, are all labour-intensive.

Gorz (1982: 124) asserts that an alternative to capitalism would entail "reducing what has necessarily to be done, whether enjoyable or not, to a minimum of each person's lifetime, and in extending as far as possible collective and/or autonomous activity seen as in itself." Juliet Schor wrote about the "overworked American." In her model of plentitude, she advocates what she terms "time wealth":

> Millions of Americans have lost control over the basic rhythm of their daily lives. They work too much, eat too quickly, socialize too little, drive and sit in traffic for too many hours, don't get enough sleep, and feel harried too much of the time. The details of time scarcity are different across socioeconomic groups; but as a culture we have a shared experience of temporal impoverishment.
>
> *(Schor 2010: 103)*

What Schor says of US culture applies more or less equally to Australian culture, despite the stereotype of Aussies being a laid-back people. Despite the fact that Australian workers pioneered the eight-hour workday, albeit at a time when the workweek was six days rather than five days, in the mid-19th century, many full-time employed Australians are now working over eight hours a day. When academics at the University of Melbourne ask each other how they are doing, a common refrain is "busy."

It is difficult or impossible to say what would be the optimal workweek. To some degree, this would vary from individual to individual. Marx characterized humans as "Homo Faber" or "Man the worker," but he was thinking of un-alienated labor where work and play are intricately interwoven, as is often the case in foraging societies. He envisioned a society where one would be able "to hunt in the morning, fish in the afternoon, rear cattle in the evening, criticise after dinner" (Marx and Engels 1978: 160). In a sense working life would never totally end so long as a person has the mental and physical capacity and desire to engage in it. Thus, people should be given the option of phasing into "retirement" rather than simply going from full-time employment to full-time retirement. Work under socialism and particularly under communism will in essence contribute to human development and allow people to achieve their full potential. While some of this occurs in capitalist societies, it is restricted to a privileged few and has become increasingly hard to attain, as many academics, for instance, who feel themselves under the demands of an increasingly corporatized university structure can testify.

The Need for a Steady-State Economy

A growing number of neo-Marxian scholars as well as non-Marxian scholars have been questioning the economic growth paradigm (Higgs 2014). On the non-Marxian side, E.F. Schumacher, who maintained that "small is beautiful," argues that "infinite growth of material consumption in a finite world is an impossibility" (quoted in Bjerg 2016: 14). Leading eco-socialist, John Bellamy Foster (1999: 114), asserts that economic growth has been the "chief reason for the rapid acceleration of the ecological crisis in the postwar period." On this matter, he even quotes Fred Cairncross, environmental editor for *The Economist*, who stated:

> Many people hope that economic growth can be made environmentally benign. It never truly can. Most economic activity involves using energy and raw materials: that, in turn, creates waste that the planet has to absorb. Green growth is therefore a chimera.
>
> *(Quoted in Foster 2002: 57)*

Not all Marxian scholars reject the growth paradigm per se. Some time ago, Barry Commoner (1972) maintained that an increase in economic activity does not necessarily translate into more environmental pollution. In other words, "what happens to the environment depends on how growth is achieved" (Commoner 1972: 141). He maintained that the US soil conservation program during the Depression helped to restore nutrients in the soil and thereby economic growth. Conversely, the "theory of socialist economics does not appear to require that growth should continue definitely" (Commoner 1972: 281). David Pepper (1993: 197) argues: "an ecological-communist utopia requires the development of productive forces … Eco-socialist growth must be a rational, planned development for everyone's equal benefit, which would therefore be ecologically benign". On the other hand, Paul Burkett (2006: 5–6) maintains:

> Historically, ever since Malthus, Marxists have been suspicious of any theory that posits purely natural limits to production and development. The reason is obvious: such theories tend to embody a conservative bias against all efforts to improve the human condition by fundamentally transforming class and other power relationships, or even by redistributing wealth and income.

In reality, this type of response has been an overreaction that has placed socialists out of touch with serious ecological considerations. A concerted redistribution of the world's resources would ensure an adequate living standard for everyone on the face of the planet, but this would entail a full discussion of how much is enough and, with the elimination of poverty, the recognition that the global population would begin to dwindle, thus placing less strain on the ecosystem.

Large sectors of countries in the Global South need to undergo growth that is directed to meeting "real needs such as access to water, for, health care, education, etc." (Foster 2011: 32). Ultimately issues of growth and de-growth and development and underdevelopment are intricately interwoven with the redistribution of resources and a drastic restructuring of the social relations of production, which would be an integral component of creating democratic eco-socialism.

Herman Daly and John Cobb (1990: 71) draw a distinction between growth and development: "'Growth' should refer to quantitative expansion in the scale of the physical dimensions of the economic system, while 'development' should refer to the qualitative change of a physically non-growing economic system in dynamic equilibrium with the environment." Growth entails utilizing more and more resources as part and parcel of the capitalist treadmill of production and consumption. Daly and Cobb (1990: 72) go on to argue:

> Any physical subsystem of a finite and non-growing earth must itself also eventually become non-growing. Therefore, growth will become unsustainable eventually, and the term "sustainable growth" would then be self-contradictory.

Development would entail providing people with adequate food, clothing, shelter, healthcare, and education. Noam Chomsky (2012: 84) maintains that the notion of growth is problematic if it entails "constant attacks on the physical environment that sustains life—like, for example greenhouse emissions, destruction of agricultural land, and so forth." Instead he advocates growth that leads to simpler lives, more livable communities, and maximization, or what he terms "growth in a different direction" (Chomsky 2012: 84).

Beyond a certain point, in terms of individual consumption, more food, clothing, and shelter are superfluous and certainly environmentally unsustainable. How much healthcare is necessary would depend on each individual's physical and mental state, which are not only interwoven but highly variable. From a critical health anthropological perspective, health can be defined as "access to and control over the basic material and nonmaterial resources that sustain and promote life at a high level of individual and group satisfaction" (Baer et al. 2013: 5). In a socialist society or a society seeking to construct socialism, there would be greater emphasis placed on preventive healthcare than curative health care.

Renewable Energy Sources, Energy Efficiency, and Green Jobs

A shift to renewable energy sources, particularly solar, wind, geothermal energy, and possibly ocean wave energy constitutes a significant component of climate change mitigation. Foster identifies two types of ecological revolution: an eco-industrial revolution and an eco-social revolution. The eco-democratic phase instigated by a broad-based radical movement would entail the following policies or goals:

- A reduction in carbon emissions in countries in the Global North of 8–10% per annum;
- A moratorium on overall economic growth combined with income and wealth redistribution, resource conservation, and economic waste reduction;
- A shift from military spending to planetary defense;
- The creation of an alternative energy infrastructure entailing a drastic shift to solar, wind, and other renewable energy resources;
- The closure of coal-fired power plants and a ban on coal-fired power plants and
- Unconventional fossil fuels, such as tar sand oil;
- A carbon fee and dividend system that would redistribute 100% of the revenue to the population on a per capita basis;
- Global initiatives to assist countries in the Global South to achieve human sustainable development;
- The implementation of environmental and climate justice to ensure that the poor, women, Indigenous people—particularly those in the Global South—do not bear the brunt of climate change;
- A new international climate regime modeled on the egalitarian and eco-centric principles delineated in the People's Agreement of the World People's Conference on Climate Change convened in Bolivia in 2010.

(Foster 2022: 78–79)

Foster (2009: 12–13) calls for a radical form of ecological modernization which

[D]raws on alternative technologies when necessary, but emphasizes the need to transform the human relations to nature and the constitution of society at its roots within the existing social relations of production. This can be accomplished only through a process of sustainable human development. This means moving decisively in the direction of egalitarian and communication forms of production, distribution, exchange, and consumption, and thus breaking with the logic of the dominant social order.

The eco-socialist phase would entail "establishing more egalitarian conditions and processes for governing global society" accompanied by the "requisite ecological, social, and economic planning" (Foster 2015: 8). Foster (2015: 8) believes that continuing ecological degradation and socioeconomic hardships will increase the numbers of an already-existing "environmental proletariat," drawing on so-called middle-class people and "some of the more enlightened elements of the ruling class" who forsake their class interests for the sake of humanity and the planet.

While acknowledging their potential usefulness, various scholars have observed that renewable energy sources are not a panacea for mitigating climate change (McCluney 2005; Trainer 2007). The creation of solar power plants, solar panels, wind farms, and geothermal plants will require the consumption of large amounts of non-renewable energy and non-renewable resources (Sarkar and Kern 2008: 20–21). In other words, renewable energy sources along with other green technologies incur *embodied* energy, the accumulated energy required to process and manufacture a product and to transport it to the site where it is utilized. Owen (2012: 187) observes that solar power plants and wind farms will require in part "vast installations, covering huge tracts of land," but theoretically this requirement can go hand-in-hand with the installation of photovoltaic panels on top of houses, apartment blocks, factories, and office buildings. In essence, nonrenewable resources, such as steel and aluminum, which may rely on coal to be processed, would be required to manufacture the infrastructure for renewable energy sources such as solar and wind power. Compared with other renewable energy sources, geothermal electricity production requires a relatively high level of water consumption. The magnetic generators used in the latest wind turbines require neodymium, a rare earth metal.

A crucial question is how much energy, regardless of the source, humanity needs. Given the demands of global capitalism to continually expand, under a business-as-usual scenario, humanity will need more and more energy in order to feed the treadmill of production and consumption and population growth. Energy utilization varies widely from region to region and from country to country. In a steady-state economy, energy requirements could theoretically level out or even eventually decline.

Ultimately, the deeper question that renewable energy enthusiasts seldom ask is "how much energy is needed in the first place," particularly in developed countries (Chambers 2011: 176). A large-scale transition to solar, wind, and other renewable energy sources will need to be coupled with a decline in per capita levels of consumption among the affluent of the world while allowing the poor to draw on these new energy sources to achieve access to basic resources. Obviously, some people in the world, particularly the poor in the Global South, desperately need access to more energy but many affluent people in both the developed and developing worlds need to reduce their energy consumption, often drastically, in order to achieve environmental sustainability and a safe climate.

A shift to eco-socialism will entail the creation of "green jobs," ones that are not only environmentally sustainable but also cater to people's social, educational, and healthcare needs. Ironically, global capitalism does create many jobs and working-class people often feel threatened by the loss of their jobs if there is a shift away from fossil fuels to renewable energy sources, although the latter will most definitely create new jobs, but also move us away from an endless treadmill of production and consumption, both of which do create jobs, both in terms of producing things and then selling them. The advertising and retail industries in themselves create numerous jobs that to a greater or lesser extent allow workers to at least meet their subsistence needs. Onaran (2010: 30) calls for "public expenditures in labour-intensive services like education, child care, nursing homes, health, and community and social services." The creation of green jobs must be accompanied by a "just transition," which means retraining displaced workers from obsolescent and environmentally destructive industries and enterprises to environmentally sustainable ones.

Sustainable Public Transportation and Travel

In his book *Ecotopia*, Ernest Callenbach (1975: 35) describes a fictional place situated in northern California, Oregon, and Washington state that has transcended cars. Aside from the question of whether such as place could exist in the modern world, as Peter Newman (2009: 108) observes, "[t]he biggest challenge in an age of radical resource-efficiency requirements will be a way to build fast rail systems for the scattered car-dependent cities." Environmentalists and other social activists began to challenge the pollution, health hazards, traffic congestion, urban sprawl, and fragmentation of social life resulting from motor vehicles and highways in the 1960s and 1970s, a period of social ferment on many fronts around the world (Golten et al. 1977). According to Ladd (2008: 133),

> Anti-freeway activists joined lovers of city life, conservationists (soon to be much more numerous and known as environmentalists), urban politicians, and a growing number of transportation planners in promoting a revival of mass transit during the 1960s. The car, they believed, was reaching the limits of its usefulness, even in the suburbs.

Belsky (2012: 54) delineates the following components of a sustainable transport system:

- Subsidies for public transport, cycling, and affordable housing close to public transport;
- Modernization of roads with real-time traffic management and operations;
- Road space protected for pedestrians, cyclists, public space;
- Bus rapid transit or rail in high-demand corridors;
- Public transport-oriented development;
- Stronger governance structures for transport and land use policy, planning, and management;
- More equitable access for the poor, disabled, young, and old.

However, his scheme still leaves too much wiggle room for roads. Modern cities have evolved following, in large part, the dictates of capital with its need for manufacturing, financial, commercial, distribution, and communication centers, as well as state bureaucracies. In cities where significant public transportation infrastructure exists but has not been developed and upgraded sufficiently to discourage car use, the growing proliferation of cars also reduces the efficiency of road-based forms of public transportation.

Of historic note, in 1926, Los Angeles had one of the largest mass transit systems in the world. To efficiently get around LA today, however, you must drive a car. The change dates to 1945 when a company known as National City Lines purchased the city's trolley system and then ripped up the tracks to make way for a new fleet of buses. The principal investors in National City Lines were General Motors, Firestone Tire and Rubber Co., Standard Oil of California, and Phillips Petroleum, entities with a clear vested interest in the success of the car. Subsequently, these companies were found guilty of forming a transportation monopoly.

Across the world, in Melbourne, Australia, buses and trams are slowed greatly by congestion caused by (often single occupancy) cars and a lack of on-road priority. Fortunately, a few Melbourne trams operate as light-rail conveyances in some motor vehicle-free stretches between the city or central business district and various suburbs.

Despite the existence of massive corporate support for the ongoing use of motor vehicles, there have been some counterhegemonic efforts to resist the auto-mobilization of society by emphasizing the need for people to rely on other forms of transportation. Unfortunately, many of the efforts to make cities greener have benefited the affluent much more than the poor and working-class people. A new urbanism that seeks to make cities more liveable and environmentally sustainable has emerged around the world and has begun to permeate urban planning. Various cities, including Singapore, Hong Kong, Zurich (Switzerland), Copenhagen (Denmark), Freiburg (Germany), Vancouver, Toronto (Canada), and Boston (United States), are encouraging residents to rely more on public transportation, including trains, trams, and buses. While the United States remains a highly car-dependent country, there are some signs that public transit has started to exhibit a modest upswing (Lane 2013).

Sustainable transportation would entail many other measures, such as limiting the use of cars as much as possible, making them smaller and more energy efficient, and even banning four-wheel drives or sports utility vehicles (SUVs), except in special circumstances (such as in the outback and rugged mountainous areas) and drastically limiting air travel. Electric cars are often offered as a more environmentally sustainable form of transportation. This might be the case if they derive their power from renewable sources of energy but not necessarily if they derive their energy from coal-fired power plants. Furthermore, electric cars will not solve congestion problems and the need to build and maintain roads, which requires an enormous amount of concrete, the manufacture of which produces carbon dioxide emissions. While shifting from cars to public transit, particularly intercity and suburban trains and trams or light-rail systems, would serve to diminish GHG emissions, these modes of transportation are not a panacea.

While there is much discussion about high-speed passenger rail substituting for cars or plane travel, according to Todorovich and Burgess (2013: 145), various studies indicate that the "direct benefits from high-speed rail in terms of overall energy and emissions may be modest." Additionally,

> [T]hese analyses also neglect the indirect impacts in terms of land-use and city-entering, which may be large and are difficult to measure and attribute ... [T]hese indirect benefits may be more important than any direct reduction in energy utilization as passenger choose high-speed rail over alternative travel modes.
>
> *(Todorovich and Burgess 2013: 145)*

Much thought is being given to the best form of public transportation, such as train, tram, or bus, in urban areas, depending on the situation. Furthermore, there is the issue of connecting small towns and rural areas with cities. In Europe, many villages are relatively well-connected to urban areas, but this is not generally the case in North America and Australasia.

Thus, given that public transportation is often infrequent or nonexistent, most people have become dependent on cars to connect them with commercial centers, family, friends, and acquaintances. Measures will need to be taken to connect rural communities to urban communities and to provide public transportation, perhaps in the form of regularly scheduled minibuses in rural areas. Furthermore, it would be possible to reinstate passenger rail service that serviced rural communities in both North America and Australia at a time in the past when their respective populations were considerably less urban than they are today.

Many cities also are increasing provisions for cycling and walking, including Canberra, the Australian national capital. Copenhagen has created bicycle right-of-ways and has fostered an ethos of respect for cyclists. Schiermonnikog, a national park off the northwest coast of the Netherlands, does not allow visitors to drive their private automobiles on the island, thus meaning that "[c]yclists, local buses, taxis, pedestrians and the occasional car of one of the island's inhabitants share the streets without any problems" (Peters 2006: 128). In terms of countries, Switzerland, admittedly a small country, has the most extensive public transport system in the world, one that connects urban and rural areas, including steep mountainous terrain. Zurich reintroduced trams at a time when they were being closed down in many German cites (Welzer 2012: 113).

In capitalist societies, "time is money," and this dictates rapid movement between places. Conversely, in a more leisurely paced world based on eco-socialist principles, people might find slower train travel—although faster than presently exists in most parts of North America and Australia—to be a time to slow down by reading, chatting with fellow passengers, enjoying the passing countryside, reflecting, and even sleeping. A more sustainable form of vacationing or holidaying would entail trips much closer to home, by train or bus, if possible, rather than to distant places either by plane or car. Cheap package holidays by airplane would become things of the past. In addition to minimizing flying, a Simpler Way would also entail a disposal of or minimizing the use of private motor vehicles and reliance on alternative modes of transportation, including walking and cycling. Many of us, particularly those of us in the Global North would do both humanity and other life forms on the planet a favor if we, as Katharine Alvord (2000) advises, were to divorce our cars.

Sustainable Food production and Forestry

Hannah Reid (2014: 45–46) provides a succinct definition of "sustainable agriculture":

> Sustainable agriculture is a method of farming based on providing for human needs for food, income, shelter and fuelwood. It also builds an understanding of the long-term effect of our activities on the environment. It integrates practices for plant and animal production with a focus on pest predator relationships, moisture and plants, soil health, and the chemical and physical relationship between plants and animals on the farm. Such agroecological approaches consider not just productivity but also system stability, sustainability and issues relating to equity and fairness.

A shift in food production away from heavy reliance on meat, particularly livestock, to organic farming, vegetarianism, and even veganism would be more environmentally sustainable and an important form of climate change mitigation. At the present time, however, humanity by and large is heading in the opposite direction as the rising middle classes in the Global South world develop an insatiable appetite for meat and dairy foods.

Animal production requires massive amounts of water and petroleum for growing, feeding, transportation, and processing. Livestock production releases methane, a powerful GHG, into the atmosphere. While ocean fishing stocks are being depleted, humanity continues to increase its consumption of fish. Many deep-sea fish are in danger of extinction due to overfishing. State-of-the-art technology allows fishermen to trawl fish from deep waters, often totally unregulated by international laws. Aquaculture production or fish farming has been on the increase globally, accounting now for about 50% of global fisheries, and relies on "wild fish further down the food chain in order to feed fish in fish farms, thus further depleting wild fish stocks" (Steffin et al. 2015: 10).

A drastic reduction in current forms of meat consumption and dairy production would greatly decrease emissions from food production and help to resolve health problems as well. Weis (2013: 12) argues that there is a strong need to deindustrialize livestock production and shift away from meat diets in general as part of an effort to create a "more sustainable, just, and humane world." However, the culling of kangaroos, wild horses, camels, and rabbits in a country like Australia could theoretically serve as a more sustainable form of meat production. Rearing livestock on grass would be more environmentally sustainable because it "helps [to] preserve native grasses and control erosion, and it eliminates the need for pesticides," but only if it is done in such a way that overgrazing is avoided (Nierenberg 2013: 36). Conversely, some leftists of different political stripes historically "have called for the total abolition of animal exploitation and adoption of worldwide vegetarianism" (Gunderson 2011: 415).

Small-scale organic farming tends to be more fuel efficient than industrial agriculture, which relies heavily on petroleum, chemical fertilizers, and pesticides. All farming requires water, but livestock production requires much more water than does crop growing. Given that about 80% of the world's water is used for agriculture, organic cultures "use much less water than agribusiness, and eating locally grown foods shrinks export-oriented agribusiness and keeps water in the country" (Piper 2014: 224). Much water is used to grow grains, including corn, which in turn feeds livestock and poultry but is also used in processed foods.

There is a strong need to shift towards agro-ecology, which relies on the extensive knowledge of farmers of local ecosystems and seeks to transcend dependence on chemical, oil-based agriculture. Crops such as maize, wheat, sorghum, millet, and vegetables can be grown in forested areas that "provide shade, improve water availability, prevent soil erosion, and add nitrogen—a natural fertilizer—to soils" (Nierenberg 2013: 194). Ducks are used in Japan for pest control in rice paddies, thus preventing the application of expensive chemical fertilizers and pesticides. Agroforestry blends trees and shrubs with perennial crops and the production of cattle, poultry, and other animals. Permaculture, which is a contraction for "permanent agriculture," a term coined by Australians Bill Mollison and David Holmgren, seeks to "integrate concepts from organic farming, sustainable forestry, no-till management, and the village design techniques of Indigenous peoples" (Bates and Hemenway 2010: 52). A shift towards vegetarianism could reverse deforestation for cattle production in the Amazon Basin with most of the meat not being consumed by Latin Americans but by Europeans and North Americans.

There is an urgent need to expand urban farming that already exists in many parts of the world, particularly the Global South. An estimated 90% of the farm produce consumed in Havana, Cuba, is grown either in the city or in its immediate hinterlands. Halwell and Nierenberg (2007: 572) maintain: "For cities confronted with growing waste disposal problems—which includes virtually all cities—the strongest environmental argument for local farming is the opportunity to reuse urban organic waste that would otherwise end up in distant, swollen landfills." Laws prohibiting farming in cities need to be repealed. Much urban farming can be done on rooftops, perhaps coupled with strategic placement of solar panels. A danger with urban farming is that it has the potential to serve as a reformist reform rather than a nonreformist or system-challenging reform in that it may fill in the gaps left by neoliberal state retrenchment, unless it is part of an effort to drastically restructure society, which it has the potential to do given that at some level it has rejected the capitalist food production system.

Despite the horror stories associated with the enforced collectivization of agriculture in the Soviet Union during the Stalinist era, Sarkar asserts that the notion of collective agriculture needs to be revisited for a number of reasons, including economies of scale, particularly if it is based on decentralized planning rather than centralized planning that would not account for regional variation within a country. Furthermore,

> In the transition period, and more so in the low-level steady state, the quantity of machines and equipment available to agriculture as a whole would also go down, so that several family farmers (on leased land), if that were the future system, would have to share each item. They would have to buy them jointly or hire them, when needed, for some state or community organization. In either case, they would have to work out a time plan for using them.
>
> *(Sarkar 1999: 227)*

Collective farming would have to find a happy medium between economies and diseconomies of scale, the latter of which were often present in the former Soviet Union.

Resisting the Culture of Consumption and Adopting Sustainable and Meaningful Consumption

Obviously, all humans need to consume a certain amount of food, clothing, and shelter in order to sustain themselves. Capitalism, however, converts "wants" into "needs" through voluminous and alluring advertisements and as a compensation for alienation in the workplace and everyday social life. From an eco-socialist perspective, Magdoff and Foster (2010: 26) argue that

> [an] economic system that is democratic, reasonably egalitarian, and able to set limits on consumption will undoubtedly mean that people will live at a significantly lower level of consumption than what is sometimes referred to in wealthy countries as a "middle class" lifestyle (which has never been universalized even in those societies).

Unfortunately, at least in the Global North, resistance to the culture of consumption remains confined to small niche groups.

Sustainable Trade

Over the past two centuries, global production has resulted in a tremendous cross-border trade of goods and services. While this increased international trade has been enhanced by free trade agreements and lower transport costs, it relies heavily on oil and contributes to GHG emissions in moving goods around the world by ship or airplane as well as trucks and trains. Furthermore, while countries in the Global South, the People's Republic of China in particular, are often criticized for their increasing GHG emissions, an appreciable amount of this is due to the fact that developed countries are importing cheap resources and manufactured goods from developing countries. International aviation and marine fuels are exempt from international taxation schemes. The global food system has undergone a tremendous rise in "food miles"—a measurement of the distance that food travels from the site of production to the site of consumption.

Vandana Shiva (2008: 104) maintains that humanity "should be reducing food-miles by eating diverse, local, and fresh foods, rather than increasing carbon pollution through the spread of corporate industrial farming, non-local food supplies, and processed and packaged food." In a similar vein, Sarkar (1999: 219) advocates the general principle of "as far as possible, produce locally or regionally" in order to avoid the emissions miles so often associated with long-distance trading. He admits that even in an eco-socialist world, some long-distance trade will be inevitable. For example, in the case of his home country of India, salt can only be produced in coastal regions, thus necessitating its exportation to regions in the interior. Sarkar (1999: 219) maintains: "Our needs and the required goods would have to be kept under control, so that long-distance trade can be kept within sustainable limits worldwide."

There is a need for the greening of shipping, which would rely on solar and hydrogen energy—powered ships, sailing ships and "kite sails" (Simms 2009: 94–95). Also given that large quantities of products are now shipped by airplane and truck, there is a strong need to revisit railroads and waterways as less energy-intensive modes of shipping.

Sustainable Settlement Patterns and Local Communities

As they have grown, cities have gobbled up precious farmland and natural areas. Overall cities are energy-intensive places on a number of counts, including in the operation of office buildings, industries, residences, shopping centers, recreational facilities, restaurants, educational institutions, hospitals, residences, transportation, highways, parking lots, airports, etc. Mike Davis (2010: 41) observes:

Heating and cooling the urban built environment alone is responsible for an estimated 35 to 45 per cent of current carbon emissions, while urban industries and transportation contribute another 35 to 40 per cent. In a sense, city life is rapidly destroying the ecological niche—Holocene climate stability—which made its evolution into complexity possible.

In reality, the ecological and carbon footprints of cities vary considerably between the Global North and the Global South, but also within cities, depending on their residential patterns (e.g., McMansions versus slum dwellings) and modes of transportation (e.g., a city with an excellent public transport system versus a highly car-dependent one). The ecological and carbon footprints of cities extend far beyond their boundaries because they rely on resources from a large hinterland that literally encompasses much of the world. Various proponents of "sustainable cities" who maintain that increasing urban density contributes to environmental sustainability downplay the "historical connections between density and economic growth" (Clement 2011: 296). Theoretically cities have the potential of becoming much greener than they are presently. Even during the early 20th century, various socialists and anarchists pioneered efforts to make cities more livable both socially and environmentally, such as the Karl Marx-Hof in Red Vienna, Austria, and the Bauhaus housing experiments in Germany (Davis 2010: 43).

A new urbanism that seeks to make cities more livable and environmentally sustainable has emerged around the world and has even permeated urban planning in some cities. While the new urbanism has been applied to some degree in many towns and cities, it needs to make a much stronger effort to be socially inclusive and counteract gentrification, which marginalizes low-income people. Conversely, in an eco-socialist world, there would be no poor people and differences in income and wealth would not be nearly as great as is currently the case in capitalist societies. The development of green cities constitutes a highly imaginative endeavor, one that will require drawing insights from numerous disciplines and fields, including architecture, building construction, urban planning, transportation planning, and last but not least the social sciences. A green or sustainable city should include medium-density housing, easy access to public transportation, and minimal reliance on automobiles. Walkability should be a core feature of the green city, which would allow people to walk as much as possible to their work sites, parks, recreational centers, theaters, shops, and eating places, while contributing to a democratized streetscape. Some psychologists have developed the notion of *ecopsychology*, which stresses the need for people, including urban dwellers, to have contact with the natural environment (Brown 2009: 162). This emergent field examines how our alienation from nature contributes to tendency to destroy our environment as well as the contribution of this alienation to depression and addiction.

Eco-communities, which increasingly are found in urban and rural areas of both the Global North and the Global South, constitute pre-figurative social experiments that potentially are part of developing more sustainable settlement patterns (Cooper and Baer 2019). Urban eco-villages can reduce car dependence or eliminate it completely if closely situated to good public transportation. Furthermore, in that an eco-village is "truly a village, it has its own internal economy that enables members to work close to home, thereby reducing their transportation needs" (Liftin 2014: 44).

The central business districts of cities around the world now have numerous high-rise office and even apartment buildings. While there has been quite a bit of discussion on how to make buildings more environmentally sustainable through the use of green roofs and walls, fitted glazing, solar panels, more efficient lighting, "under the pressure of the accelerating debate around sustainability has the idea begun to grow that perhaps not more office buildings but rather fewer buildings, more intelligently and intensively used" (Duffy 2008: 11). Of course, in an eco-socialist world, many buildings presently occupied by financial institutions, marketing firms, and department stores that facilitate the capitalist patterns of production and consumption would not be needed and could be converted to serve more socially beneficial purposes.

Cities should be easily interconnected with other cities via trains rather than automotive or plane transportation. Also, there is the question as to what the optimum maximum population of a city is. Some cities have become so incredibly large that it almost defies the imagination, with the world now having numerous cities with over ten million people and some with over 30 million people. Obviously, there is no easy answer to this question because it depends on the national context and notions of population density.

Density can have both positive and negative aspects. On the upside, by bringing large numbers of people together, higher density makes public transit possible while helping small shops, restaurants, and other amenities to thrive. Further, it entices people to walk rather than drive and reduces stress on the environment. On the downside, the rigid distribution of housing into different price levels advances segregation and inequality. As has been stressed, sound environmental behavior and social justice go hand-in-hand.

Demilitarization

Militarism and warfare are ubiquitous on the world scene and constitute significant sources of GHG emissions and drivers of climate change, often overlooked and understated. In the cases of the world's largest military apparatus, Belcher et al. (2019) refer to its "hidden carbon costs," noting:

> This is particularly true of oil, but also other material acquisitions like food, machinery, and other apparently mundane materials (e.g., sand,

concrete, and water). As the US military continues to carry out the everywhere war in some of the least accessible corners of the globe, its supply chains require logistical sophistication like never before.

This suggests that the climate movement must join forces with a much-needed reinvigorated peace movement. In a lecture presented at the Old South Church in Boston on April 11, 2019, Noam Chomsky (2020) argued that humanity faces three major threats, namely climate change, a nuclear holocaust, and the undermining of democracy, exemplified by the Trump presidency that fortunately was in its last days at the time. By and large, the peace movement has been most pronounced in the Global North and historically has waxed and waned, but has experienced a pronounced lull since exuberant periods in the 1980s during the last days of the Cold War and then again in 2001–2003 with the invasions of Afghanistan and Iraq on the part of a US-led coalition of military powers, which included the United States and Australia. The peace movement also needs to be expanded in the Global South. While it is partially anti-systemic in that it "opposes 1st world imperialism and excessive militarization of 1st world societies," as Starr (2000: 54) observes, "few versions of the peace movement put forth critiques of capitalism." However, a revitalized peace movement would need to be more explicitly anti-systemic in its calls for denuclearization and demilitarization.

Conclusion

The radical transitional reforms that we have delineated constitute loose guidelines for shifting human societies or countries towards democratic eco-socialism. We do not purport that our suggested guidelines are comprehensive because undoubtedly others could be added to the list. Anthropologist David Graeber, a self-espoused anarchist, asserts that he is less interested in specifically what kind of alternative economic system needs to be created than in how people decide to get there. He suggests the following five steps for getting to a viable free society, one that avoids economic and ecological disasters:

- A debt jubilee or a general cancellation of the debts of ordinary people around the world;
- Working less given the grim reality that the "current pace of the global work machine is rapidly rendering the planet uninhabitable";
- Shifting work from the creation of "ever more consumer products and ever more disciplined labor" to work that focuses on caretaking and teaching;
- Critiquing and shifting away from the bureaucratization of social life;

- Reclaiming the notion of *communism* by shifting away from viewing it as the "absence of private property arrangements" to the "original definition: 'from each according to their abilities, to each according to their needs.'"

(Graeber 2013: 283–295; emphasis in the original)

The question remains whether societies and a world system based on eco-socialist principles would allow room for markets. In our vision of an eco-socialist society, new left parties that rise to power in democratic elections would in time nationalize the means of production, particularly the multinational corporations and even national corporations that today exert a powerful influence, both directly and indirectly, on governments and mainstream political parties. We can see room for small businesses, such as restaurants, cafés, bookshops, clothing stores, experimental theaters, and so forth, preferably worker-owned and worker-managed, but possibly owned by independent entrepreneurs or families. Obviously, markets would not completely disappear but would be much diminished compared to their strong presence in capitalist societies. In contrast to reciprocity and redistribution as modes of economic exchange that are characteristic of Indigenous societies, markets involve the exchange of goods and services with the use of money, predate capitalism by millennia and even existed in rudimentary form in Indigenous societies, became more elaborate in pre-capitalist state societies, but became particularly pronounced and hegemonic with the full-scale development of capitalism. In Schweickart's (2016: 6) conception of a socialist-oriented *economic democracy*, small businesses would "provide jobs for large numbers of people, and goods and services to even more," as they presently do. Unfortunately, most small businesses today often pay their employees low wages and provide them with minimal social benefits, if any. Such practices would have to be seriously curbed under eco-socialism. Ultimately, an eco-socialist society and world system, in order to prevent the emergence of large corporations, would have to create legislation that would have to restrict the size of businesses, lest the capitalist tendency for the bigger businesses to take over the smaller businesses gains ascendancy.

References

Althusser, Louis. 2014. *On the Reproduction of Capitalism: Ideology and Ideological State Apparatuses.* London: Verso

Alvord, Katarine. 2000. *Divorce Your Car! Ending Your Love Affair with the Automobile.* Gabriola, BC: New Society Publishers

Amin, Samir. 2013. China 2013. *Monthly Review* 64(10): 14–33

Arrighi, Giovanni, Hopkins, Terence K., and Wallerstein, Immanuel. 1989. *Anti-Systemic Movements.* London: Verso

Baer, Hans A., Singer, Merrill, and Susser, Ida, 2013. *Medical Anthropology and the World System: A Critical Perspective* (3rd edn). New York: Praeger

Bates, Albert and Hemenway, Toby. 2010. From agriculture to permaculture. In *2010 State of the World: Transforming Cultures—From Consumerism to Sustainability.* Linda Starke and Lisa Mastny, eds. pp. 47–53. New York: W.W. Norton & Company

Belcher, Oliver, Bigger, Patrick, Neimark, Ben, and Kenelly, Cara. 2019. Hidden carbon costs of the "everywhere war": Logistics, geopolitical ecology, and the carbon boot-print of the US military. *Transactions of the Institute of British Geography* 45(1): 65–80

Belsky, Eric S. 2012. Planning for inclusive and sustainable urban development. In *State of the World 2012: Moving Toward Sustainable Prosperity.* Linda Starke, ed. Pp. 38–65. Washington, DC: Island Press

Bjerg, Ole. 2016. *Parallax of Growth: The Philosophy of Ecology and Economy.* London: Polity

Brown, Lester R. 2009. *Plan B 4.0: Mobilizing to Save Civilization.* New York: W.W. Norton & Company

Burkett, Paul. 2006. *Marxism and Ecological Economics: Toward a Red and Green Economy.* Leiden: Brill

Callenbach, Ernest. 1975. *Ecotopia: The Notebooks and Reports of William Weston.* New York: Bantam Books

Chambers, Neil B. 2011. *Urban Green: Architecture for the Future.* New York: Palgrave Macmillan

Chomsky, Noam. 2012. *Occupy.* Brooklyn, NY: Zucotti Park Press

Chomsky, Noam. 2020. The third threat: The hollowing out of our democracy. In *Internationalism or Extinction: Noam Chomsky.* Charles Derber, ed. Pp. 86–98. New York: Routledge

Clement, Matthew Thomas. 2011. The town-country antithesis and the environment: A sociological critique of the "real utopian" project. *Organization & Environment* 24: 292–311

Cochran, D. 2021. Why the Russian Revolution actually owes its success to nonviolent resistance. *Waging Nonviolence.* https://wagingnonviolence.org/2021/07/why-the-russia n-revolution-actually-owes-its-success-to-nonviolent-resistance/#:~:text=The%20Russia n%20Revolution%20of%201917,landlords%20or%20soldiers%20their%20officers

Commoner, Barry. 1972. *The Closing Circle: Confronting the Environmental Crisis.* London: Cape

Cooper, Liam and Baer, Hans A. 2019. *Urban Eco-Communities in Australia: Real Utopian Responses to the Ecological Crisis or Niche Markets?* Singapore: Springer

Daly, Herman and Cobb, John B. Jr. 1990. *For the Common Good.* London: Green Point

Davis, Mike. 2010. Who will build the Ark? *New Left Review* 61: 29–46

Diesendorf, Mark. 2014. *Sustainable Energy Solutions for Climate Change.* Sydney: UNSW Press

Duffy, Frank. 2008. *Work and the City.* London: Black Dog Publishing

Foster, John Bellamy. 1999. *The Vulnerable Planet: A Short History of the Environment.* New York: Monthly Review Press

Foster, John Bellamy. 2002. *Ecology against Capitalism.* New York: Monthly Review Press

Foster, John Bellamy. 2009. *The Ecological Revolution: Making Peace with the Planet.* New York: Monthly Review Press

Foster, John Bellamy. 2011. Capitalism and degrowth: An impossibility theorem. *Monthly Review* 62(8): 26–33

Foster, John Bellamy. 2013. James Hansen and the climate-change exit strategy. *Monthly Review* 64(9): 1–19

Foster, John Bellamy. 2015. Marxism and ecology: Common forms of a Great Transition. *Great Transition Initiative*, October. https://www.greattransition.org/

Foster, John Bellamy. 2022. *Capitalism in the Anthropocene: Ecological Ruin or Ecological Revolution.* New York: Monthly Review Press

Gilbert, Richard and Perl, Anthony. 2010. Transportation in the post-carbon world. In *The Post-Carbon Reader: Managing the 21st Century's Sustainability Crisis.* Richard Heinberg and Daniel Lercy, eds. pp. 347–360. Healdsburg, CA: Watershed Media

Golten, T.E. and B.R. Golten, eds. 1977. *The End of the Road: A Citizen's Guide to Transportation Problem Solving.* Washington, DC: National Wildlife Federation

Gorz, André. 1973. *Socialism and Revolution.* Garden City, NY: Anchor

Gorz, André. 1982. *Farewell to the Working Class: An Essay on Post-Industrial Socialism.* London: Pluto Press

Gorz, André. 1994. *Capitalism, Socialism, Ecology.* London: Verso

Graeber, David. 2013. *The Democracy Project: A History, a Crisis, a Movement.* New York: Spiegel & Grau

Gunderson. Ryan. 2011. The metabolic rifts of livestock and agriculture. *Organization & Environment* 24: 402–422

Halwell, Brian and Nierenberg, Danielle. 2007. Farming and cities. In *State of the World 2007: Our Urban Future.* Linda Starke, ed. pp. 48–66. New York: W.W. Norton

Higgs, Kerryn. 2014. *Collision Course: Endless Growth on a Finite Planet.* Cambridge, MA: MIT Press

Hoffman, John. 1984. *The Gramscian Challenge: Coercion and Consent in Marxist Political Theory.* Oxford: Basil Blackwell

Klein, N. 2007. *The Shock Doctrine: The Rise of Disaster Capitalism.* New York: Knopf

Kovel, Joel. 2007. *The End of Nature: The End of Capitalism* (2nd edn). London: Zed Books

Ladd, Brian. 2008. *Autophobia: Love and Hate in the Automotive Age.* Chicago: University of Chicago Press

Lane, Bradley. 2013. Public transportation as a solution to oil dependence. In *Transport Beyond Oil: Policy Choices for a Multimodal Future.* John L. Renne and Billy Fields, eds. pp. 107–125. Washington, DC: Island Press

Liftin, Karen T. 2014. *Eco-Villages: Lessons for Sustainable Community.* London: Polity

Loewy, Michel. 2015. *Ecosocialism: A Radical Alternative to Capitalist Catastrophe.* Chicago: Haymarket Books

Lohmann, Larry. 2006. Carbon trading: A critical conversation on climate change, privatization and power. *Development Dialogue* 28

Magdoff, Fred. 2013. Global resource depletion: Is population the problem? *Monthly Review* 64(8); 13–28

Magdoff, Fred and Foster, John Bellamy. 2010. What every environmentalist needs to know about capitalism. *Monthly Review* 61(10): 1–30

Marx, Karl. 1975. Critical notes on the King of Prussia and social reform. In *Early Writings.* Harmondsworth: Penguin

Marx, Karl and Engels, Friedrich. 1978. The German ideology. In *Marx-Engels Reader* (2nd edn). New York: W.W. Norton and Company

McCluney, Ross. 2005. Renewable energy limits. In *The Final Energy Crisis.* Andrew McKillop and Sheila Newman, eds. pp. 153–175. London: Pluto Press

Newman, Peter. 2009. Transportation opportunities: Towards a resilient city. In *Opportunities Beyond Carbon: Looking for a Sustainable World.* John O'Brien, ed. Pp. 98–115. Melbourne: Melbourne University Press

Newman, Peter. 2010. Responding to oil vulnerability and climate change in our cities. In *Managing Climate Change: Papers from the Greenhouse 2009 Conference.* Imogen Jubb, Paul Holper, and Wenjce Cai. Melbourne: CSIRO Press

Nierenberg, Danielle. 2013. Agriculture: Growing good—and solutions. In *State of the World 2014: Is Sustainability Still Possible?*Linda Starke, ed. pp. 190–200. Washington, DC: Island Press

Onaran, Oezlem. 2010. The crisis of capitalism in Europe, West and East. *Monthly Review* 62(5); 18–33

Owen, David. 2012. *The Conundrum: How Scientific Innovation Increased Efficiency, and Good Intentions Can Make Our Energy and Climate Problems Worse.* Melbourne: Scribe

Pepper, David. 1993. *Eco-Socialism: From Deep Ecology to Social Justice.* London: Routledge

Pérotin, V. 2017. What do we really know about workers' co-operatives?Co-operatives UK. https://web.archive.org/web/20170113124521/ https://www.uk.coop/sites/defa ult/files/uploads/attachments/worker_co-op_report.pdf

Peters, Peter Frank. 2006. *Time, Innovation and Mobilities: Travel in Technological Culture.* London: Routledge

Piper, Karen. 2014. *The Price of Thirst: Global Water Equality and the Coming Chaos.* Minneapolis: University of Minnesota Press

Reid, Hannah. 2014. *Climate Change and Human Development.* London: Zed Books

Robinson, William I. 2014. *Global Capitalism and the Crisis of Humanity.* Cambridge: Cambridge University Press

Rogers, Chris. 2014. *Capitalism and Its Alternatives.* London: Zed Books

Ross-Nazzal, J. and students. 2018. Saving capitalism: The New Deal era, 1933–1940. In *Our Story: An Ancillary to US History. From Pre Contact to Post Factual America.* J. Ross-Nazall, ed. PressBooks. https://pressbooks.pub/ourstory/

Ryle, Martin. 1988. *Ecology and Socialism.* London: Radius

Sarkar, Saral. 1999. *Eco-Socialism or Eco-Capitalism? A Critical Analysis of Humanity's Fundamental Choices.* London: Zed Books

Sarkar, Saral and Kern, Bruno. 2008. *Eco-Socialism or Barbarism: An Up-to-Date Critique of Capitalism.* Cologne: Initiative Eco-Socialism

Schor, Juliet. 2010. *Plentitude: The New Economics of True Wealth.* Melbourne: Scribe

Schweickart, David. 2016. Economic democracy: An ethically desirable socialism that is economically viable. The Next System Project. https://nextsystemproject.org

Shiva, Vandana. 2008. *Soil Not Oil: Environmental Justice in the Age of Climate Crisis.* Boston, MA: South End Press

Shue, Henry. 2014. *Climate Justice: Vulnerability and Protection.* Oxford: Oxford University Press

Sim, Stuart. 2010. *The Carbon Footprint Wars: What Might Happen If We Retreat from Globalization?*Edinburgh: Edinburgh University Press

Simms, Andrew. 2009. *Ecological Debt: Global Warming and the Wealth of Nations* (2nd edn). London: Pluto Press

Starr, Amory. 2000. *Naming the Enemy: Anti-Corporate Movements Confront Globalization.* New York: Zed Books

Steffin, Will, Broadgate, Wendy, Deutsch, Lisa, Gaffney, Owen, and Ludwig, Cornelia. 2015. The trajectory of the Anthropocene: The Great Acceleration. *Anthropocene Review* 2(1): 1–18

Stilwell, Frank. 2000. *Changing Track: A New Political Economic Direction for Australia.* Sydney: Pluto Press

Tanuro, Daniel. 2010. *Green Capitalism: Why It Can't Work*. London: Merlin Press

Todorovich, Petra and Burgess, Edward. 2013. High-speed rail and reducing oil dependence. In *Transport Beyond Oil: Policy Choices for a Multimodal Future*. John L. Renne and Billy Fields, eds. pp. 141–160. Washington, DC: Island Press

Trainer, Ted. 2007. *Renewable Energy Cannot Sustain a Consumer Society*. Sydney: UNSW Press

Wall, Derek. 2010. *The Rise of the Worldwide Ecosocialist Movement*. London: Pluto Press

Wallis, Victor. 2009. Economic/ecological crisis and conversion. *Socialism and Democracy* 23(2): 94–101

Weis, Tony. 2013. *The Ecological Footprint: The Global Burden of Industrial Livestock*. London: Zed Books

Welzer, Harald. 2012. *Climate Wars: Why People Will Be Killed in the Twenty-First Century*. Patrick Camiller, trans. London: Polity

Williams, Chris. 2010. *Ecology & Socialism: Solutions to Capitalist Ecological Crisis*. Chicago: Haymarket Books

Wlodzimierz, Wesolowski. 1967. Marx's theory of class domination. In *Marx and the Western World*. Nicholas Lobdowicz, ed. pp. 103–131

Wolff, Richard. 2012. *Democracy at Work: A Cure for Capitalism*. Chicago: Haymarket Books

Wright, Olin Erik. 2010. *Envisioning Real Utopias*. London: Verso

EPILOGUE

Humanity is at a crossroads. Like all other living things that have inhabited this planet, eventually we humans will become extinct, if not due to developments of our own making, then due to natural events over which we have no control. Yet it seems that as a species we can exert some degree of agency or control over our fate during this pivotal century. Although global capitalism has produced an array of impressive technological innovations—many of which began outside of the private sector in universities or government programs—it is a system fraught with contradictions, including an incessant drive for economic growth and accumulation, increasing social inequalities, authoritarianism, militarism, depletion of natural resources, and environmental degradation, including global warming and associated climatic changes. It has become increasingly apparent that human societies will have to adapt to the best of their ability to the reality of climate change using a variety of strategies. Clearly there will be a need for all manner of technological innovations, reliance on renewable energy sources, more efficient forms of heating and cooling, the development of buildings that are more energy efficient, the redesign of cities to control their energy demands and heat outputs, restoration of degraded environments, the use of sustainable agriculture, reforestation, protection of biodiversity, less reliance on private motor vehicles and airplanes, and less reliance on the factory farming of animals. As important as these and other strategies will be in mitigating climate change, on their own they are insufficient if they are not ultimately part of a longer-term effort to transcend global capitalism and its ravenous demands.

The heart of our dilemma is that, on the one hand, as "the global situation of the environment is becoming worse and worse, and the threats are becoming closer and closer," people's recognition of the need for structural change grows. On the other hand, "time is running out, because in some years—no one can say how much—the damage may be irreversible" (Loewy 2006: 307).

DOI: 10.4324/9781003469940-13

It is our belief that anthropologists and other social scientists can play a critical role in contributing their analytical skills and insights to humanity's struggle for a better future—one in which we as a species learn to live in reasonable harmony with one another and with nature. From our perspective, the anthropology of climate change and the critical anthropology of climate change in particular, have the potential to help to create a world system based on social justice, deep democracy, environmental sustainability, and a safe climate. Perhaps more than any other issue, climate change allows critical social scientists to identify the contradictions of the existing capitalist world system and to contemplate the creation of an alternative world system based on eco-socialist principles and to apply them in keeping with the notion of praxis—the merger of theory and action.

Anthropology as an academic discipline that seeks to make holistic cross-cultural and temporal observations must collaborate with the other social and behavioral sciences, the humanities, and the natural sciences, particularly climate science and earth science. Anthropology has been very good at looking at human societies in the present and in the past, the immediate past and even the distant past. With a few exceptions, none of the social sciences have been good at successfully looking at the future with new, insightful imaginings, but the seriousness of anthropogenic climate change and the related ecological emergency, as well as the widening disparities around which societies are structured demand a very serious consideration of future scenarios.

The social science of carefully envisioning the future based on current trends and developments, or futurology as it is called, sometimes traces its origins to the influential and highly esteemed writings of H.G. Wells (which, nonetheless, were burned by the Nazis in Berlin during World War II). While others saw him as a science fiction author, in his own assessment he was more of a social critic and historian. His most famous novel, published just before the turn of the 20th century, *The War of the Worlds*, portrays the arrival on Earth of imperialist aliens with high-technology weaponry and evil designs on the planet. The story, in fact, was a futurist recasting of Well's brother's account of the British empire's real-world genocide committed against the Indigenous Tasmanians of Australia. In Well's tale, the invaders indiscriminately kill people with deadly heat rays, or what we now known as lasers. Thus, it is human biology and our pathogens that ultimately defeat the invaders. The message of Well's novel, according to some reviewers, is that advanced technology does not necessarily confer invincibility. Wells also envisioned the possibility of future utopias and plotted alternative ways and means of achieving them. His success in predicting future events and technologies provides a model for the potential of thinking about the future in the social sciences.

More than any other issue, climate change does perhaps allow critical social scientists to illuminate the contradictions of the existing capitalist world system and to contemplate the creation of an alternative world system based on eco-socialist principles. Anthropology is a discipline that has the potential to reveal the "big picture" of the world's interlocked economic and social system while drawing on close observations of "small pictures" of the many different societies and cultures around the world. While anthropologists and other critical social scientists are too small a group to act as a vanguard in the struggle against climate change and global capitalism, they and their students can form alliances with anti-systemic movements, including the labor, anti-corporate globalization, social justice, peace, Indigenous and ethnic rights, women's rights, the environmental, and climate justice movements. The time for this vital task is now.

Reference

Loewy, Michael. 2006. Eco-socialism and democratic planning. In *Coming to Terms with Nature*. Leo Panitch and Colin Leys, eds. pp. 294–309. London: Merlin Press

INDEX

Note: Page numbers in *italics* indicate figures.

Abbott, Tony 50
ablation 19
Abramoff, Rose 194
Abrupt Climate Change Scenario, An 215
academic air travel 168–9
achieving democratic capture of state 240–41
acidification 106–7
adaptive processes 31–3
adopting sustainable consumption 257–8
aerosols 16
affective orientation 44–5
Agarwal, Anil 38
Agrawal, Arun 33
AI *see* artificial intelligence
air quality 22–3, 103, 165
"airworlds" 37
Akther, Hasina 75
Albanese, Anthony 152, 156, 214
Albritton, Robert 223
algal bloom 105
Ali, I. 136
Ali, Tariq 222–3
alterative world system 221–5; democratic eco-socialism 224–5; democratic socialism 222–3; eco-socialism 223–4
alternative forms of flying 173–8
Alternative für Deutschland 3
alternatives to Anthropocene and Capitalocene 68–9

Althusser, Louis 240–41
Altvater, Elmar 66
Alvarez, Alex 216
Alvord, Katherine 255
Amazon rainforest 25
ambient air pollution 106–7
American Anthropological Association 12, 58, 178
American Enterprise Institute 145
American Petroleum Institute 79, 82–3
Amin, Samir 221, 244
Andean Altiplano 133–4
Anderson, E.N. 32
Anderson, James 185, 212, 226
Anderson, Kevin 44, 174–5
Andina, Acción 150
Angus, Ian 1, 65–8
Anthropocene 58
Anthropocene or Capitalocene? 65–7
Anthropocene era 58–72; alternatives to Anthropocene and Capitalocene 68–9; Anthropocene vs. Capitalocene 65–8; contested concept of Anthropocene 59–63; postcolonial historical take on Anthropocene 63–5; recognition of human–climate nexus 58–9; reservations about "Capitalocene" 69–70
Anthropocene vs. Capitalocene 65–8
anthropocentrism 8–9, 64
anthropogenic deforestation 32
anthropogenic disruption 28

anthropogenic impact on Earth 14–17
anthropologists and air travel 169–71
Anthropology of Climate Change 1–2
Anthropology and Contemporary Human Problems 210
anthropology and engagement with climate change 11–14
anti-corporate resistance 142–3
anti-fracking 52
anti-imperialism 67
anti-industrialization 142–3
anti-systemic movement 199–202
antibiotic era 37
antimicrobial resistance 101–2
archaeology 13, 31–2
Arctic Climate Impact Assessment project 32
Armelagos, George G. 37
Aronoff, Kate 158
Arrenhuis, Svante 11
Arrighi, Giovanni 241
artificial intelligence 69
asceticism 62
asylum 124
atmospheric aerosol loading 27
atolls 130–31
atomic bomb 60
attribution 115
Australia 151–7, 212, 225–30, 255–6
authoritarianism 63–4
avian viruses 63
Aviation Justice 178
avoiding ecological disaster 261–2
awareness of outside world 8

Baba, M. 124
"background noise" 126
bad Anthropocene 61–2
"bad" ozone *see* ground-level ozone
Badiou, Alan 223
Baer, Hans 169, 176–7, 226
Bajau people 8
Bangladesh 73–5, 129
bankruptcy 138
"barefoot scientists" 196
Barrios, Roberto 159
Baskin, Jeremy 219
Bauman, Zygmunt 168
Beaufort Gyre 21
Beck, Ulrich 217
beels 36
Belcher, Oliver 260
Belsky, Eric S. 252
benign environmental practices 40
Berker, Lars E. 188–9

Berkhout, Frans 60–61
Berstein, Aaron 89
Bezos, Jeff 150
Biden, Joe 90, 158
Big Switch, The 154
Bill Gates Problem, The 149
Bill and Melinda Gates Foundation 147
bio "recruitment" 25
biochemical flows 27
biodegradable plastic 151
biofuel production 112–13, 148, 154, 168
biofuel soot 16–17
biomass 16–17, 152–3
biomedicine 47
biomes 21
biosphere integrity 27
biotic attrition 25
Bjørkdahl, Kristian 174
black carbon 16–17, 91
Black Friday 48
Black, Joseph 10
black mold 105
blame for poor 82–6
BlueSkyModel 169
Bodely, John 210
Bolsanaro, Jair 213–14
Bond, Patrick 196
Bookchin, Murray 226
boomtowns 51
Boswell, Terry 224
boundary positions 26–8
Boyer, Dominic 44–5
BP 80, 176, 200–201
Bradbrook, Gail 189–90
Bramwell, Simon 189–90
Branson, Richard 203
Brazil 195
Break Through 62
Breakthrough Energy 147, 150
Breakthrough Institute 62, 65, 145
Brecht, Bertolt 87
Bridger, Rose 155
Brookings Institute 145
brown hydrogen 152
Brown, Lester 144
Brown, Margot 150
Brundtland Commission 38
Buechler, Steven M. 201
bunker ammonia 153, 156
Burgess, Edward 254
Burgmann, Verity 198
Burkett, Paul 244, 248

Cairncross, Fred 248
Callenbach, Ernest 252

CAN *see* Climate Action Network
Capitalinian sub-type 68
Capitalocene era 58–72; *see also*
  Anthropocene era
capitalophilanthropy 109
Carbon Brief 6
carbon capture 144
carbon dioxide 10–11, 14–15, 21, 33, 59,
  80, 89, 107, 142–3
Carbon Disclosure Project 79
Carbon Majors Database 79
Carbon Pricing Mechanism 15
carbon reduction 13
carbon trading 187
Carter, Neil 143–4
Castañeda, H. 123
Castells, Manuel 184, 201
catastrophic convergence 214
Caton, Steven C. 37
causation of climate change 10
CDP *see* Carbon Disclosure Project
Centre for Policy Research 6
Chagnon, Napoleon 169–70
Chakrabarty, Dipesh 63–5
challenges to planetary health 115
changing world 7
Chase-Dunn, Christopher 199, 202, 224
cheap labor 66, 73–4, 98
chemicalization 106
Chevron Corporation 79–80, 82, 84–5
chicken farms 63
"Chicken Little" dilemma 3, 84
child morbidity 136
China 195
chlorination in water treatment 16
chlorofluorocarbons 15
chloroform 16
cholera 102, 105–6, 128–9
Chomsky, Noam 232, 249, 261
Christoff, Peter 217–18
Chthulucene era 68–9
Chu, Steven 147
City in the Sky 168
civil disobedience 194
Clark, Brett 68, 146
Clark, Duncan 167
clean energy 145, 147, 157
Climate Accountability Institute 80
climate action movements 186–90; 350.
  org 187; Extinction Rebellion 189–90;
  School Strike 4 Climate 187–9
Climate Action Network 185–6
climate action vs. climate justice 183–207;
  anti-systemic movement 199–202;
  clash of interests 203; climate justice

movement 191–4; the climate
  movement 186–90; climate movements
  184–5; from climate movement to
  climate justice movement 197–9;
  Global South climate movement
  195–7; tendencies in international
  climate movement 185–6; turmoil and
  youth 183–4
climate activism 3
climate apartheid 138
climate capitalism 49
Climate, Capitalism and Communities 51–2
Climate Change Act 2008 43
climate change and anthropology 6–29;
  anthropogenic disruption 28; changing
  world 7; engaging climate change
  11–14; fundamental planetary changes
  17–27; impact of anthropology 14–17;
  misnamed planet Earth 7–10; short
  history of climate science 10–11;
  unheard voices in climate change
  science 6–7
climate change denialism 3, 82, 160, 185,
  212–13
Climate Change and Displacement 126
Climate Crisis and the Global New Deal
  232
climate engineering 219
climate feedback loop 15–16
Climate Gangsters vs. Climate Justice 191
Climate of History in a Planetary Age
  63–4
climate hotspots 36, 127
climate justice 173–4, 183–207; *see also*
  climate action vs. climate justice
Climate Justice 203
climate justice movement 191–4
climate migration 12–13, 125–30; and
  health 128–30; nexus of 125–8
Climate Models and CO2 Warming 83
climate movement as anti-systemic
  movement 199–202
climate movement to climate justice
  movement 197–9
climate movements 184–5
climate refugees 7, 120–41; climate
  migration and health 128–30; climate
  migration nexus 125–8; decrease in
  GHG emissions 137–8; fleeing drought
  and rain 134–5; floods of starvation
  120–21; high altitude communities
  133–4; island nations at growing risk
  130–33; refugees of war, conflict,
  oppression 122–5; super flooding
  135–7

climate science 6–29; *see also* climate change and anthropology
climate wars 214–16
"climate worlds" 42
Climate Worlds project 42
climate-change syndemics 130
climatology 11–12
Clinton, Bill 175–6
Club of Rome 208
coal gasification 152
coastal erosion 126
Coates, Ta-nehisi 100–101
cobalt 97–8
Cobalt Red 97
Cobb, John 249
Coca-Cola 201
coffee production 62–3
Cold War 199, 261
collaboration 53, 113–14
collapse of major civilizations 13
collective allegiance 40
collectivization 257
Collins, Edward 81
colonial dispossession 100
colonial strategy of planetary health 108–115
Commoner, Barry 248
Communarian sub-type 68
conflicting anthropological perspectives 30–57; critical anthropology of climate change 45–52; cultural ecological anthropological perspective 31–7; cultural interpretive perspective 37–45; integrated critical understanding 52–4; understanding course of human–climate nexus 30–31
confronting zoonoses 101
congested drainage 73, 129
Connell, Raewyn 177
Connolly, William 213
Conquergood, D. 124
Cons, Jason 36
consumerism 38
consumption culture 33, 257–8
containing climate change 142–63; *see also* ecological modernization
contested concept of Anthropocene 59–63; early, middle, late Anthropocene 60–61; good Anthropocene 61–2; patchy Anthropocene 62–3
continuing research on climate change 36–7
cooling down 52
Copernicus, Nicolaus 9

coral bleaching 51
corporate environmentalism 160
Corporate Watch 191
cosmopolitanism 217
cosmovision 42–3
counter-revolution 212
COVID-19 pandemic 17, 54, 63, 86, 102–4, 106, 155, 165–6, 177–9, 203
cracks in the system 199–202
Crate, Susan 34–5, 50–51
Crist, Eileen 66
critical anthropological perspective 30–57; *see also* conflicting anthropological perspectives
critical anthropology of climate refugees 120–41; *see also* climate refugees
critical anthropology of future 208–236; eco-socialism 221–5; future global catastrophe 208–9; how long can humanity survive 231–3; playing with time 209–210; radical perspectives in dialogue with eco-socialism 225–31; reflexive modernization 217–21; road to dystopia 209–216
critical health anthropological perspective 95–119; *see also* planetary health
critical medical anthropology 47
critique of ecological modernization 157–9
Croeser, Eve 185
cross-cultural knowledge 39–40
Crumley, Carol 14, 31–2
Crutzen, Paul 59–60, 65
crystalline ice 18–20
cultural ecological perspective 30–57; *see also* conflicting anthropological perspectives
Cultural Ecology 32
cultural interpretive perspective 30–57; *see also* conflicting anthropological perspectives
cultural models 39
cyanobacteria 105
cyborgs 69

Daly, Herman 249
"dance of global warming" 44
Darian-Smith, Eve 213–14
dark doldrums 143
"dark heat" 10–11
Dark Mountain project 213
Darwinian evolution 30
Davis, Heather 100
Davis, Mike 258–9
Dawson, Ashley 231
De Lucia, Vito 203

de Moor, Joost 193
de-growth 229–30
debris tailing 88
decarbonization 53, 64
decolonization 110, 112–13
decrease in GHG emissions 137–8
deferral 243
defining planetary health 99–102
deforestation 27, 32, 103, 137
deglobalization 156
degree of transgression 27
delegitimization 50
demilitarization 260–61
democratic capture of state 240–41
democratic eco-socialism 222,
    224–5
democratic socialism 222–3, 238
demonic duo *see* interface of climate
    change and inequality
dengue fever 37, 104, 128, 136
depletion of resources 38
depletion of stratospheric ozone 27
depreciation 84
depression 129, 259
Derber, Charles 184
destabilization 12
destruction 23–4
Dhaka 73–5, 128
diarrhea 74, 128–9
die-offs 25
Dietz, Matthias 195
diphtheria 102
direct air capture 156
dislocation 130–33
displacement 63, 106, 124–7, 129
disturbing Anthropocentric perspective
    7–10
Doerr, John 149–50, 160
Doherty, Peter 171–2
domino effect of risk 125
Donn, William L. 83
doomism 63, 175
"Doomsday Glacier" *see* Thwaites
    Glacier
Douglas, Mary 14, 38, 40
"down under" ecological modernization
    151–7
Dreamers 100–101
drivers of climate change 86–90
drought 2, 126, 134–5, 138, 210–211
Dry Corridor 78
Dubash, Navroz 5
Duharte, Adrian Santiago Franco 174
Dulles, Allan 98
Dunkelflaute 143

dynamics of geosphere-biosphere
    interactions 27

E7 144
early Anthropocene 60–61
Earth, Atmospheric, and Planetary
    Sciences 22
Earth Day 89
Earth democracy 48
Earth Policy Institute 144
Earth System Science Center 50
"Earth, Wind, and Fire" 146
Earth's otherness 64
Earthshot project 150–51
Eckersley, Robyn 217–18
eco-anarchism 48, 221, 223, 226–8
eco-anxiety 129
eco-authoritarian regimes 211–14
eco-equity 91
eco-fascism 212–13
eco-feminism 225, 228
eco-Marxism 36–8, 65–6
eco-socialism 48, 67–8, 142–3, 173,
    221–32, 237–63; as alternative world
    system 221–5; avoiding ecological
    disaster 261–2; fighting climate change
    237–8; radical perspectives in dialogue
    with 225–31; transitioning to 238–61
ecocide 63, 100
ecological anthropology 35
ecological civilization 224, 229
ecological modernization 49, 142–63,
    220–21, 250–51; anti-corporate
    resistance 142–3; corporate
    environmentalism 160; critique of
    157–9; "down under" 151–7;
    ecomodernism 145–6; in the
    mainstream 143–5; other proponents of
    149–51; rich proponents of 146–9;
    technical fixes 143
ecological resilience 35
"ecology of capture" 36
ecomodernism 62, 145–6
ecopsychology 259
ecosocial perspective on climate change
    45–52
Ecosocialism in a Nutshell 223–4
ecosyndemics 130
Ectopia 252
effects of greenhouse gases 14–17
egalitarianism 76
El-Hinnawi, Essam 125
electric batteries 95–8
electrolysis 152
electrolyte derangement 105

"elephant in the sky" *see* flying and climate crisis
elite polluters 11, 78–81, 107–8
emancipation 115
embedded energy 168, 251
emergency aid 12–13
emic views 39
emissions taxes 242–3
encountering effects of GHGs 14–17
End to Nature, The 187
energy consumption 7
energy efficiency 249–52
energy superpower 151–4
Engels, Friedrich 70, 241
environment and spread of infectious disease 102–6
environmental anthropological perspective 30–57; *see also* conflicting anthropological perspectives
environmental degradation on humanity 14
environmental determinism 31
Environmental Justice Foundation 122–3
Environmental Protection Agency 85
environmental refugees 125
environmental suffering 95–8
environmental sustainability 38, 158–60, 209–210
environmental violence 106–8
environmentally irresponsible behavior 179–80
epistemic disobedience 113
eradicating poverty 100, 202
Eriksen, Thomas Hylland 51–2, 157
erosion 21
"escalator to extinction" 25
Escrigas, C. 176
ethnocentrism 9–10
European Wind Energy Association 144
existential dread 129
existing vulnerabilities 12–13
exoplanets 8
expansive egalitarianism 41
exploitation 63, 68–9, 87, 100, 110, 255
externality 144
extinction 26
Extinction Rebellion 187, 189–90, 194, 198
extreme weather events 23–4, 134, 186
Exxon 80, 82–3

F4F *see* Fridays 4 Future
Facing the Anthropocene 65–7
Fagan, Brian 14
Falck, Christiane 88

famine 134
Faria, Pedro 79
"Father of Capitalism" 1
Federal Trade Commission 79
feral proliferations 62–3
Ferris, Elizabeth 126
fighting climate change 237–8
Figueres, Christiana 150
Finan, Timothy J. 36
fine particulate matter 106
Finkel, Alan 155–7
fire weather 23
firn snow 19
Fiske, Shirley 61
fleeing drought/rain 134–5
flight shame activism 173–5
flooding 1, 73–5, 106, 120–21, 134–8, 160
flying and climate crisis 164–82; academic air travel 168–9; alternative forms of flying 173–8; anthropologists and air travel 169–71; increase in flight numbers 166–8; irresponsible behavior 179–80; morality of flying 164; other academics and air travel 171–3; staying grounded 165–6; teleconferencing 178–9
Foer, Jonathan 9
Fogarty International Center 13
Fonda, Jane 189
food production sustainability 255–7
Foot, Eunice Newton 11
foraging to farming 60, 76
forestry 255–7
Fortress World 203, 210, 216
fossil fascism 212
fossil fuels 3, 16–17, 32, 59–60, 63, 66, 78–82, 86, 89, 95–6, 137, 142–3, 148
Foster, John Bellamy 68, 146, 158, 223, 225, 248, 250, 257
Fotopoulos, Takis 226
Fourier, Jean-Baptiste Joseph 10–11
fracking 52, 61, 89
fragility of nature 40–41
Frankfurter Allgemeine Zeitung 188
free-market authoritarianism 213
freedom of choice 40
Freeman, Benjamin 25
frequent flying 169–73, 179–80
fresh water change 27
Fridays 4 Future 188–9
Friends of the Earth 186
Fuel Revolution 60
Fuentes, Augustin 107–8
Fugelli, Per 111
full solutions 143

fundamental planetary changes 17–27; intensifying wildfires/hurricanes 22–4; melting icecaps/glaciers 18–22; rising temperatures 17–18; species shifts/loss of biodiversity 24–6; transgressing planetary boundaries 26–7
future scenarios 208–236; *see also* critical anthropology of future

Gaia hypothesis 69, 212
Galilei, Galileo 9
Ganzi Daily 96
Garnaut Reviews 153
Garnaut, Ross 151–4, 156–7
Garrelts, Heiko 195
Gates, Bill 146–9, 219, 221
Gatrell, P. 122
genocide 63, 100
gentrification 259
geocentrism 9
geoengineering 219–21
Georgia Institute of Technology 25
Gergis, Joelle 172–3
GHGs *see* greenhouse gas emissions
Giacomini, Terran 194
"gift of God" 49–50, 81
Giroux, Henri 176
glacial melting 18–22, 135
glaciers 18–19
Glikson, Andrew 60
Global Anti-Aerotropolis Movement 178
global average temperatures 2
global capitalism 38, 40, 47–9, 51–4, 69–70, 137–8, 165, 176, 201
global catastrophe 32, 208–9
Global Climate Coalition 160
Global Climate Science Communications Plan 82
global democracy 221, 224–5
global economy 34, 60
Global Forum on Cities conference (1994) 43
Global Marsall Plan 144
global monopoly capitalism 68
Global Plan for Solving Our Climate Crisis Now 150, 160
global resilience capacity 27
Global South climate movement 6, 195–7; Brazil 195; China 195; India 196–7; Kiribati 195–6; sub-Saharan Africa 186
global threats 12
global warming 2–3, 8, 15–17, 19–22, 25, 32, 37, 40–41, 51, 64, 69, 77–9, 83–4
Global Warming Potential 14–15

Glover, Andrew 171
Gonzalez, Guillermo 97
good Anthropocene 61–3, 146
"good" ozone *see* stratospheric ozone
"good victims" 123
Gore, Al 143–4, 147, 150, 186
Gorz, André 229, 237–8, 247
Graeber, David 261
Gramsci, Antonio 63
Grand Challenge Symposia 61
grappling with reliance on flying 164–82; *see also* flying and climate crisis
grassroots fight against climate change 86–90
grassroots resistance to mining 196–7
Great Acceleration 59, 68
Great Barrier Reef 51
Great Extinction 64
Great Transition 69
green capitalism 49, 156
Green Deal 3, 44, 218, 232
green electromobility 98
green hydrogen 153
green jobs 249–52
Green Left 226
Green politics 62, 68
Green Premiums 148
Green Revolution 87, 137, 148
greenhouse gas emissions 3, 11, 13–18, 21, 37–49, 51, 65, 76–86, 89–91, 128–30, 137–8
Greenpeace 186, 189, 196
greenwashing 79, 151–7
grid-group model 40, *41*
Griffith, Saul 154–7
ground-level ozone 15–16, 106
Guatemala 78
Guterres, António 103
GWP *see* Global Warming Potential

half-earth socialism 230–31
Half-Earth Socialism 230
Hallam, Roger 189–90
Halwell, Brian 256–7
Haman, Inas 105
Hamid, S. 136
Hansen, James 242
Haraway, Donna 68–9
Harms, Arne 196
Harper, Kristin N. 37
Harrell-Bond, B.E. 124
Harrison, Fay 110
Hartley, Daniel 66
Harvey, David 146, 202
Hastrup, Kirsten 36–7, 42

healing rituals 76
health and climate migration 128–30
Heart Grown Bitter, The 124
heat domes 18
heat stroke 18
heat waves 2, 18, 25, 135, 198–9, 210–211
heat-related death 2
heat-trapping 16
Heede, Richard 80–81
Heffernan, C. 130
Helm, Dieter 143
Herrington, Gaya 208–9
Herschel, William 10–11
Hickel, Jason 229
hierarchicalization 50
High, Mette 49
high-altitude communities 12, 133–4
Hindenburg 173
Historical Ecology 31–2
HIV/AIDS 102–3, 106
"hoax" of climate change 11
"hockey stick hypothesis" 175
Hoffman, Susanna 52
Holloway, John 200
Holmgren, David 256
Holocene era 17, 65, 76–7
homelessness 230
Homo sapiens 17–18
Hornborg, Alf 35, 45, 59, 66, 220
Hughes, David McDermott 142–3
Human Choice and Climate Change 38
human creation of inequality 76–7
Human Ecology 35
human exceptionalism 8–10
human extinction 208–9
human migration 125–8
human social organization 40
human suffering 12, 95–8, 106
human-centrism 10
human-induced climate change 24, 34
human–environment relations 31–7;
    continuing research on climate change
    36–7
human–nature nexus 58–9, 63
hurricanes 2, 23–4, 106, 129, 132
hydrochlorofluorocarbons 15
hydrofluorocarbons 15
hydrogen power 151–2
hypovolemic shock 105

ICARUS 33
ICE *see* Informed Citizens for the
    Environment
ice ages 17–18, 60
ice discharge 19

ice floes 19–20
ice shelves 18–19
IEP *see* Institute for Economics & Peace
IKEA 201
IMF *see* International Monetary Fund
impact of human practice on nature 14
imperialism 62
imported luxury goods 78–9
Imposing Aid 124
Inconvenient Truth, An 144
increase in flight numbers 166–8
increasing social equality 244–5; *see also*
    social inequality
India 196–7; resistance to coal mining
    196–7
Indigenous communities 10, 21, 35, 42–3,
    53, 59, 76–7, 88–90, 110–115, 134–5,
    158–9, 169–71, 192, 227, 238–9
Indigenous Environmental Network 89
Indigenous Resistance Against
    Carbon 89
Indus Basin Irrigation System 137
industrial pollution 47, 106–7
Industrial Revolution 15, 17–18, 59,
    66, 100
Informed Citizens for the Environment 84
infrared radiation 10–11, 16–17
Initiative on Climate Adaptation
    Research and Understanding through
    the Social Sciences *see* ICARUS
Institute for Economics & Peace 127
institutionalization of power 76
Intangible Drilling Costs Deduction 83
integrated critical understanding of
    climate change 52–4
intelligent sabotage 199
intensifying wildfires 22–4
interface of climate change and
    inequality 77–8
International Air Transport Association
    166
International Day for Biological
    Diversity 103
International Monetary Fund 77–8, 86,
    202, 229, 240
Inuit 34–5, 42
IPCC *see* UN Intergovermental Panel on
    Climate Change
irrigation 31
island nations at growing risk 12, 52,
    130–33, 196

Jackson, Tim 218–19
Jacobin 146
Jasanoff, Sheila 209

Jerez, Barbara 97
Jevons paradox 142, 165–6, 168
Johanson, Donald 170
Jones, Rhys 112
Jore, Solveig 104

Kahn, Laura H. 101
Kalmus, Peter 194
Kaplan, Bruce 101
Kara, Siddharth 97–8
Keith, David 219
Kempton, Willet 38
Kennedy, John F. 150
Kenya 134–5
Kerry, John 150
"king tides" 131
Kingsnorth, Paul 213
Kiribati 195–6
Klein, Naomi 189, 243
Klepp, Silja 195
Knights, Sam 190
Knox, Hannah 43–4
Koch, Max 178
Koenig, Claudia 195
Konisky, David 186
Kovel, Joel 245
Kropotkin, Peter 226
Kurtti, T. 104
Kyoto Protocol 82, 186–7

Ladd, Brian 252
Lahsen, Myanna 50–51
Lancet Commission 106, 108–9
Lancet, The 108–9
land system change 27
landfill 15
late Anthropocene 60–61
Latouche, Serge 229
Latour, Bruno 58, 61–2, 65
laughing gas *see* nitrous oxide
Le Pen, Marine 3, 213
League of Nations 122
Leahy, Terry 227–8
legitimization 50
Leigey, Vincent 229
leishmaniasis 128
Lemos, Maria 33
Leppert, Amanda 159
Lewis, S.L. 60
Limbaugh, Rush 85
liminal zones 124
limitations of reflexive modernization 220–21
Lipset, David 89
lithium 95–8

"lithium triangle" 95–6
Little, Andrew 114
living at risk 73–5
lobbying 81–2
local communities 258–60
local knowledge 39
Lohmann, Larry 242
Loizos, Peter 124
Lorde, Audre 109
loss of biodiversity 24–6, 63, 103, 131
loss of mythological symbols 35
Lovelock, James 69, 212
Lovins, Amory 147, 150
low-carbon cities 43, 85, 146
Lumumba, Patrice 98
luxury emissions 38
Lynas, Mark 211

McAdams, Jane 126
McBurney, Stuart 165
McDonald's 201
McElroy, Ann 34
McGranahan, Carole 110
McKibben, Bill 187, 189, 198
Macmillan, Alexandra 112
Magdoff, Fred 245, 257
Mahan Forests coal mining 196–7
mainstream ecological modernization 143–5
malaria 102, 128–9, 136, 220
Malinowski, Bronislaw 170
Malm, Andreas 66, 187, 190, 197–9, 212
malnutrition 129, 134, 136
Malpass, David 3
Mann, Michael 80, 175–6
Marcantonio, Rice 107–8
marginalization 110, 134–5
Marx, Karl 65, 70, 237–8, 241, 247
Marxist political economy 223–4
Maslin, M.A. 60
mastery of fire 60
Mathews, Andrew 58
Matsutake Worlds Research Group 10
Matthews, Freya 228
Mazou, Raouf 123
Mazzocchi, Tony 232
Mead, Margaret 13, 170
meaningful consumption 257–8
meaningful work 246–7
medieval feudalism 66
melting icecaps 17–22, 36–7, 76, 84–5
Mendes, Paulo 52, 157
mental health 129
Merkel, Angela 188
Mesozoic era 9

methane 15, 21, 59, 152
methyl tert-butyl ether 16
middle Anthropocene 60–61
Milei, Javier 3
Miles, Tiya 101
Miliband, Ralph 222
militarism 260–61
Milito, Erik 83
Miller, Todd 216
Milton, Kay 31, 39–42
misnamed planet Earth 7–10
mitigation efforts 43, 78, 142–3, 154, 212–13
Mokbul, Ahmad 75
Mollison, Bill 256
"monster" of ecomodernism 62
Moonshot project 150
moonsoon 73, 135
Moore, Jason 65–7
Morales, Evo 192
morality of flying 164
Moran, Emilio 33–4
Morrison, Scott 213–14, 232
Mother Earth 110–112, 115, 192–3, 228
Mother Nature 43, 100, 111–12
movement of ocean currents 20–21
multispecies ethnography 9–10
Munderloh, U. 104
mushrooms 10

Namugerwa, Leah 183–4
Narain, Sunita 38
NASA 8, 18
National Audubon Society 25–6
natural disasters 164
natural gas production 80
nature/culture dualism 64–5
Necrocene era 68–9
need for steady-state economy 248–9
Nelson, Anitra 229
neocolonialism 230
neoliberalism 61, 124, 159
net zero carbon 44, 147–8, 157, 210–211
Nevins, Joseph 174–5
"new normal" 44
new progressive parties 240–41
New York Times 84
Newell, Peter 218
Newman, Peter 252
nexus of climate migration 125–8
Nierenberg, Danielle 256–7
Nita, Marta 186
nitrous oxide 15, 21, 148
noise pollution 165

non-renewable energy sources 95–8, 144
nonanthropocentrism 64
nonreformist reforms 237–8
nonsecular cosmology 63
nonviolence 194
nonzoonotic disease 102–3
Nordhaus, Ted 62, 145, 232
Notpla 150–51
novel entities 27
Novocene era 69
nuclear weapons technology 68
Nuttall, Mark 32, 34–5
Nyberg, Daniel 160

Obama, Barack 85, 123, 147, 194
ocean acidification 27
ocean noise pollution 107
offshore drilling 83
Oil Change International 89
O'Lear, Shannon 69
Onaran, Oezlem 252
One Health model 101–2, 115
Oppenheim, Alan 83
optimism of will 63
Orban, Viktor 3
Ord, Tony 220
original affluent societies 76
Orlove, Ben 33, 37
Orr, Yancey 52
other academics and flying 171–3
overcrowding 73–4
overheating 51–3
Overseas Development Institute 86
overwintering 25
Oxford Climate Journalism Network 6
ozone depletion 27, 38

Pakistan 135–7
paleoecology 35
Parenti, Christian 66, 214
Paris Agreement (2015) 2, 137, 183, 220
Parker, Martin 170
Parkinson, C. 176
partial solutions 143
Partij voor de Frijheid 3
Partisan Power 145
patchy Anthropocene 62–3
Patterson, Matthew 218
Pendergrass, Drew 230–31
Peni, Emmanuel 88
People and Nature 33–4
Pepper, David 248
Perceptions of Climate Change in India 44
perfluorocarbons 15
permafrost 21–2

pessimism of intellect 63
petro-colonialism 89, 155
Phillips, Leigh 146
phosphorous 61
photosynthesis 27
Pillai, Priya 197
Pitron, G. 159
Plan B 144
Plane Stupid 178
planetary health 95–119; challenges for 115;
    as colonial strategy 108–115; definition
    99–102; electric batteries and suffering
    95–8; environmental violence 106–8;
    spread of infectious disease 102–6
Plantationocene era 68–9
plastic pollution 103–4
playing with time 209–210
PM2·5 106
Polemocene era 69
political ecology 46
Pollex, Jan 188–9
Pollin, Robert 232
pollution 11, 47, 78–81, 103–4, 106–8,
    128–9
poor sanitation 73–4, 106, 120–21, 128–9,
    136
Popper, Karl 30
population dynamics 39
possibility of eco-authoritarian regimes
    211–14
post-Fordism 209
postcapitalism 65
postcolonial historical take on
    Anthropocene 63–5
potential for climate wars 214–16
poverty 73–5, 78, 100, 132–3, 136–7, 202,
    214
Powalla, Oliver 196
*Principles of Tackling Climate Change in
    Manchester* 44
private ownership of means of production
    1–2
Provenzale, Antonello 59–60
public ownership of means of production
    224–5, 243–4

"quarry nation" 51
Question of Resilience 27, 36

racism 9, 63–4
radical perspectives in dialogue with
    eco-socialism 225–31; de-growth
    229–30; eco-anarchism 226–8;
    eco-feminism 228; half-earth socialism
    230–31; indigenous voices 230

radioactivity 106
Radonic, L. 107
Rahman, Amar 125
Rahman, Atiq 74
Rahman, Md. Ashiqur 36
Rainbow Alliance 150
Randall, Doug 215–16
Rassemblement national 3
Ravanneid, Aase 44
raw sewage 107
Ray, Celeste 39, 40
Rayner, Steve 14
real utopias 225, 238
rebound effect 142, 165–6, 168
recognition of climate change 10–11
recognition of human–nature nexus 58–9
recycling 49, 155, 217, 224–5, 247
redistribution of wealth 147–8
Redlinger, Robert 85
Redvers, Nicole 111
reflexive modernization 217–21;
    geoengineering 219–20; limitations of
    220–21
reformist reforms 237–8
Refugee Convention (1951) 122
refugees of war, conflict, oppression 122–5
Reid, Hannah 255
Reid, Papaarangi 112
release of carbon dioxide 10
reliance on flying 164–82; *see also* flying
    and climate crisis
relocation 35
Renaghan, Quinn 193–4
renewable energy sources 61, 63, 65, 85,
    142–4, 147, 151, 158–9, 173–4, 220–21,
    249–52
reservations about "Capitalocene" 69–70
*Reset: Restoring Australia after the
    Pandemic Recession* 152
resilience of Earth system 26–7
resistance from below 86–90
resistance to coal mining in India 196–7
resisting culture of consumption 257–8
resource capitalism 145
respiratory problems 16, 103, 105, 108
restoration 12–13
rethinking era of climate change
    production 58–72; alternatives to
    Anthropocene and Capitalocene 68–9;
    Anthropocene as contested concept
    59–63; Anthropocene vs. Capitalocene
    65–8; postcolonial historical take on
    Anthropocene 63–5; recognition of
    human–climate nexus 58–9; reservations
    about "Capitalocene" 69–70

rhizomic sociality 10
Ribot, Jesse 33
rich powerful people addressing
    ecological crisis 142–63; *see also*
    ecological modernization
rich proponents of ecological
    modernization 146–9; Bill Gates 147–9
Richardson, Katherine 27
rights of nature 43
Risher, Dana 193–4
rising temperatures 7, 17–18, 76–7, 210–211
road to dystopia 210–216; eco-
    authoritarian regimes 211–14; Mark
    Lynas 211; potential for climate wars
    214–16
Robinson, Mary 203
Robinson, William I. 240
Rockefeller Foundation 108–9
Rockefeller, John D. 79, 109
Rockström, Johan 35, 53
Rogers, Chris 244
Roncoli, Carla 39
Roosevelt, Franklin D. 237
Rosewarne, Stuart 156
Ross-Nazzal, J. 237
Ruddiman, William 60
RWE 52
Ryle, Martin 241

safe climate 157
Sagan, Carl 14
Sahlins, Marshall 76
salinity 20, 132
Salleh, Ariel 160, 228
Saloa, Kelesoma 132
Sarkar, Saral 246, 258
Save the Sepik 88
saving life on Earth 2
Schapira, Allan 136
Schellenger, Michael 62
School of Advanced Research 31
School Strike 4 Climate 187–9
Schor, Juliet 247
Schroeder, Patrick 195
Schumacher, E.F. 248
Schwab, Tim 149
Schwartz, Peter 215–16
Schweickart, David 262
Scott, James 87
sea ice melting 18–22
sea level rises 130–34
seeking to address ecological crisis
    142–63; *see also* ecological
    modernization
Segebart, Doerte 195

self-centredness 10
self-determination 89, 113
self-interest 1
sequestration of carbon 144
settlement patterns 258–60
semi-periphery countries 45–6
sexism 9, 108, 228
Shaw, Christopher 157
Shell 80, 200–201
Shellenberger, Michael 145
Shiva, Vandana 258
short history of climate science 10–11
shortening working week 246–7
Sila Nanotechnologies 97
Singer, M. 130
situating humans within environment
    37–45
Six Degrees 211
slave labor 66, 68
Smith, Adam 1
Smith, Dan 216
Smithsonian Institution 61
SMR *see* solar radiation management
social disruption 193–4
social equality 146, 244–5
social inequality 52–3, 73–94, 137–8;
    human creation of inequality 76–7;
    interface of climate change and
    inequality 77–8; living at risk 73–5;
    resistance from below 86–90;
    transforming physical environment 91;
    unequal distribution of power 78–81;
    welfare for wealthy, blame to the poor
    82–6
social justice 54, 142–3
"socialism of wind" 142–3
"socialist lies" 3
socialist planning 245–6
sociocultural configurations 39–40
socioeconomic gap between rich and
    poor 48–9, 76–7
solar energy 61, 145, 153, 155, 158
solar heat 21
solar radiation management 219–20
South Sudan 120–21
sovereignty 89, 113
Spangenberg, Joachim H. 69
spatiotemporal patterns of geosphere-
    biosphere interactions 27
species extinction 2
species shifts 24–6
Speth, James Gustave 38
Spiegel, Samuel 89
spread of infectious disease 2, 12, 33, 37,
    101–6, 128

Standard Oil Company 79
Starr, Amory 261
starvation floods 120–21
staying grounded 165–6, 173
steady-state economy 248–9
Steffin, Will 60
Stensrud, Astrid 51
Stern, Nicholas 48, 81
Steward, Julian 31
Stilwell, Frank 221, 245
Stoermer, Eugene 59
storm surges 24, 131–2
Stoutenberg, Jenny 125–56
stratospheric ozone 15–16, 27; depletion of 27
Strengers, Yolande 179
stress on habitats 26
structural violence 108
sub-Saharan Africa 196
subalterns 52, 110, 147
subsistence agriculture 132–3
suicidal behaviors 129
sulfur hexafluoride 15
Sundarbans 36, 129
sunrise industries 144, 156
superflooding 135–7
Superpower Transformation 152
survival emissions 38
survival of humanity 231–3
Susser, Ida 222
sustainable agriculture 255–7
sustainable global population 245
sustainable public transportation/travel 252–5
sustainable trace 258
Sutton, Mark 32
Symons, Jonathan 145–6
systemic adaptation 37
Szolucha, Anna 52

Talukdar, Ruchira 197
Tanuro, Daniel 197, 243–4
tar sands oil 89–90
Te Oranga o te Taiao 113
Teafa, Lily 132
technical fixes 143
teleconferencing 177–9
Teller, Edward 219
tendencies in international climate movement 185–6, *185*
Terrio, S. 123
Tesla electric cars 95–6
theoretical perspectives on climate change 30–57; *see also* conflicting anthropological perspectives

theory of human needs 38
theory of wants 38
Thinking Like Climate 43
350.org 185, 187
Thunberg, Greta 4, 173, 183, 187, 189
thunderstorms 23
Thwaites Glacier 20
Todd, Zoe 100
Todorovich, Petra 254
towards the future 1–5
Townsend, Patrick K. 34
toxic leakages 96–7, 108
toxication 106
traditional ecological knowledge 34, 42, 110–115, 134
traditional worldview 42–3
Trainer, Ted 158, 226–7
transforming physical environment 91
transgressing planetary boundaries 26–7, 99–102
Transition of a Sustainable and Just World, The 226
transitioning to eco-socialism 238–61; demilitarization 260–61; democratic capture of state 240–41; emissions taxes 242–3; increasing social equality 244–5; meaningful work/shortening working week 246–7; public onwership of means of production 242–4; renewable energy 249–52; resisting culture of consumption 257–8; steady-state economy 248–9; sustainable food production/forestry 255–7; sustainable global population 245; sustainable public transport/travel 252–5; sustainable settlement patterns 258–60; sustainable trade 258; workers'democracy/socialist planning 245–6
trauma of conquest 97
travelling less by air 173–8
Treaty of Waitangi (1840) 114
trichloroethylene 16
Tropic of Chaos 214
tropical cyclones 23–4
Trump, Donald 3, 83, 85, 90, 184–5, 212, 261
Trusiewicz, Eric 150
Tsing, Anna 10, 62–3
tuberculosis 37, 102
Tulele Peisa 131
tundra warming up 18–22
turmoil and youth 3, 183–4
Turner, Terisa 194
Tuvalu 130–33

two genres of climate movement 183–207; *see also* climate action vs. climate justice
Tyndall Centre for Climate Change Research 44, 174
Tyndall, John 11
type 2 diabetes 128

ultimate climate change mitigation strategy 237–63; *see also* eco-socialism
ultraviolet light 16
UN Climate Change conference 147, 183, 188, 191–3
UN High Commissioner for Refugees 122–3, 126
UN Intergovermental Panel on Climate Change 2, 13, 33–4, 53, 113, 131, 133, 167, 172, 209
underground aquifiers 8
understanding climate change 10–11
understanding course of human–climate nexus 30–31
unequal distribution of power 78–81
unethical behavior 97–8
UNHCR *see* UN High Commissioner for Refugees
unheard voices in climate change science 6–7
United Nations 73–4, 100, 103, 122
"universal statements" about the world 30
unregulated production 91
urbanization 37
Urry, John 145
"utopian fantasy" 3
utopistics 210

van Helmost, Jan Baptist 10
Vardy, Mark 44–5
vector-borne disease 33, 37, 101, 104
vertical wind shear 23
vested economic interests 39
Vettese, Troy 230–31
Victor, David 13
Vieira-da-Silva, L.M. 109
Viluni Sakha 34
Virchow, Rudolf 101
Vivekananda, Janani 216
VOCs *see* volatile organic compounds
volatile organic compounds 16
vulnerability to climate change 34, 73, 108

walkability 259
Wall, Derek 158, 160, 238, 244

Wallerstein, Immanuel 210, 232–3
Wallis, Victor 200, 239
Walmart 201
Wang, Zhihe 223–4
War Against the Common 1
Ward, Peter 61
wastewater treatment 15, 89
water shortages 63
water treatment 16
water vapor 16
water-borne disease 74, 101, 128, 134, 136
waterlogging 129
Waterworlds project 36–7
Watt, James 59
Wealth of Nations 1
wealthy proponents of ecological modernization 149–51
Weapons of the Weak 87
Weik, Elkie 170
welfare for wealthy 82–6
Welzer, Harald 214
West Nile disease 104–5
whakapapa 114
Whelan, Tenise 150
white gold 96
White, Kandi 89
White, Leslie C. 60
White, Tim 170
WHO *see* World Health Organization
Who Owns the Wind? 142
Whyte, Kyle 100
Wijkman, Anders 35, 53
WildAid 150
Wilde, Parke 176
Wilders, Geert 3
wildfires 2–3, 22–3, 198–9, 211, 213, 232
Wilson, E.O. 230
wind energy 61, 142, 145, 155, 158–9
Wirth, Michael K. 79
Wolf, Eric 46, 87
Wolff, Richard 245
workers' democracy 245–6
World Bank 3, 77, 100, 136–7, 202, 229, 240
World Commission on Environment and Development 38
World Economic Forum 79, 173, 189, 202, 240
World Health Organization 102, 105, 108–9, 136
World Heritage Site 90
World Urban Forum 183
World War I 122

World War II 59, 68, 122, 244
World Wildlife Fund 201
worldwide vegetarianism 255
Wright, Christopher 160
Wright, Eric Olin 222, 225, 238
WWF *see* World Wildlife Fund

XR *see* Extinction Rebellion

youth-led grassroots global climate
    activism 3–4
Yushin, Gleb 97–8

zero emissions 167–8
Zetkin Collective 212
Žižek, Slavoj 223
zoonotic disease 102–3, 105